Bacteriophages as Drivers of Evolution

Stephen T. Abedon

Bacteriophages as Drivers of Evolution

An Evolutionary Ecological Perspective

Springer

Stephen T. Abedon
Department of Microbiology
The Ohio State University
Columbus, OH, USA

ISBN 978-3-030-94308-0 ISBN 978-3-030-94309-7 (eBook)
https://doi.org/10.1007/978-3-030-94309-7

This Springer imprint is published by the registered company Springer Nature Switzerland AG
The registered company address is: Gewerbestrasse 11, 6330 Cham, Switzerland

In loving memory of my aunt, Sheila Tobias, 1935–2021, who was my inspiration to become serious about writing.

Preface

When I was much younger, now 40 or so years past, I felt a desire to study evolutionary biology. What I imagined 'studying evolutionary biology' to be, however, was not what I would now think of as *doing* evolutionary biology, i.e., as an academic researcher. Nevertheless, after four years of diligently studying as an undergraduate almost anything but evolutionary biology, I wanted to at least be doing something that I considered to be *fun*. Thus, I chose Ecology and Evolutionary Biology as my minor in graduate school. Oddly, in actually attempting to take up the study of evolutionary biology, I found myself attracted to things much more ecological. Far too many years later, I would come to realize that my true passion was for what can be described as evolutionary ecology. Consequently, today, I find myself not an evolutionary biologist, and certainly not an ecologist, but perhaps what might best be described as an evolutionary ecology *aspirant*.

Meanwhile, I have an undergraduate degree in Biochemistry, and a graduate concentration in Molecular Genetics, though to this day I am not really quite sure how the latter came to be. My research emphasis, however, has almost never been on the biochemical nor molecular genetic underpinnings of organisms. I like to think that I nevertheless still retain a biochemist's appreciation for not ignoring potentially useful details, molecular or otherwise, along with an affinity for molecular genetics, though the latter is especially as it originally existed as a much more organismally oriented endeavor, dating back to before I was born. Less than a year after choosing to study Ecology and Evolutionary Biology as a minor, and after an odd veering off to attempt to study nutritional biochemistry toward a master's degree, I discovered bacteriophages along with microbiology. For me, ultimately, phages have been a means by which I have been able to combine my early training in biochemistry with my later aspirations toward evolutionary ecology, and ecology. Ultimately, this all has resulted in my writing this monograph, on the evolutionary ecological impact of bacteriophages especially on bacteria.

Bacteriophages, or phages, are the viruses of domain *Bacteria*. They play important roles in bacterial ecology, serving as predators, parasites, commensals, and even as mutualistic symbionts. Stemming from these ecological roles, phages also impact bacteria evolutionarily. Phages, in particular, can contribute to bacterial evolution by each of the four basic evolutionary forces, those of mutation, genetic migration, natural selection, and genetic drift, plus non-random mating. Phages can increase bacterial mutation rates such as by disrupting bacterial genes (as in the course of prophage integration), phages can play numerous important roles in the movement of genetic material between bacteria (via gene transduction), phages can select for certain bacterial genotypes over others (resulting in antagonistic coevolution), and phage action can result in random impacts on gene (allele) frequencies in bacteria, giving rise to genetic drift. Phages, also through transduction, can bias 'matings' between different bacteria, resulting in the noted non-random mating. That is, phages may increase rates of gene exchange between some bacteria but not between others within bacterial species. Phage-phage interactions at both population and community levels also can result in modifications of allele frequencies within *phage* populations.

Beginning with these various concepts, in this monograph I emphasize an exploration of the many facets of especially bacterial evolution that are impacted by the interactions of bacteria with phages, though also and to a far lesser degree, the evolutionary impact of phages on organisms other than bacteria. There are six units in total. These include (Part I) an introductory unit, (Part II) a unit on phages as they impact the various evolutionary forces except genetic migration and natural selection, (Part III) a unit that focuses on the phage impact on genetic migration, (Part IV) a unit that can be viewed as an extension of the phage impact on genetic migration though focusing on phage latent (lysogenic) infections, (Part V) a unit on acquired resistance by bacteria to phages as can be viewed as a first installment of looking at phages as effectors of natural selection, and (Part VI) an extension of the previous unit, this one titled simply, "Natural selection". It is at the end of latter unit that antagonistic coevolution is looked at in some detail. For many individuals, other than transduction, such coevolution might be viewed as perhaps even the *only* way that phages in fact drive evolution. It is not.

Overall, my intention has been to supply a relatively math free overview of this field. This in part is because I am not a trained mathematician, though I have been known to dabble a bit in relatively simple mathematical methods. It is also because I tend to prefer logical explanations to mathematical ones, and because I am hoping to not have excessive mathematical calculations limit my audience. Especially, I think that broadening people's perspectives on just what phage-driven evolution might consist of is important and, except for those who are really into mathematical explanations, I don't think that presenting equations necessarily is required to reach that goal. Having said all of that, however, it is important to realize that you can't get completely away from numbers, particularly probabilities, nor even equations (e.g., in calculating the fitness costs of phage resistance), but I have done my best to not lean too heavily on these. I'm also very much a 'phage guy', interested in phages intrinsically rather than as model systems for studying, e.g., basic principles of ecology and evolutionary biology, or for that matter using phages to study common molecular aspects of living things. Thus, this is not a book on how phages have or can contribute to mathematical biology, ecology, evolutionary biology, evolutionary ecology, or molecular aspects of biology—I won't be mentioning, for example, either Red Queens, pan genomes or Green Beards, nor even adaptive landscapes or Prisoner's Dilemmas—but instead this book is about how an understanding of phages can provide a better appreciation of the ways that phages can impact other organisms evolutionarily. Infused in that emphasis also are my interests in the history of ideas as well as the related issue of not claiming for myself ideas for which others hold precedence. Though I will be the first to admit that being aware of everything that has come before is really hard to accomplish, and certainly I have most likely utterly failed in many areas in this regard, nevertheless I have a strong predilection toward celebrating especially the ideas of others, with often a delight in discovering who has already said what.

The genesis of this monograph otherwise followed a still yet unusual path. First is my developing and teaching a course on microbial evolution, which by my choice involved a lot of consideration of microbial evolutionary ecology rather than molecular evolution. Thus, I was primed. Second is that I had reached a sufficient level of frustration with editing multi-authored monographs that I had vowed, never again! Thus, I was circumscribed. Lastly, Springer Nature in February of 2018 asked me to edit a volume on "Phages and Their Hosts as Drivers of Evolution". Thus, there was opportunity. I replied, though, that this was far too broad a subject to cover in a single volume and that I would rather write a single-authored monograph than edit a multi-authored one. I also proposed shortening the title to just, "Bacteriophages as Drivers of Evolution". This, in my opinion, was just too nice a title to waste, and hence—following long hours sitting at a computer—truly this monograph was born. As an aside, the phrase, "Drivers of evolution", in quotation marks, in fact yields nearly 1500 hits in a Google Scholar search, and a "Drivers of Evolution" AND (phage OR bacteriophage) search yields about 150 hits.

My target audience ranges from those with only an interest in phages, particularly in terms of their potential to impact bacteria, to those with an interest explicitly in evolutionary ecology involving phages. To substantially paraphrase Dr. 'Bones' McCoy, "Damn it Jim, I'm a teacher, but a scientist, too." For the former audience, I provide necessary background material so that basic evolutionary ecological ideas, I hope, are understandable, e.g., just what exactly is antagonistic coevolution? For readers who are already experts, including of phage-bacterium evolutionary and ecological interactions, I provide a number of insights that, so far as I have been able to tell, have not been proffered elsewhere, or at least are not easily found or recognized in the literature. This includes new thinking not previously published on by myself. Indeed, the whole point of a project such as this, so far as I am concerned, is to have an opportunity to at least attempt to synthesize new ideas. I apologize in advance, and profusely, to the extent that I am merely reinventing someone else's insights without proper attribution.

Much of the goal of this monograph nevertheless is to drive home two points. The first is that antagonistic coevolution along with transduction are not the only ways that phages can drive especially bacterial evolution. The second is that it can be profitable, toward understanding phage ecology and evolutionary biology, to focus on things other than molecular details, and particularly other than information that may be gleaned exclusively from the study of nucleic acid sequences. Instead, what I present is my perspectives on microbial evolutionary ecology as applied to phages, as informed by over 30 years of pondering, pondering particularly on how individual phages can impact other organisms. Ultimately, this means that what is found in this book is there due to a combination of its having come to my attention while piquing my interest, there being enough space to include it, and also there being sufficient time for me to effectively cover it. Among those topics mostly missing are considerations of the evolution of both phage tail-like bacteriocins (tailocins as they increasing are being called) and gene transfer agents. In addition, if something were an issue predominantly involving phage evolution, or phage ecology rather than evolution, and not especially the phage impact on bacterial evolution, then I mostly have intentionally avoided covering it. Notwithstanding these limitations, I hope that you find my approach to understanding the phage evolutionary impact to be at least a tiny bit enlightening.

Columbus, OH, USA Stephen T. Abedon

Glossary of Terms

A glossary defining the following terms can be found at the end of this monograph: Abiotic; Abortive infection; Abortive infection system; Abortive transduction; Absolute fitness; Accessory gene; Acquired resistance; Adaptation; Adaptive immunity; Adsorptive host range; Adsorption; Adsorption resistance; All the world's a phage; Allele; Allele frequency; Allelic variation; Allelopathy; Amixis; Antagonist; Antagonistic coevolution; Antagonistic interaction; Antagonistic pleiotropy; Antibiotic; Archaea; Archaeal virus; Arms race; Assortative mating; Assortative transduction; Attachment; Attachment site; Auxiliary metabolic genes; Avoidance; Bacteria; Bacterial virus; Bactericidal host range; Bacteriocin; Bacteriophage; Bacteriophage exclusion (BREX); Base sequence; Biofilm; Biogeochemical cycling; Biogeography; Biological species concept; Biotic; Burst size; *Caudovirales* (order); Capsid; Capsule; Cellular arrangement; Cheating; Chronic infection/release; Circular permutation; Clone/clonal; Coadsorption; Coevolution; Coinfection; Coliphage; Colloid; Colony; Colony-forming unit; Commensal; Commensalism; Community; Community dynamics; Community ecology; Community resistance; Compensating mutation; Competition; Concatemer; Conspecific; Converting gene; Conjugation; Core gene; *cos* site; CRISPR; CRISPR-Cas system; CRISPR adaptation; Cumulative evolution; Curing; Cryptic prophage; Darwinian fitness; Defective interfering particle (DI particle); Defective prophage; Delay; Deletion mutation; Destructive infection; Determinism; Deterministic evolution; Direct reciprocal transduction; Directional selection; Disassortative mating; Disperse; Dispersion; Disruptive frequency-dependent selection; Divergent evolution; Diversifying selection; Domestication (of genes); Eclipse; Ecological species concept; Ecology; Ecophysiology; Ecosystem; Ecosystem Ecology; Ecotype; Efficiency of plating (EOP); Encounter; Endosymbiosis; Environment; Environmental DNA (eDNA); Endolysin; Envelope-level resistance; Epistasis; Eukaryote; Evolution; Evolutionary biology; Evolutionary ecology; Evolutionary fitness; Evolvability; Exploitative competition; Exploiter; Extended phenotype; Extracellular polymeric substance (EPS); Extracellular search; Extinction; Fitness; Fixation; Fixed; Focus organism; Founder effect; Free

DNA; Free Phage; Frequency-dependent selection; Gene; Gene exchange; Gene flow; Gene pool; Gene transfer agent (GTA); Generalized transduction; Genetic bottleneck; Genetic diversity; Genetic drift; Genetic hitchhiking; Genetic material; Genetic migration; Genetic sequence; Genetic structure; Genetic transduction; Genetic variation; Genomic island; Genotype; Geographical range; Group selection; Hard selection; Hardy-Weinberg; Head-tailed; Headful packaging; Helper phage; Heteroimmunity; Heterospecific; Historical contingency; Hitchhiking; Homoimmunity; Homologous recombination; Homospecific; Horizontal gene transfer (HGT); Host; Host range; Hybridization; Illegitimate recombination; *In situ*; *In vitro*; *In vivo*; Inclusive fitness; Indirect reciprocal transduction; Individual selection; Induction; Infection; Innate immunity; Insertion mutation; Insertion (prophage); Insertion site; Insertional mutagenesis; Integrase; Integration; Interspecific; Intraspecific; Irreversible attachment/adsorption; Invasive; Iterative coevolution; Iteroparity; Killing the winner; Kin selection; Lambdoid phage; Latent infection; Latent period; Life cycle; Lineage; Linkage; Locus/Loci; Lotka-Volterra; Lysin; Lysis; Lysis from within; Lysis from without; Lysis-lysogeny decision; Lysogen; Lysogen allelopathy; Lysogeny; Lysogenic conversion; Lysogenic cycle; Lysogenic host range; Lytic conversion; Lytic cycle; Lytic phage; Macroevolution; Mass action; Meiosis; Microcolony; Microevolution; Migration (genetic); Mobile genetic element; Moron (broad definition); Moron (narrow definition); Moron accretion hypothesis; Mosaic evolution; Multiplicity of infection (MOI); Muller's ratchet; Mutant; Mutation; Mutation, compensating; Mutation, deletion; Mutation, insertion; Mutation, point; Mutation rate; Mutualism; Mutualistic coevolution; Myovirus; Natural selection; Negation; Negative selection (with replica plating); Niche; Non-Darwinian evolution; Non-homologous recombination; Non-random mating; Non-reciprocal transduction; Non-synonymous substitution; Nucleotide sequence; Numerical refuge; Numerical vulnerability; Obligately lytic; Obligatorily sexual; One-step growth; Organismal ecology; Orthologous replacement; *pac* site; Packaging; Panmixis; Parasite; Parasitism; Pathogenicity island; Penetrative host range; Periodic selection; Phage; Phage-inducible chromosomal island; Phage-susceptibility type; Phage λ; Phage therapy; Phagocytosis; Phagotrophic; Phase variation; Phenotype; Physical; Plaque; Plaque-forming unit (PFU); Plasmid; Plasmid prophage; Pleiotropy; Podovirus; Point mutation; Polymorphism; Population; Population ecology; Population genetics; Population growth; Positive selection; Postzygotic barriers; Predator; Prezygotic barriers; Prey; Productive cycle/infection; Productive host range; Professionally lytic; Prokaryote; Prophage; Prophage depletion; Prophage, integrated; Prophage, plasmid; Pseudogene; Pseudolysogeny; Pseudoreversion; Random mating; Range; Receptor, adsorption; Reciprocal transduction; Recombination; Reduced infection vigor; Reductive infection; Relative fitness; Release; Reproductive success; Resistance; Resistance, acquired; Resistance, chromosomal; Resistance, extrachromosomal; Restriction; Restriction endonuclease; Restriction-modification; Resource; Reversible attachment/adsorption; Reversion mutation; Sampling error; Satellite plasmid; Satellite virus; Secondary adsorption; Secondary infection; Secondary phage; Segregation; Selection; Selective sweep; Self-sacrifice; Semelparity; Sequence space; Sex; Sexual reproduction; Single-nucleotide polymorphism (SNP); Single-step growth; Siphovirus; Sister lysogen; Site-specific recombination; Spatial structure; Specialized transduction; Spatial vulnerability; Speciation; Species; Stable polymorphism; Stabilizing frequency-dependent selection; Stabilizing selection; Stochastic; Strain (taxonomy); Strictly lytic; Subspecies; Superinfection; Superinfection exclusion; Superinfection immunity; Superspreader; Surface receptor; Survival; Symbiont; Symbiosis; Synonymous substitution; Tailocin; Tailed phage; Temperate phage; Titer; Tradeoff; Transducing particle; Transducing phage; Transduction; Transductive host range; Transductive species; Transformation; Translocation; Transposon; Treatment resistance; Unit of selection; Uptake; Victim; Virion; Virion particle; Virion progeny; Virocell; Virulence (pathogen); Virulence (phage); Virulence factor; Virus.

Acknowledgments

I know at this point that I am supposed to thank all of those people who I had discussions with over the course of writing this monograph. Unfortunately, however, most of the time spent writing this monograph I was conversing with people about a subject that has only a little to do with what I have written here, i.e., phage therapy in its various guises. One exception to that was a brief email conversation that I had with Fred Cohan about ecotypes, and I am grateful for that. Thank you, too, to Kate Hargreaves for convincing me of the utility of navigating Word documents using header styles, which was truly an invaluable skill toward writing this monograph. Unfortunately, though, much of the ideas that she and I had developed together failed to 'make the cut'. From toward the end of the writing process, I also would like to thank Sébastien Lemire for helping me to better understand *pac* sites and associated phage DNA packaging. Oh yes, and I certainly would like to thank the Ohio State University Libraries system, and staff, without which and whom both my career and this monograph would not have been possible!

Lastly, I would like to thank all of my various collaborators and mentors over the years including though not limited to Harris Bernstein, Carol Bernstein, John Spizizen, Rick Michod, Stephen Zegura, Bob Doms, Rich Calendar, Betty Kutter, John Riedl, Patrick Paddison, David Stopar, Paul Hyman, Cameron Thomas (now Cameron Thomas-Abedon), Jason Gill, Larry Goodridge, Jeff LeJeune, Mya Breitbart, Forest Rohwer, John Dennehy, Paul Turner, John Yin, Stephen Krone, Siobain Duffy, Sarah Kuhl, Catherine Loc-Carrillo, Bob Blasdel, Andrew Curtright, Ben Chan, Jason Clark, Heather Allen, Zack Hobbs, Rustam Aminov, Matt Sullivan, Diana Alves, Tena Katsaounis, Krystyna Dąbrowska, Andrzej Górski, Claudia Igler, Gary Trubl, Katarzyna Danis-Wlodarczyk, David Harper, Ben Burrowes, Malcolm McConville, and Dan Wozniak, all of whom have helped contribute to my thinking on phages through the years.

Contents

Abbreviations

CFU	Colony-forming unit
eDNA	Environmental DNA
ICE	Integrative and conjugative element
min	Minutes

Part I
Introductions

Chapter 1
Bacteriophages, a Brief Introduction

Contents

Abstract Bacteriophages, or phages for short, are the viruses of bacteria. They were discovered over 100 years ago and have been well studied ever since. In total there are more than ten described family-level phage taxa. This chapter provides a brief introduction to the general biology of bacteriophages. Introduced are the concepts of phage productive, phage reductive, and phage destructive infections, which are phage infections that either produce and release virion progeny, have deferred their production of virion progeny presumably until conditions are more conducive to successful virion production, or in which the infecting virus does not survive, respectively. The basics of adsorption of virions to host bacteria, lytic cycles versus lysogenic cycles, and also genetic transduction are discussed, where transduction is virion-mediated movement of bacterial DNA between bacterial cells. A distinction is drawn between phages that are strictly productive versus temperate. Strictly lytic phages only release their virion progeny lytically and never display lysogenic cycles whereas temperate phages display either productive phage infections or instead reductive (lysogenic) infections, depending on circumstances.

> Phages have been of interest to scientists as tools to understand fundamental molecular biology, as vectors of horizontal gene transfer and drivers of bacterial evolution, as sources of diagnostic and genetic tools and as novel therapeutic agents.—Martha R. J. Clokie et al. (2011, p. 31)

Bacteriophages are a category of viruses that are distinguishable from other viruses in terms of what host organisms they infect, which for phages are members of domain *Bacteria*. Though more formally described as bacterial viruses, upon their

S. T. Abedon, *Bacteriophages as Drivers of Evolution*,
https://doi.org/10.1007/978-3-030-94309-7_1

discovery they were said to be 'un bacteriophage obligatoire' (d'Herelle 1917). D'Herelle's use of this descriptor presumably was due to the phage ability to macroscopically clear bacterial cultures. This thereby may have given an impression of something 'eating' or 'devouring' bacteria, with the word 'phage' stemming from the Greek word, φαγειν, meaning just that (Stent 1963). A second and third category of viruses exist. These are the viruses that infect eukaryotic organisms, that is, the viruses of *domain* Eukarya, and the viruses that infect members of domain *Archaea*, the latter which like bacteria are also prokaryotes.

Also for historical reasons, viruses of archaea came to be predominantly called archaeal viruses rather than phages (Abedon and Murray 2013). Archaeal viruses nevertheless presumably fill similar niches to those of bacteriophages, ones of predominantly infecting relatively small, single-celled organisms. Thereby, at least to a first approximation, archaeal viruses may have similar impacts on the evolution of their hosts as do bacteriophages. Indeed, a majority of archaeal virus isolates morphologically resemble the majority of known bacteriophages, i.e., as so-called head-tailed viruses, which are members of the virus order *Caudovirales* (Abedon 2018; Lehman 2018). Much of what I've written here nonetheless is explicitly from the perspective of phages rather than from that of archaeal viruses.

Phages also represent the archetype of what can be described as viruses of microorganisms, as includes not just also archaeal viruses but viruses of single-celled eukaryotes as well (Hyman and Abedon 2012, 2018). Phages as drivers of evolution, particularly of bacterial evolution, to some degree thus may be similar in principle to viruses of eukaryotes serving as drivers of the evolution of eukaryotic microorganisms, or the noted archaeal viruses as drivers of the evolution of archaea. Nevertheless, and reflecting the focus of this monograph on *phages* as drivers of evolution, in this introductory chapter I provide a general overview of just phages.

1.1 Introduction to Phages and Their Biology

There are at least 11 morphologically distinct phage families (Lehman 2018) (Fig. 1.1). My preference is to differentiate these families into those with smaller, single-stranded genomes; those with medium-length, double-stranded genomes; and those with a combination of larger, double-stranded DNA genomes and head-tail morphologies, the latter being the members of virus order *Caudovirales* (Abedon 2009; Hyman and Abedon 2012). In addition to their virion characteristics (Sanz-Gaitero et al. 2019), phages also can be distinguished (Fig. 1.2) in terms of whether they lyse bacteria (Dennehy and Abedon 2021c), lysogenize bacteria and thereby exist for part of their life cycles as prophages (Howard-Varona et al. 2017; Liu et al. 2020; Łoś et al. 2021), how their infections of bacteria are initiated (Dennehy and Abedon 2021a), what receptors they use for adsorption (Silva et al. 2016; Hyman 2017), and abilities to move DNA between bacteria (genetic transduction, or simply 'transduction') (Penades et al. 2015; Chiang et al. 2019; Schneider 2021).

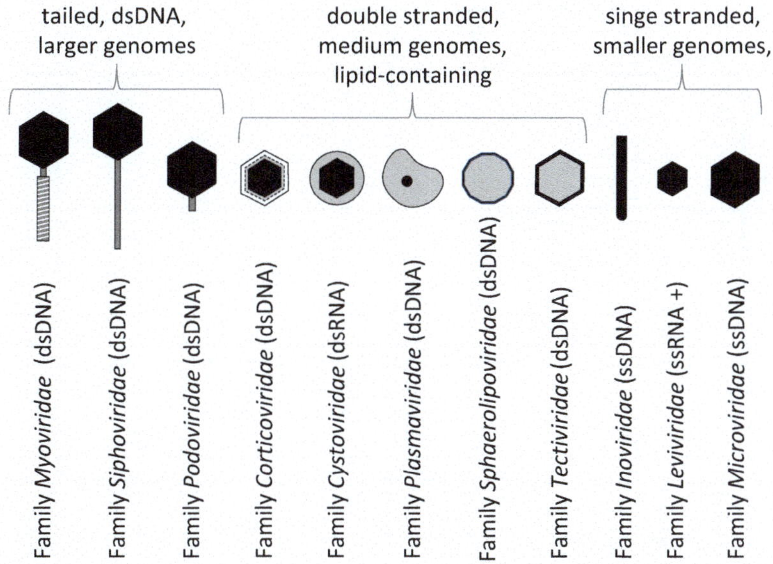

Fig. 1.1 Families of bacterial viruses with their basic virion morphologies and genome characteristics. Black indicates capsids, light gray lipids, and darker gray tails. Sizes of individual virions are not drawn to scale. The figure is adapted from those of Ackermann (2006), Hyman and Abedon (2012), and Lehman (2018). Note, though, that phages recently have been distinguished into additional taxa based on genetic rather that morphological characteristics

I generally also like to classify phage infections into ones that are 'Productive' versus 'Reductive' or 'Destructive' (Abedon 2020) (Fig. 1.3). Productive infections *produce* virions, reductive infections include lysogenic cycles, i.e., as infecting phages are "'Reduced' into prophage" (Lwoff 1953), and the term destructive here refers to a loss of phage viability upon infection (phages are *destroyed* by bacteria). This is, rather than destruction of the infected bacterium, though all lytic productive and also many phage-destructive infections result as well in loss of viability of phage-infected bacteria. For phage-destructive infections, those that are also bacteria-killing notably can be described as abortive. Reductive along with non-abortive but still phage-destructive infections by contrast do not result in a loss of bacterial viability.

1.2 Most Phages Lyse Their Hosts to Release Virion Progeny

In the course of their productive infections, we can distinguish phages into those that are lytic versus those that are not (Figs. 1.2 and 1.3). Lytic viruses release their virion progeny by lysing their host bacteria. This lysis has at least three effects:

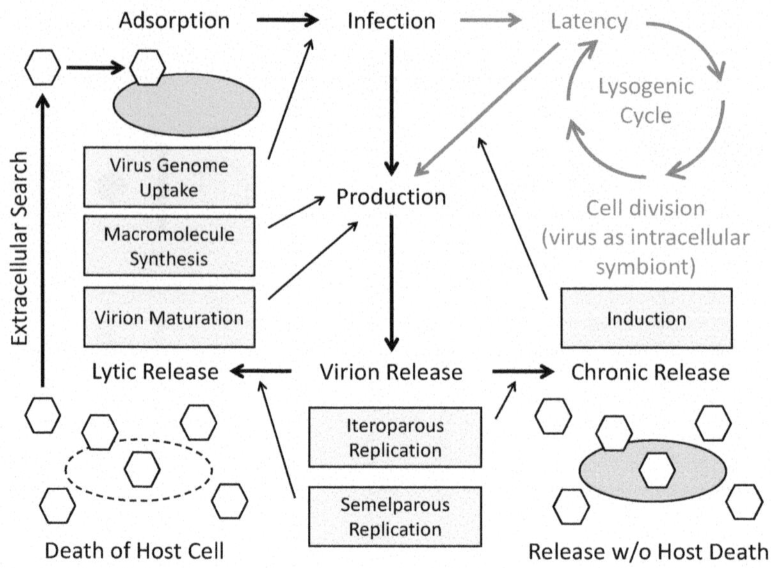

Fig. 1.2 Variations on phage life cycles. Productive cycles are shown in black while latent cycles are shown in gray. Note that chronically releasing phages do not possess icosahedral virions despite their representation as such in the figure

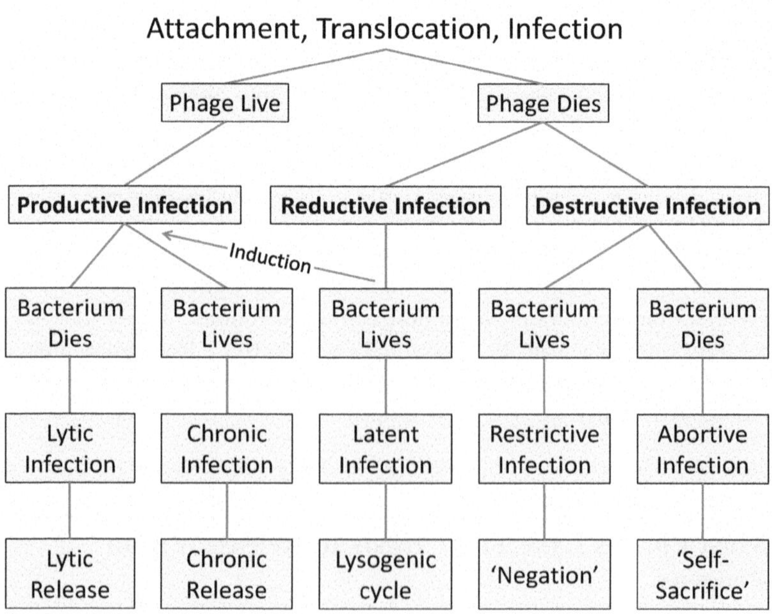

Fig. 1.3 Productive, reductive, and destructive phage infections and their various outcomes. This figure is derived from a figure by Dennehy and Abedon (2021c)

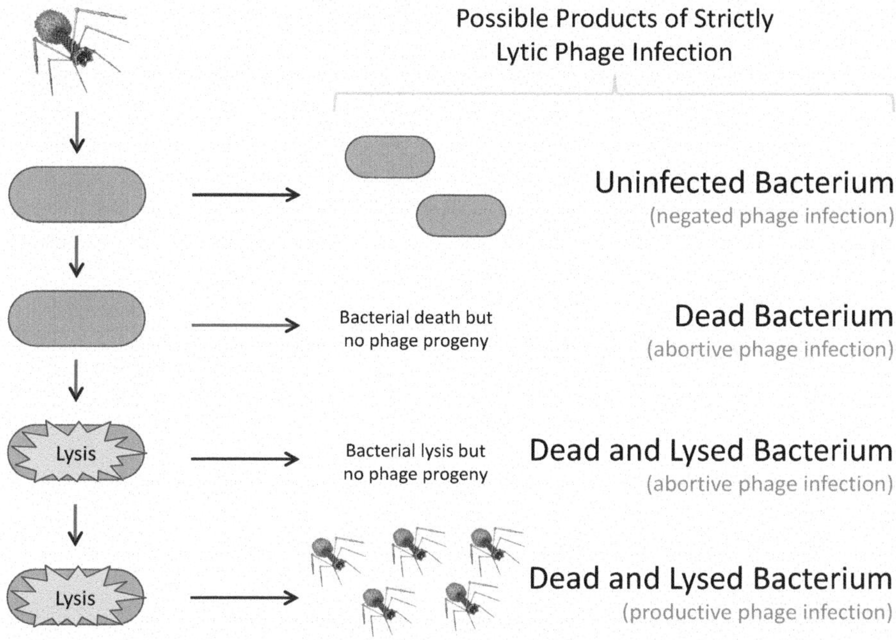

Fig. 1.4 Variations on products of infections by strictly lytic phages. Depending on the genetics of the phage, the genetics of the hosting bacterium, and environmental circumstances, not all such infections will be successful, i.e., as resulting in the production and release of new, progeny virion particles. Instead, it is possible for the host bacterium to block phage virion production, with the host either surviving and continue to reproduce (here described as a negated phage infection; Sects. 18.3 and 19.2) or not surviving and therefore not producing new, daughter bacteria (here described as an abortive or bacterial self-sacrificial phage infection; Sects. 18.4 and 20.1). This figure is derived from that of Danis-Wlodarczyk et al. (2020)

(1) termination of the phage infection (which is physiological), (2) release of virion progeny into the extracellular environment (which is ecological), and (3) a partial solubilizing and releasing of non-phage contents of those bacteria (which also is ecological) (Dennehy and Abedon 2021b, c). Along the way, physiologically as well as ecologically, and hugely important in terms of phages acting as drivers of evolution, the phage-infected bacterium is killed (Fig. 1.4).

Non-lytic phages, by contrast, release their virions chronically, without either killing or otherwise destroying or solubilizing their host bacteria. The vast majority of known phage types are lytic, however, with only two phage families containing members that instead are chronically released. Those are members of phage family *Inoviridae* (a smaller, single-stranded type of DNA virus) (Russel and Model 2006) and members of phage family *Plasmaviridae* (a medium-sized, non-tailed, indeed capsid-lacking, also DNA virus) (Maniloff and Dybvig 2006) (Fig. 1.1).

Phages that can only successfully infect lytically, rather than also lysogenically or instead chronically, we can describe as strictly lytic or obligately lytic. That latter also can be described as 'virulent'. These designations, however, come with the caveat that certain phages can be strictly lytic while successfully infecting some bacterial hosts but not strictly lytic while successfully infecting other hosts (Campbell 1961). As a consequence of that caveat, we can describe strictly lytic phages that are strictly lytic on *all* hosts as professionally lytic (Hobbs and Abedon 2016), though here we won't worry much about that distinction, instead simply assuming for the sake of convenience that all strictly lytic phages are strictly lytic while infecting all hosts. We thus can differentiate a large majority of phages into two lytically infecting categories, those that are always strictly lytic and those that instead decide each time they adsorb a bacterium to infect either lytically or sometimes instead lysogenically. The latter are introduced in the following section (Sect. 1.3). In addition, and as also may be true for archaeal viruses as well (above), it is thought that a majority of phage types are head-tailed, and all head-tailed phages both contain large, dsDNA genomes and are lytic (Fig. 1.1). These are members of the morphological families *Myoviridae*, *Siphoviridae*, and *Podoviridae*, as illustrated on the right of that figure.

1.3 Some Variations on the Concept of Lysogeny

Lysogeny is a semi-permanent association between phages and their bacterial hosts. A more general term for the associated lysogenic phage infections, as applied to non-phage viruses, is latent infection, though this latter concept is not to be confused with that of 'latent period' (Abedon 2022). The associated 'lysogenic cycles' (Bertani 1953) begin with phage virion attachment to a phage-susceptible host bacterium and then the infection of that bacterium. The resulting phage genome is then converted into what is known as a prophage and this is rather than directly giving rise to a productive infection. Prophages are able to maintain themselves within the infected bacterium either by integrating into the host chromosome or instead by serving as a replicating plasmid (Fig. 1.5). These infections, however, can be converted into productive cycles via a process known as induction (Fig. 1.2), which can either occur spontaneously or instead can be forced. It is prior to induction that the resulting lysogenized bacteria are described as lysogens.

Roughly half of tailed bacteriophages are temperate—that is, able to display latent infections as well as productive infections—and temperate phages are also found in phage families *Inoviridae* and *Plasmaviridae* (Maniloff et al. 1994; Ackermann 2006; Roux et al. 2019). That is, there are phages which are not lytic but which still form prophages. Furthermore, roughly half of bacteria have been thought to possess prophages (Paul 2008; Touchon et al. 2016), though that at least in marine environments could be an overestimation (Tuttle and Buchan 2020). These fractions also are not consistent across locations or conditions (Williamson et al. 2002; Payet and Suttle 2013; Brum et al. 2015). In any case, temperate phages are distinct from non-temperate phages and are distinct especially from strictly lytic phages in that during lysogenic cycles the phage and host bacterial genomes functionally merge over what can be very long time spans. These mergers can be associated with expression of prophage-encoded genes that are both novel and useful to the bacterial host (see the concept of lysogenic conversion, Chap. 15).

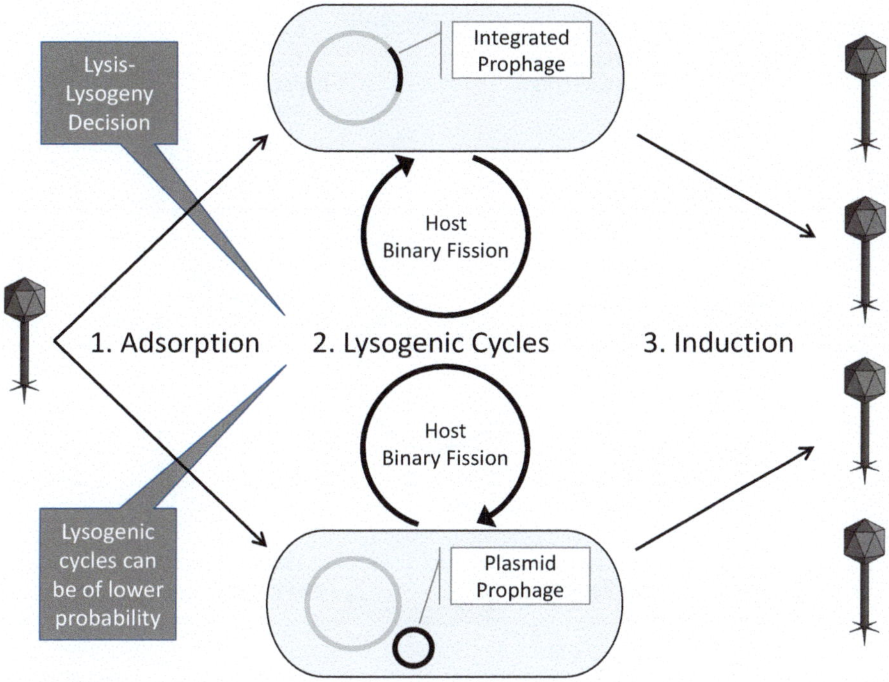

Fig. 1.5 Lysogenic cycles. Shown are lysogenic cycles associated with tailed, lytic, siphoviruses, e.g., as could be phage λ on top or phage N15 on the bottom. Not shown is induction resulting in chronic rather than lytic virion release, e.g., as associated with the temperate inovirus, CTXPhi

1.4 Phage Infections Start with Adsorption

As viruses, phages exist as extracellular virions for at least part of their life cycles, though for temperate phages this extracellular state can be temporarily avoided via their display of lysogenic cycles. Virions diffuse within aqueous or at least well wetted environments, randomly encountering materials with which they can reversibly interact. Those materials can include bacterial surfaces, associated extracellular polymeric substances, and also non-bacterial substances. The latter includes most notably mucus (Barr et al. 2013), but also can include any number of especially organic substances that may be found in their environments (Trubl et al. 2020). These interactions can be specific, e.g., as with bacterial-host molecules, or with the noted mucus, or instead can be non-specific, e.g., as with colloidal substances found in and otherwise making up soils.

Reversible attachment to a potential host bacterium can proceed to an irreversible attachment, which is then followed by phage genome translocation into the bacterial cytoplasm (Dennehy and Abedon 2021a). After genome translocation, then any lysis-lysogeny decisions must be made. That is, for temperate phages a choice must be made to proceed with a productive infection or instead to initiate a lysogenic

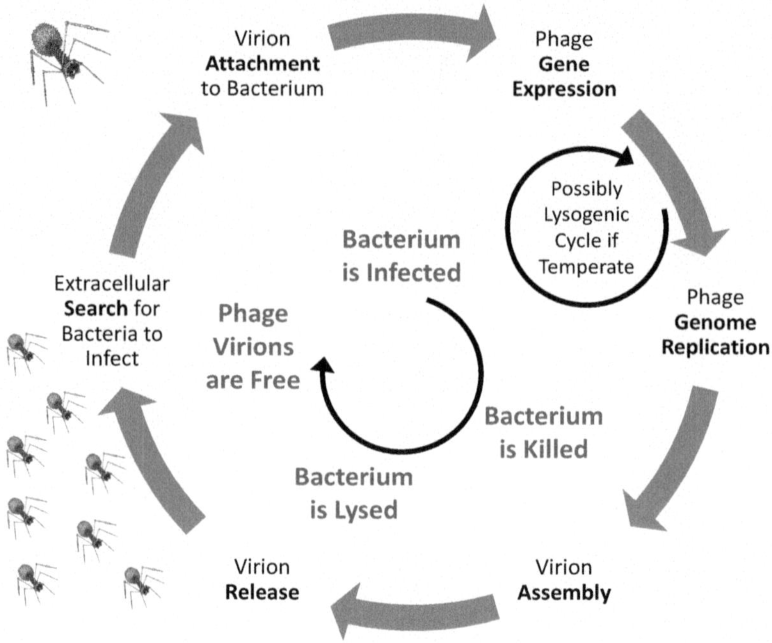

Fig. 1.6 Phage productive cycle with emphasis on a lytic cycle. Shown also is potentially a lysogenic cycle if infection is by a temperate phage. Note that bacteria often are no longer genetically alive (capable of producing new bacteria progeny) prior to their phage-induced bacterial lysis

cycle (Ptashne 2004; Little 2006; Golding 2018). In either case, expression of phage genes occurs following entrance of the phage genome into the bacterial cytoplasm, and this gene expression is either limited (lysogenic cycles; but see Chap. 15) or instead is extensive (productive cycles). In the case of productive cycles, (1) phage genomes are extensively replicated, (2) the resulting progeny genomes are packaged into virions, and (3) the virions then released into the extracellular environment (Dennehy and Abedon 2021c). A phage productive cycle, beginning with virion attachment to a bacterium, is summarized in Fig. 1.6.

1.5 Some Phages Can Transport Bacterial DNA Between Bacteria via Transduction

In the course of productive and perhaps especially lytic phage infections, it is possible—to varying degrees depending on the phage—for bacterial DNA to be packaged into virion particles (Fig. 1.7). This is either along with or instead of phage DNA also being packaged, thereby producing what is known as a transducing particle (Schneider 2021). Transducing particles are able to transfer non-phage

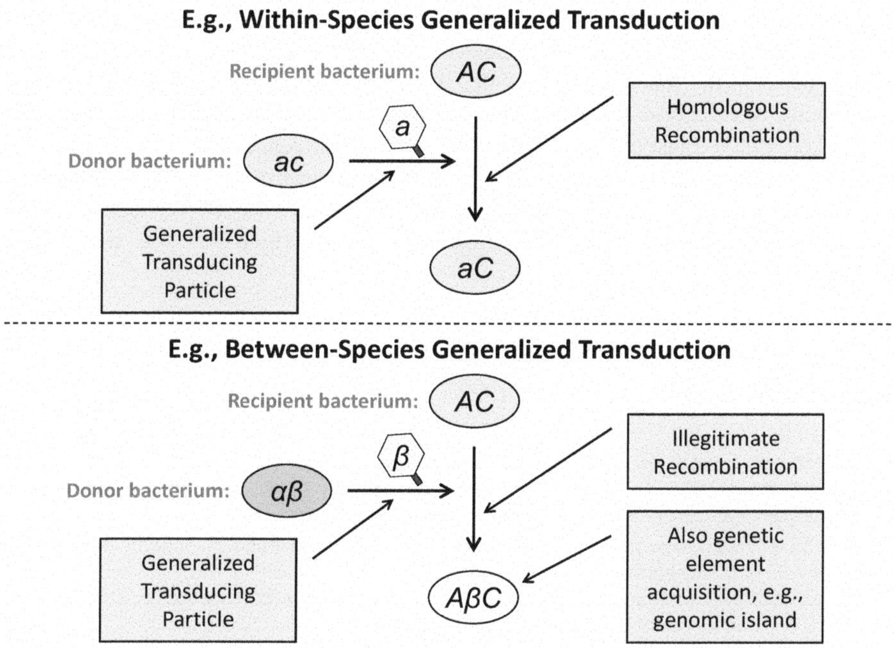

Fig. 1.7 Migration of alleles both within populations (top) and between populations (bottom). Greek letters indicate alleles coming from a different species or at least from different strains of the same species. Homologous recombination results in no net-gain in genes but instead involves a replacement of one allele with another. Illegitimate recombination (Chap. 13, Sect. 13.2) can give rise to a net gain in alleles (here shown as conversion of *AB* genotype to *ABα*). Note in any case how movement of DNA into bacterial cytoplasms is followed by different means of retaining alleles once they have been acquired (recombination, plasmid establishment, etc.)

DNA between bacteria, thereby effecting bacterium-to-bacterium horizontal gene transfer. If phage DNA is also included in the transducing particle, then the resulting mechanism is described as specialized transduction, but see also the concept of phage morons (Chap. 14). If there is no phage DNA, then the mechanism is described instead as generalized transduction (Chap. 13). With specialized transduction, or especially with the carriage of phage morons, there is at least some potential that a transducing particle can give rise to a somewhat normal productive or lysogenic cycle (Fig. 1.2). With generalized transduction, by contrast, the virion is not capable of effecting an infection, with the transducing particles thereby serving solely as carriers of non-phage DNA between bacteria. A novel additional mechanism of transduction, termed lateral transduction, was recently discovered in *Staphylococcus aureus* as mediated by certain temperate phages such as phage 80α (Chen et al. 2018). This results in substantially (10^3-fold) higher levels of transduction of large amounts bacterial DNA that, like generalized transduction, is not packaged in association with phage DNA.

This phage carriage of new DNA into bacteria, regardless of the specific mechanism, is an important driver of bacterial evolution, though one that acts, unlike with lytic infections, only when the delivering phages do not kill transduction-recipient bacteria. Such lack of killing is of greater likelihood the less phage DNA that is also being carried in the transducing particle. It also is more likely the lower the phage multiplicity of infection: the more phages that infect a bacterium at once, then the more likely that a transducing particle will coinfect with a non-transducing, lytically infecting phage—the latter typically produced by phages in the majority—thereby resulting in bacterial lysis rather than horizontal gene transfer between bacteria. Still, transduction is thought to be a dominant means by which DNA is transferred between bacteria.

References

Abedon ST (2009) Phage evolution and ecology. Adv Appl Microbiol 67:1–45
Abedon ST (2018) Viruses of domain Archaea. In: Hyman P, Abedon ST (eds) Viruses of microorganisms. Caister Academic Press, Norwich, pp 167–192
Abedon ST (2020) Phage-phage, phage-bacteria, and phage-environment communication. In: Witzany G (ed) Biocommunication of phages. Springer, pp 23–70
Abedon ST (2022) Evolution of bacteriophage latent period length. In: Dickins TE, Dickens BJA (eds) Evolutionary biology: new perspectives on its development. Springer
Abedon ST, Murray KL (2013) Archaeal viruses, not archaeal phages: an archaeological dig. Archaea 2013:251245
Ackermann H-W (2006) Classification of bacteriophages. In: Calendar R, Abedon ST (eds) The bacteriophages. Oxford University Press, Oxford, pp 8–16
Barr JJ, Auro R, Furlan M, Whiteson KL, Erb ML, Pogliano J, Stotland A, Wolkowicz R, Cutting AS, Doran KS, Salamon P, Youle M, Rohwer F (2013) Bacteriophage adhering to mucus provide a non-host-derived immunity. Proc Natl Acad Sci U S A 110:10771–10776
Bertani G (1953) Lysogenic versus lytic cycle of phage multiplication. Cold Spring Harb Symp Quant Biol 18:65–70
Brum JR, Hurwitz BL, Schofield O, Ducklow HW, Sullivan MB (2015) Seasonal time bombs: dominant temperate viruses affect Southern Ocean microbial dynamics. ISME J 10:437–449
Campbell A (1961) Conditions for the existence of bacteriophages. Evolution; international journal of organic evolution 15:153–165
Chen J, Quiles-Puchalt N, Chiang YN, Bacigalupe R, Fillol-Salom A, Chee MSJ, Fitzgerald JR, Penades JR (2018) Genome hypermobility by lateral transduction. Science (New York, NY) 362:207–212
Chiang YN, Penades JR, Chen J (2019) Genetic transduction by phages and chromosomal islands: the new and noncanonical. PLoS Pathog 15:e1007878
Clokie MR, Millard AD, Letarov AV, Heaphy S (2011) Phages in nature. Bacteriophage 1:31–45
d'Herelle F (1917) Sur un microbe invisible antagoniste des bacilles dysentériques. C R Acad Sci Ser D 165:373–375
Danis-Wlodarczyk K, Dąbrowska K, Abedon ST (2020) Phage therapy: the pharmacology of antibacterial viruses. In: Coffey A (ed) Exploitation of bacteriophages for biocontrol and therapeutics. Caister Academic Press, Norwich
Dennehy JJ, Abedon ST (2021a) Adsorption: phage acquisition of bacteria. In: Harper D, Abedon ST, Burrowes BH, McConville M (eds) Bacteriophages: biology, technology, therapy. Springer, New York, pp 93–117
Dennehy JJ, Abedon ST (2021b) Bacteriophage ecology. In: Harper D, Abedon ST, Burrowes BH, McConville M (eds) Bacteriophages: biology, technology, therapy. Springer, New York, pp 253–294

Dennehy JJ, Abedon ST (2021c) Phage infection and lysis. In: Harper D, Abedon ST, Burrowes BH, McConville M (eds) Bacteriophages: biology, technology, therapy. Springer, New York, pp 341–383
Golding I (2018) Infection by bacteriophage lambda: an evolving paradigm for cellular individuality. Curr Opin Microbiol 43:9–13
Hobbs Z, Abedon ST (2016) Diversity of phage infection types and associated terminology: the problem with 'lytic or lysogenic'. FEMS Microbiol Lett 363:fnw047
Howard-Varona C, Hargreaves KR, Abedon ST, Sullivan MB (2017) Lysogeny in nature: mechanisms, impact and ecology of temperate phages. ISME J 11:1511–1520
Hyman P (2017) Phage receptor. Reference module in life sciences. Elsevier
Hyman P, Abedon ST (2012) Smaller fleas: viruses of microorganisms. Scientifica 2012:734023
Hyman P, Abedon ST (2018) Viruses of microorganisms. Caister Academic Press, Norwich
Lehman SM (2018) Bacteriophage diversity. In: Hyman P, Abedon ST (eds) Viruses of microorganisms. Caister Academic Press, Norwich, pp 145–165
Little JW (2006) Gene regulatory circuitry of phage λ. In: Calendar R, Abedon ST (eds) The bacteriophages. Oxford University Press, Oxford, pp 74–82
Liu Z, Deng Y, Ji M, Sun W, Fan X (2020) Prophages domesticated by bacteria promote the adaptability of bacterial cells. Biocell 44:157
Łoś J, Zielińka S, Krajewska A, Michalina Z, Malachowska A, Kwaśnicka K, Łoś M (2021) Temperate phages, prophages and lysogeny. In: Harper DR, Abedon ST, Burrowes BH, McConville M (eds) Bacteriophages: biology, technology, therapy. Springer, New York, pp 119–150
Lwoff A (1953) Lysogeny. Bacteriol Rev 17:269–337
Maniloff J, Dybvig K (2006) Mycoplasma phages. In: Calendar R, Abedon ST (eds) The bacteriophages. Oxford University Press, Oxford, pp 636–652
Maniloff J, Kampo GJ, Dascher CC (1994) Sequence analysis of a unique temperature phage: mycoplasma virus L2. Gene 141:1–8
Paul JH (2008) Prophages in marine bacteria: dangerous molecular time bombs or the key to survival in the seas? ISME J 2:579–589
Payet JP, Suttle CA (2013) To kill or not to kill: the ballance between lytic and lysogenic viral infection is driven by trophic status. Limnol Oceanogr 58:465–474
Penades JR, Chen J, Quiles-Puchalt N, Carpena N, Novick RP (2015) Bacteriophage-mediated spread of bacterial virulence genes. Curr Opin Microbiol 23:171–178
Ptashne M (2004) Genetic switch: phage lambda revisited. Cold Spring Harbor Laboratory Press, Cold Spring Harbor, NY
Roux S, Krupovic M, Daly RA, Borges AL, Nayfach S, Schulz F, Sharrar A, Matheus Carnevali PB, Cheng JF, Ivanova NN, Bondy-Denomy J, Wrighton KC, Woyke T, Visel A, Kyrpides NC, Eloe-Fadrosh EA (2019) Cryptic inoviruses revealed as pervasive in bacteria and archaea across Earth's biomes. Nat Microbiol 4:1895–1906
Russel M, Model P (2006) Filamentous bacteriophages. In: Calendar R, Abedon ST (eds) The bacteriophages. Oxford University Press, Oxford, pp 146–160
Sanz-Gaitero M, Seoane-Blanco M, van Raaij MJ (2019) Structure and function of bacteriophages. In: Harper D, Abedon ST, Burrowes BH, McConville M (eds) Bacteriophages: biology, technology, therapy. Springer
Schneider CL (2021) Bacteriophage-mediated horizontal gene transfer: transduction. In: Harper DR, Abedon ST, Burrowes B, McConville M (eds) Bacteriophages: biology, technology, therapy. Springer, New York, pp 151–192
Silva JB, Storms Z, Sauvageau D (2016) Host receptors for bacteriophage adsorption. FEMS Microbiol Lett 363:fnw002
Stent GS (1963) Molecular biology of bacterial viruses. WH Freeman and Co., San Francisco, CA

Touchon M, Bernheim A, Rocha EP (2016) Genetic and life-history traits associated with the distribution of prophages in bacteria. ISME J 10:2744–2754
Trubl G, Hyman P, Roux S, Abedon ST (2020) Coming-of-age characterization of soil viruses: a user's guide to virus isolation, detection within metagenomes, and viromics. Soil Sys 4:23
Tuttle MJ, Buchan A (2020) Lysogeny in the oceans: lessons from cultivated model systems and a reanalysis of its prevalence. Environ Microbiol 22:4919–4933
Williamson SJ, Houchin LA, McDaniel L, Paul JH (2002) Seasonal variation in lysogeny as depicted by prophage induction in Tampa Bay, Florida. Appl Environ Microbiol 68:4307–4314

Chapter 2
A Closer Overview of Phage Infections

Contents

Abstract Phage life cycles can be differentiated into adsorption, infection, and release stages. Not all adsorptions progress to infections, however, with infection defined here as beginning at the point of phage nucleic acid entry into the bacterial cytoplasm. A basic distinction that can be made between different types of successful, not phage-destructive infections is whether they do or do not lead directly to the production and release of virions (virion-productive versus reductive infections) and then, if virions are produced and released, how that release takes place. In terms of these latter, productive infections, release can occur either lytically or chronically. This chapter takes a closer look at some of the biology of lytically virion-releasing phage infections as well as chronically virion-releasing phage infections, particularly as relevant to understanding phages as drivers of bacterial evolution. Additional and greater emphasis, however, is placed on gaining an appreciation of the biology of lysogenic cycles including from ecological and evolutionary perspectives.

> When a virulent phage infects a bacterium susceptible to it, the cell dies, produces more phage, and lyses. When a temperate phage infects a sensitive cell, the result is sometimes the same as with a virulent phage. However, in a fraction of the infected cells, a different pathway is followed.—Allan Campbell (1961, p. 153)

In this chapter, we take a closer look at the different types of phage infections, particularly phage infections that are successful, i.e., ones that are not phage-destructive (Fig. 1.3). Specifically, these are lytic, chronic, or lysogenic infections and this is rather than infections that are either infecting-phage 'negating' or in which bacteria display self-sacrificing in order to destroy the infecting phage. The latter are

S. T. Abedon, *Bacteriophages as Drivers of Evolution*,
https://doi.org/10.1007/978-3-030-94309-7_2

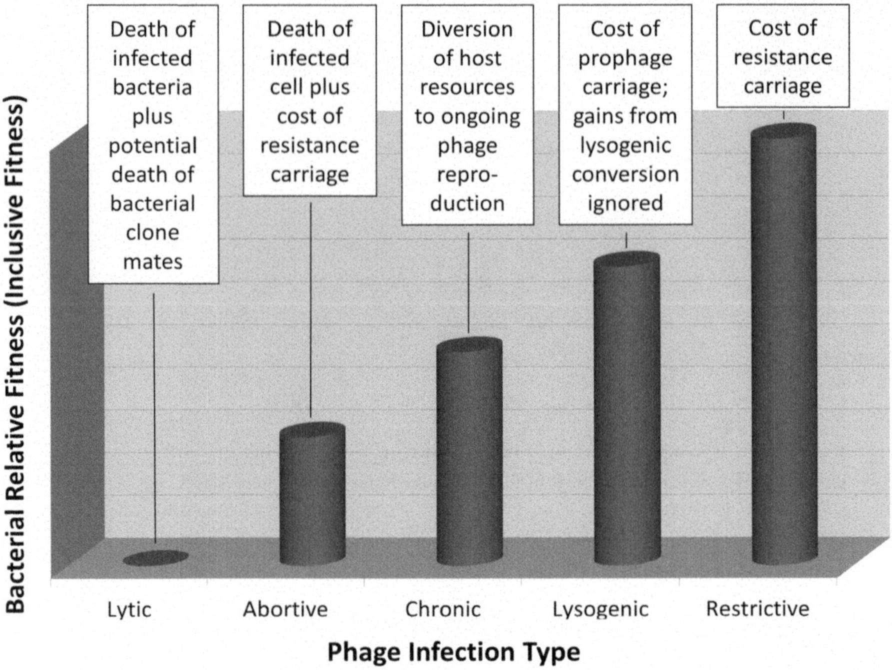

Fig. 2.1 Consideration of the costs of various phage infection types on bacteria. Costs are indicated in terms of bacterial Darwinian fitness, though this is provided in terms of qualitative differences rather than with specific quantitative differences. That is, the higher the cost of a given infection type, or resistance mechanism carriage, then the lower a bacterium's relative fitness. Furthermore, relative fitnesses are presented including in terms of inclusive fitness, meaning that what is being considered is not just the costs to individual bacteria but, if present, costs also to entire clonal bacterial populations. By way of example, a lytic, productive infection not only kills the infected bacterium, but generates phages which can go on to kill clonally related bacteria found within the same environment. The primary argument, then, is that bacteria allowing lytic phage population growth can have a large negative impact also on bacterial inclusive fitness, assuming that bacterial numbers are sufficient to support substantial phage population growth, whereas blocking phage population growth either with or without the infected cell dying ("abortive" versus "restrictive" infection, respectively) is preferable. Not shown is adsorption resistance (interference with phage adsorption), which can have varied costs depending on how a phage's receptor molecule has been modified along with how important that molecule is to the displaying bacterium within a given environment. Also not indicated is transduction, which occasionally should be able to actually increase a recipient-bacterium's fitness

discussed instead in Chaps. 18, 19, and 20. Also among unsuccessful phage infections are those that are initiated by substantially defective phages, including phages effecting transduction (see Chaps. 10 and 13 for the latter along with as discussed in the previous chapter, Chap. 1). To bring us briefly back to the theme of this monograph, which is on the impact of phages on evolution rather than strictly on considerations of phage biology, see Figure 2.1 for an approximation of the relative impacts of these different phage infection types on bacterial evolutionary fitness (the latter a concept introduced in Chap. 3, Sect. 3.3).

2.1 Lytic Cycles Kill Bacteria but also Produce New Phage Virions

The primary as well as most familiar means of phage productive infection is the lytic cycle. Lytic cycles are so named because they end with a phage-induced lysis of the phage-infected bacterium. Technically, this lysis can be described as a lysis from within (Young 1992) to distinguish it from much less common, as well as less relevant to us here, lysis from without (Abedon 2011). The reason that lysis from within is called lysis from within is that phage proteins involved in this process are generated *within* the phage-infected bacterium, and then act starting from there on the bacterial cell envelope, i.e., digesting the bacterial cell wall from *within*. Lysis from without, by contrast, involves the action of cell-envelope degrading enzymes as supplied or applied from outside of the bacterium (Danis-Wlodarczyk et al. 2021).

The processes involved in phage lytic cycles can be quite complex, and will vary in details—often by substantial amounts—from phage type to phage type. The important components, however, are that phage genomes, following phage adsorption, reach the bacterial cytoplasm in the course of the process of genome translocation or uptake. Phage gene expression then occurs, and the metabolism of the hosting bacterium is directed to varying degrees toward producing new phages. Then, after an often well-defined interval, phage virions are released in the course of breakdown of the host-cell envelope, i.e., in the course of phage-induced bacterial lysis from within (Dennehy and Abedon 2021).

Lysis terminates the phage infection, terminates the production of new virions by the phage infection, terminates cell metabolic activities, releases phage virions into the extracellular environment so that they might disseminate to infect new bacteria, initiates the process of decay of the phage-infected bacterium, releases soluble or colloidal materials previously found within the confines of the bacterial cell (including bacterial enzymes as well as phage-encoded enzymes), and, as part of the latter, releases as well any DNA that was still found within the bacterium. The degree to which this released, naked DNA—thus becoming environmental DNA or eDNA—is of bacterial origin depends on the phage, as different lytic phages break down the bacterial chromosome to varying degrees in the course of lytic infections. This host DNA degradation is exemplified especially by the action during *Escherichia coli* infections of the large-genomed myovirus, bacteriophage T4, though phage T4 infections do not result in complete degradation of all host DNA (Kutter et al. 2018). By contrast, genome degradation is not seen during infections by the large-genomed *Bacillus subtilis* myovirus, phage SPO1 (Yehle and Ganesan 1972). See also the results of Keen et al. (2017) as discussed in Chap. 12 (Sect. 12.3.2).

Generally, here when considering generic phages, it will often be lytic phages and lytic cycles that I have in mind. This is unless I am explicitly referring to lysogenic cycles—though typically as still caused by lytic phages—or instead referring to chronic infection cycles. Thus, though at times I will generalize to describe 'productive' cycles, for most phages these productive cycles will be lytic even if that is not stated explicitly—I just don't want to be categorically ignoring chronic infection cycles! Nonetheless, what's most relevant here about lytic cycles, i.e., in terms of their impact on evolution rather than on ecology, are (1) that they result in the death of bacteria and thereby typically will strongly select for the evolution in bacteria of phage resistance, (2) that they can be associated with the generation of transducing particles of various kinds, (3) that upon lysis they can release the noted eDNA, and (iv) that lytic cycles (versus chronic virion release) are the typical consequence of induction of prophages to terminate lysogenic cycles, the latter because most temperate phages are also lytic phages rather than chronically releasing phages.

2.2 Some Vagaries of Lysogenic Cycles

Campbell (1996) introduces the concept of temperate phages as follows (p. 2041):

> The defining characteristic of temperate phages is their ability to enter a prophage state, where most of the phage genes are unexpressed and the prophage replicates as part of the host genome. Classically, the presence of the prophage was manifested by the occasional breakdown (either spontaneous or induced) of latency, so that every culture of a prophage-bearing (lysogenic) strain contains some free infectious phage liberated by lysis of a few cells in the culture.

We can extend these ideas in four different directions, which I do as follows. See Fig. 2.2 for summary.

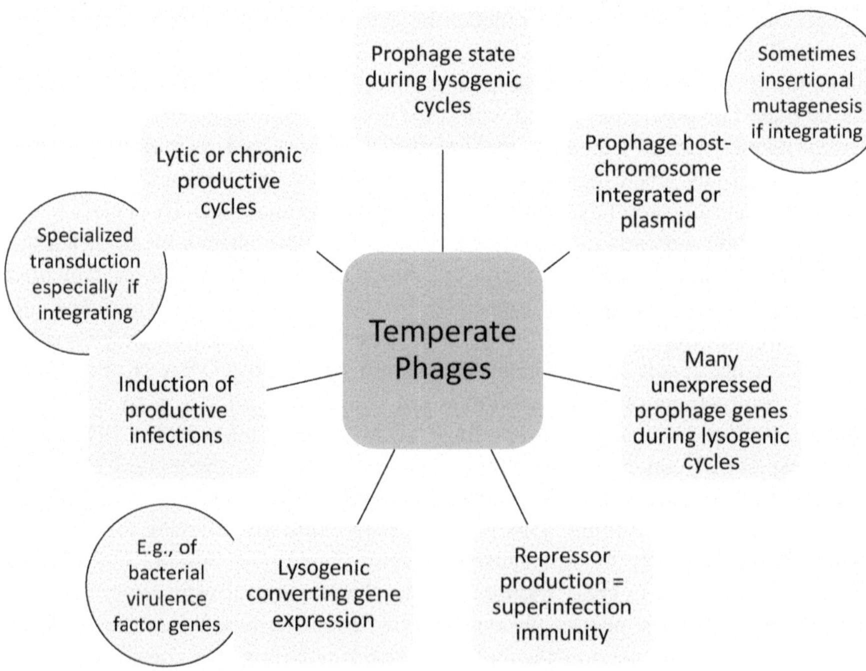

Fig. 2.2 Many aspects of temperate phages and some variations

The first consideration, and as already noted, is that not all temperate phages are also lytic phages (Yamada et al. 2007). For example, phage CTXPhi, the phage that encodes cholera toxin, is a filamentous phage of family *Inoviridae*, which chronically releases progeny virions in the course of its productive cycle rather than lytically. It also happens to be a temperate phage with a prophage that integrates into its host's chromosome, *Vibrio cholerae* (McLeod et al. 2005). Nonetheless, the very term, 'lysogeny', is derived from the potential for bacterial lysogens to occasionally release lytic phage virions, with 'lysogen' presumable being a portmanteau combining 'lysis' and 'generating'. Though technically phage CTXPhi latent infections are not lysis generating, nevertheless they contain phage CTXPhi prophages and are generally described as lysogenic.

Second on this list of extensions of Campbell's description of temperate phages, not all prophages insert into their host chromosome but some instead exist as plasmids during lysogenic cycles (Lehnherr 2006; Ravin 2006; Gilcrease and Casjens 2018) (Fig. 1.5). Such plasmid prophages thus deviate from the textbook description of lysogenic cycles as involving prophage genomes being replicated in conjunction with the bacterial chromosomes within which they are covalently associated. This places limitations on the potential for certain temperate phages to effect specialized transduction, since a phage cannot accidentally pick up bacterial chromosomal DNA during excision following induction if excision by definition never occurs. Presumably being a plasmid rather than integrated also would add an additional step to the domestication of prophage genes by bacteria (Chap. 17) since for a phage gene to become a bacterial chromosomal gene (sans the rest of the prophage), it must first become part of the bacterial chromosome, which is a given following prophage integration but not necessarily a given for prophages that exist instead as plasmids. On the other hand, presumably a plasmid prophage can much more easily become just a plasmid than can prophages which normally are segregated solely as integrated into the bacterial chromosome.

Third, of those relatively few phage genes that are expressed during lysogenic cycles, most notable are those encoding phage repressors, which are proteins that serve to maintain the lysogenic state (Ptashne 2004; Little 2006). Thus, a functional prophage at a minimum exists as the prophage genome (DNA) in combination with prophage repressor gene expression and activity. Phage repressor proteins are also directly responsible for prophage expression of superinfection immunity (Sects. 4.4, 8.1.3, and 16.1 of Chaps. 4, 8, and 16).

Lastly, prophages can express additional genes during lysogenic cycles, ones that can serve to additionally modify, i.e., 'convert', the lysogen's phenotype from that seen absent prophage presence. This gives rise to what is known as lysogenic conversion (Chap. 15). Generally, the concept of converting genes and transduction, including specialized transduction, are not considered to be identical, as converting genes are not thought to be near-term accidental acquisitions of DNA by prophages whereas specialized transduced genes are indeed considered to be near-term accidental acquisitions by prophages (Chap. 14, Sect. 14.2 but also Chap. 13, Sect. 13.1.1).

2.3 Looking at Lysogenic Cycles from Ecological Perspectives

Lysogenic cycles begin at the point of phage adsorption and then continue over the course of cycles of bacterial binary fission. Some fraction of temperate phages upon adsorption enter into lysogenic cycles, versus entering into productive cycles, with what that fraction is depending on the phage, the host, and various conditions. It is certainly an unfortunate exaggeration to assume that most temperate phage adsorptions result in lysogenic cycles, as it is the opposite that generally or at least often is true, with infections by temperate phages resulting in lysogenic cycles for example only about 10% of time (Sinha et al. 2017); see also (Avlund et al. 2009). Thus, for many or most temperate phages, one should think of them as productively infecting phages, e.g., lytic phages, that instead only occasionally infect lysogenically. And this certainly is rather than the unfortunately common but very much incorrect 'lysogenic phage' moniker that one so often hears and reads (Hobbs and Abedon 2016).

Contrasting that virion-centered perspective, temperate phages are often discovered as prophages within lysogens, rather than as free phages (Howard-Varona et al. 2017; Paez-Espino et al. 2019). Hence there exists a discovery bias toward viewing temperate phages from a perspective of lysogenic cycles rather than from a perspective of productive cycles, or instead as free phages. This issue of the frequency of lysogeny versus 'lyticality', however, is more ecological rather than it is evolutionary and hence not of primary concern here. Furthermore, even if lysogenic cycles aren't always strictly the center of a temperate phage's existence, nevertheless in terms of temperate phages versus non-temperate phages as drivers of evolution, clearly lysogenic cycles and their characteristics can be of interest, though temperate phage propensities to generate transducing particles are highly relevant as well (Chap. 13). As evidence for the latter, though certainly not a random sampling, Schneider (2021) in her historical review of transduction focusses solely on temperate phages as generating generalized transducing particles, i.e., phages Mu, P1, and P22 (phage λ is discussed as well, though is not a generalized transducing phage, presumably due to its utilization of *cos* sites for packaging; Chap. 13, Sect. 13.1.2). Potentially non-temperate phages that don't degrade their host's DNA, especially if they display headful packaging, can transduce as well, and that includes even a modified version of phage T4 (Wilson et al. 1979).

An additional ecological consideration is that temperate phages basically serve as the 'perennials' among lytic phages, where the strictly lytic (non-temperate) phages instead are the 'annuals' (see Fig. 2.3 for additional analogies). That is, temperate phage 'life spans' as infections can encompass multiple 'growing seasons' rather than just a single 'growing season', e.g., as may be measured approximately in terms of bacterial host doubling times. This analogy can explain a number of ecological characteristics of lysogenic cycles. For example, as they are not obligated to produce virions every infection, then latently infecting temperate phages presumably can optimize their prophage induction, and thereby virion production and release, for more ideal environmental conditions. They also can produce multiple 'clutches' of

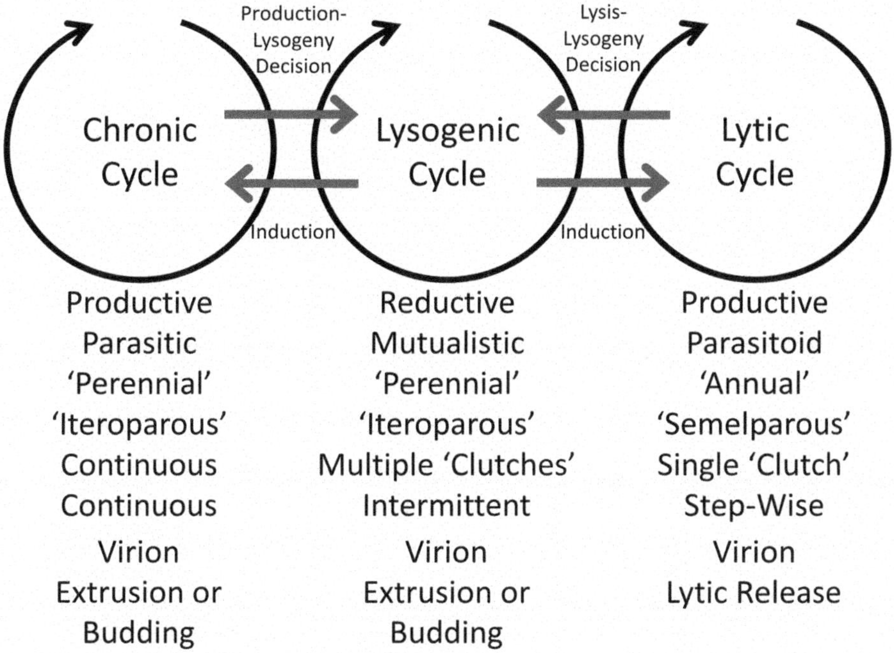

Fig. 2.3 Various perspectives on chronic versus lysogenic versus lytic cycles. A 'clutch' is the product of a single reproductive episode

phages (individual induced productive cycles) per initial adsorption, i.e., as equivalent to per seed sown, rather than just a single clutch (or burst as these are more familiarly described for phages). That resulting 'iteroparity' (Abedon 2022) occurs because of the potential, in the course of bacterial binary fission, for prophages to increase the number of individual bacteria that they are infecting without first going through productive cycles, versus the 'semelparity' of many annual plants as well as obligately lytic phage infections (Gill and Abedon 2003). On the other hand, as temperate phages can display both productive and lysogenic cycles, they also can take advantage, as a population, of any benefits associated with an 'annual' life cycle, which basically means that, like a weed, a temperate phage will retain a potential for more rapid increases in population size than can occur via lysogenic cycles alone.

2.4 Looking at Lysogenic Cycles from Evolutionary Perspectives

There are a number of evolutionary concepts that are associated with lysogenic cycles. Perhaps the most obvious is that, as noted, integrated prophages can effect specialized transduction. Prophages, upon insertion into bacterial chromosomes, also can be inherently mutagenic (insertional mutagenesis; Chap. 5), though some temperate phages are more prone to so mutagenizing bacteria than others. Prophages furthermore serve as symbiotic and indeed potentially mutualistic partners with their bacterial hosts, presumably resulting in coevolution toward improving the 'fit' between bacterial metabolism and the prophage. Prophages particularly can supply the noted lysogenic converting genes (Chap. 15), with phages thereby potentially serving as especially efficient agents of horizontal gene transfer of those genes. Converting gene expression and bacterial metabolisms also can coevolve. It furthermore is possible for temperate phages to acquire host genes, such as during previous infections (Sects. 4.4 and 16.1), which then can serve to replace or augment existing host genes, which also is an aspect of lysogenic conversion (Campbell 1996).

As noted above, because prophage genes often reside on bacterial chromosomes, they in principles may be coopted by the bacterial host toward less phage-associated functions, i.e., the process of prophage gene 'domestication'. The latter is seen with the formation of defective or cryptic prophages (Campbell 1996), i.e., where the prophage decays while certain genes remain, thereby with prophage genes essentially becoming bacterial genes (Chap. 17). Prophages otherwise can be fully or partially deleted from bacteria, what is known as being 'cured' (Stent 1963), thereby potentially leaving behind only random bits of temperate phage genome, or no prophage sequence at all. A nice further quotation from Campbell (1996) is that (p. 2045),

> Cryptic [or defective] prophages have been aptly described as genetic debris that clutters the bacterial genome (Strathern and Herskowitz 1975). From the cell's perspective, all prophages might be considered junk DNA; cryptic prophages appear to be junk from the phage's perspective as well. Although the available information is somewhat anecdotal, perhaps the most noteworthy feature of cryptic prophages is their prevalence.

The prevalence of prophage DNA, of course, is something that has become even more obvious upon routine sequencing of bacterial genomes (Roux et al. 2015; Howard-Varona et al. 2017), though distinguishing defective from non-defective prophages (Fig. 2.4) is not necessarily bioinformatically trivial (Song et al. 2019). Lastly, defective prophages may or may not be inducible, and if inducible may or may not result in the release of particles resembling virions (Fig. 2.4).

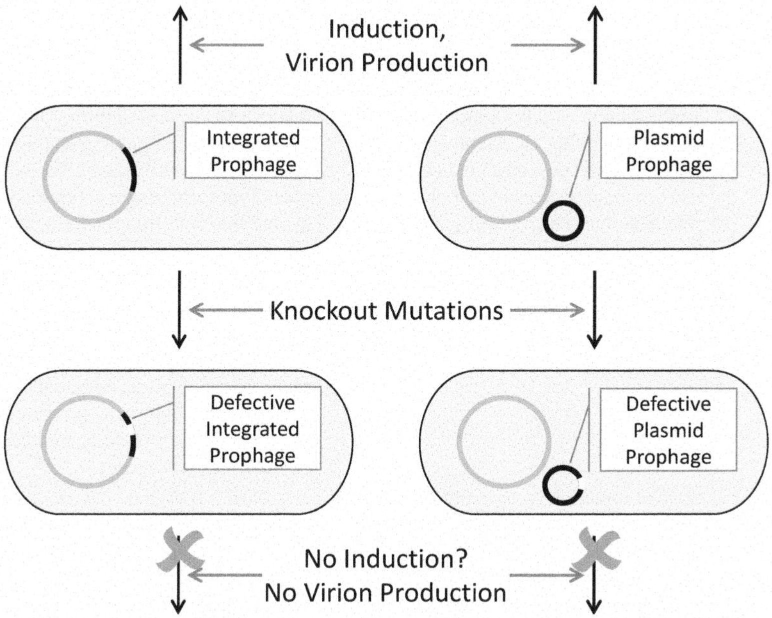

Fig. 2.4 Prophages can be found integrated into existing bacterial DNA, i.e., such as the bacterial chromosome, as follows a site-specific recombination event. Alternatively, prophages can exist as plasmids. Defective prophages possess mutations that knockout especially the ability these prophages to subsequently display productive infections, as potentially can occur due to defects that interfere with any part of the prophage induction process, but also can involve interference in any other aspect of the resulting phage productive cycle

Phage genes also can recombine with host genes. The resulting recombinant bacteria, however, are less likely to persist given DNA acquisition from strictly lytic phages since bacteria subject to lytic infections are, of course, less likely to survive versus bacteria subject to lysogenic infections. Furthermore, due to a mechanism known as superinfection immunity (Sects. 4.4, 8.1.3, and 16.1 of Chaps. 4, 8, and 16), non-prophage temperate phage genomes can be found in relatively inert states within bacterial cytoplasms. Meynell (1969) suggests (p. 113; emphasis mine), though without attribution, that non-replicative plasmid DNA "is believed to be diluted out during subsequent bacterial divisions *just as superinfecting phage segregates in a strain rendered immune by lysogenization by a related phage*." Presumably this retention of a not-integrated, not-replicating phage genome can allow for recombination of aspects of it with the bacterial chromosome (Chap. 17, Sect. 17.3). These otherwise inert phage genomes also in principle can recombine with existing prophages, therefore resulting in modification of temperate phages, and also prophages can recombine with productively infecting phages, thereby modifying those phages as well (Hendrix and Casjens 2008). Indeed, Campbell (1996) concluded his review by noting that (p. 2045), "It is plausible that the most common confrontation allowing generation of new recombinant types

Table 2.1 Lysogeny and its Basic Evolutionary Impacts

Concept	Description	Impact on bacterial evolution
Prophage	Genetic form taken by a phage upon establishment of a lysogenic cycle. Prophages can either be established in the course of insertion into the bacterial chromosome or instead can persist as plasmids that are physically separate from the bacterial chromosome	Prophages supply to bacteria new genes that can be expressed during lysogenic cycles, thereby supplying superinfection immunity or lysogenic conversion phenotypes. Prophage genes also can evolve into becoming strictly bacterial genes rather than phage genes
Integration	This is the result of insertion of the prophage genome directly into bacterial DNA	As prophages genetically decay, the genes of integrated prophages that are useful to bacteria can become bacterial genes, as those genes are already found physically within the bacterial chromosome. Alternatively, the process of integration in some cases can disrupt existing bacterial genes
Lysogenic cycle	State of infection by a prophage, as more generally can be described as a latent phage infection. Latent phage infections can give rise to productive phage infections either spontaneously or upon exposure to inducing agents	It is during lysogenic cycles that prophages both become genetically part of the host bacterium and can supply new functions to their host bacteria, such as via lysogenic conversion
Induction	Process by which a latent phage infection is converted to a virion-productive phage infection, resulting in either lytic or chronic virion release, depending on the phage type	As a default we should assume that induction would be selectively detrimental to the host bacterium as it should directly lower host fitness, particularly given lytic virion release. It is possible, however, that some, especially population-level benefits may be associated with induction such as allelopathic effects against competing bacteria

is infection by an active phage of a strain harboring a related cryptic [or defective] prophage." For summary of concepts associated with lysogeny and their impacts on bacterial evolution, see Table 2.1.

2.5 Chronic Cycles Release New Phage Virions Without Killing Host Bacteria

Chronic release of phage virions is seen in only two phage families, family *Inoviridae* and family *Plasmaviridae*. The former is the familiar filamentous phage virion, with phage M13 a prominent example. As plasmaviruses are so rare, inoviruses will be the primary emphasis of this section. All inoviruses possess ssDNA genomes for which the final stages of encapsidation occur over the course of their release from bacteria. Notably, that release neither terminates the phage infection nor kills the infected bacterium. Also, infected bacteria continuously release virions into the extracellular environment over time. Indeed, from Horiuchi (1983, p. 403): "The infected cell is not killed, but continues to grow at a reduced rate while secreting about 500 progeny phage per generation... the infected cell becomes resistant to superinfection... Due to the unique filamentous morphology of the virion, there is little, if any, constraint on the length of DNA to be packaged." Of interest, some evolved strains of phage M13 are able to actually improve the fitness of the bacterial hosts they are infecting (Shapiro and Turner 2018).

Just as with lytic phages, some inoviruses can also infect latently. These, for consistency with the phage literature, I also describe as lysogenic, even though no 'lysis generation' should be observed. Thus, in many cases where lysogenic cycles are relevant, then potentially those lysogenic cycles may be associated with chronically releasing phages rather than lytically releasing phages. For example is the above-noted phage CTXPhi, the phage that encodes cholera toxin. Keep in mind that the lysogenic cycles of these phages still remain distinct from their productive cycles, i.e., virion production occurs upon prophage induction rather than while still existing as prophages. That is, it is not correct to equate lysogenic cycles with chronically releasing productive cycles, and indeed chronically releasing phages that do not display latent infections are not temperate phages (e.g., again phage M13). It is also not reasonable to suggest that the interactions of chronically releasing phages such as phage M13 with bacteria should be viewed as forms of predation.

Presumably, given extracellular breakdown of Inovirus capsids, then phage eDNA may be released. Perhaps these viruses can in some manner also transduce bacterial DNA between cells other than in terms of lysogenic conversion. By and large, however, most of the emphasis of this monograph is on phages that are lytically rather than chronically released, whether associated with strictly lytic or instead with temperate phages that release their virions lytically.

References

Abedon ST (2011) Lysis from without. Bacteriophage 1:46–49

Abedon ST (2022) Evolution of bacteriophage latent period length. In: Dickins TE, Dickens BJA (eds) Evolutionary biology: new perspectives on its development. Springer

Avlund M, Dodd IB, Semsey S, Sneppen K, Krishna S (2009) Why do phage play dice? J Virol 83: 11416–11420

Campbell A (1961) Conditions for the existence of bacteriophages. Evol Internat J Organ Evol 15: 153–165

Campbell AM (1996) Cryptic prophages. In: Neidhardt FC (ed) *Escherichia coli* and *Salmonella* cellular and molecular biology. ASM Press, Washington, DC, pp 2041–2046

Danis-Wlodarczyk KM, Wozniak, DJ, Abedon ST (2021) Treating bacterial infections with bacteriophage-based enzybiotics: in vitro, in vivo and clinical application. Antibiotics 10:1497

Dennehy JJ, Abedon ST (2021) Phage infection and lysis. In: Harper D, Abedon ST, Burrowes BH, McConville M (eds) Bacteriophages: biology, technology, therapy. Springer, New York, pp 341–383

Gilcrease EB, Casjens SR (2018) The genome sequence of *Escherichia coli* tailed phage D6 and the diversity of *Enterobacteriales* circular plasmid prophages. Virology 515:203–214

Gill JJ, Abedon ST (2003) Bacteriophage ecology and plants. APSnet feature. https://www.apsnet.org/edcenter/apsnetfeatures/Documents/2003/BacteriophageEcology.pdf

Hendrix RW, Casjens SR (2008) The role of bacteriophages in the generation and spread of bacterial pathogens. In: Hensel M, Schmidt H (eds) Horizontal gene transfer in the evolution of pathogenesis. Cambridge University Press, Cambridge, UK, pp 79–112

Hobbs Z, Abedon ST (2016) Diversity of phage infection types and associated terminology: the problem with 'lytic or lysogenic'. FEMS Microbiol Lett 363:fnw047

Horiuchi K (1983) Co-evolution of a filamentous bacteriophage and its defective interfering particles. J Mol Biol 169:389–407

Howard-Varona C, Hargreaves KR, Abedon ST, Sullivan MB (2017) Lysogeny in nature: mechanisms, impact and ecology of temperate phages. ISME J 11:1511–1520

Keen EC, Bliskovsky VV, Malagon F, Baker JD, Prince JS, Klaus JS, Adhya SL (2017) Novel "superspreader" bacteriophages promote horizontal gene transfer by transformation. MBio 8: e02115–e02116

Kutter E, Bryan D, Ray G, Brewster E, Blasdel B, Guttman B (2018) From host to phage metabolism: hot tales of phage T4's takeover of *E. coli*. Viruses 10:387

Lehnherr H (2006) Bacteriophage P1. In: Calendar R, Abedon ST (eds) The bacteriophages. Oxford University Press, Oxford

Little JW (2006) Gene regulatory circuitry of phage λ. In: Calendar R, Abedon ST (eds) The bacteriophages. Oxford University Press, Oxford, pp 74–82

McLeod SM, Kimsey HH, Davis BM, Waldor MK (2005) CTXϕ and *Vibrio cholerae*: exploring a newly recognized type of phage-host cell relationship. Mol Microbiol 57:347–356

Meynell GG (1969) Exclusion, superinfection immunity and abortive recombinants in I+ × I+ bacterial crosses. Genet Res 13:113–115

Paez-Espino D, Roux S, Chen IA, Palaniappan K, Ratner A, Chu K, Huntemann M, Reddy TBK, Pons JC, Llabres M, Eloe-Fadrosh EA, Ivanova NN, Kyrpides NC (2019) IMG/VR v.2.0: an integrated data management and analysis system for cultivated and environmental viral genomes. Nucl Acids Res 47:D678–D686

Ptashne M (2004) Genetic switch: phage lambda revisited. Cold Spring Harbor Laboratory Press, New York

Ravin NV (2006) N15: the linaer plasmid prophage. In: Calendar R, Abedon ST (eds) The bacteriophages. Oxford University Press, Oxford, pp 448–456

Roux S, Enault F, Hurwitz BL, Sullivan MB (2015) VirSorter: mining viral signal from microbial genomic data. PeerJ 3:e985

Schneider CL (2021) Bacteriophage-mediated horizontal gene transfer: transduction. In: Harper DR, Abedon ST, Burrowes B, McConville M (eds) Bacteriophages: biology, technology, therapy. Springer, New York, pp 151–192

Shapiro JW, Turner PE (2018) Evolution of mutualism from parasitism in experimental virus populations. Evoltion 72:707–712

Sinha V, Goyal A, Svenningsen SL, Semsey S, Krishna S (2017) *In silico* evolution of lysis-lysogeny strategies reproduces observed lysogeny propensities in temperate bacteriophages. Front Microbiol 8:1386

Song W, Sun HX, Zhang C, Cheng L, Peng Y, Deng Z, Wang D, Wang Y, Hu M, Liu W, Yang H, Shen Y, Li J, You L, Xiao M (2019) Prophage hunter: an integrative hunting tool for active prophages. Nucl Acids Res 47:W74–W80

Stent GS (1963) Molecular biology of bacterial viruses. WH Freeman and Co., San Francisco, CA

Strathern A, Herskowitz I (1975) Defective prophage in *Escherichia coli* K12 strains. Virology 67: 136–143

Wilson GG, Young KKY, Edlin GJ, Konigsberg W (1979) High-frequency generalized transduction by bacteriophage T4. Nature (London) 280:80–82

Yamada T, Kawasaki T, Nagata S, Fujiwara A, Usami S, Fujie M (2007) New bacteriophages that infect the phytopathogen *Ralstonia solanacearum*. Microbiology 153:2630–2639

Yehle CO, Ganesan AT (1972) Deoxyribonucleic acid synthesis in bacteriophage SPO1-infected *Bacillus subtilis*. I. Bacteriophage deoxyribonucleic acid synthesis and fate of host deoxyribonucleic acid in normal and polymerase-deficient strains. J Virol 9:263–272

Young R (1992) Bacteriophage lysis: mechanisms and regulation. Microbiol Rev 56:430–481

Chapter 3
Evolutionary Biology Basics

Contents

Abstract This chapter introduces some basic concepts of evolutionary biology, particularly as aimed toward the non-specialist and as are relevant to appreciating the evolutionary ecology of phages as drivers of evolution, especially the evolution of bacteria. Emphasis is placed on understanding the four basic mechanisms by which the frequencies of alleles can be modified within populations over time, that is, the four basic drivers of evolutionary change. These are mutation, genetic drift, migration, and natural selection. Mutations are changes in nucleic acid sequences, genetic drift is the consequence of sampling error, migration is the movement of alleles from one population into another, and natural selection represents biased as well as deterministic changes in allele frequencies over time. Natural selection generally has a strong environmental component and also is the emphasis of evolutionary ecology as a scientific discipline, as natural selection serves as the primary means by which adaptations increase in frequency within populations. Discussed as well in this chapter are the concepts of Darwinian fitness, relative fitness, and absolute fitness.

> . . .it is only in very recent years that the study of phage evolution has attracted the attention of more than a handful of individuals.—Roger W. Hendrix (2008, p. 177)

In this monograph, I consider the numerous ways that bacteriophages can impact the evolution of other organisms, particularly the evolution of their bacterial hosts. This impact, I argue, spans the four phenomena that are known to impact allele

S. T. Abedon, *Bacteriophages as Drivers of Evolution*,
https://doi.org/10.1007/978-3-030-94309-7_3

frequencies within populations, where change in allele frequencies over time is the standard population geneticist's definition of evolution. These phenomena, in the order that they are considered here, are (1) mutation, (2) genetic drift, (3) migration—of genes, such as seen with horizontal gene transfer, i.e., genetic migration—and (4) natural selection. Appreciating the idea that bacteriophages can drive evolutionary change requires gaining some familiarity with both what phages consist of (Chaps. 1 and 2) and how evolutionary processes are thought to take place (this chapter). Though our concerns here are primarily with evolutionary processes, it is important to always recognize that evolution occurs within an ecological context, and for phage-driven evolution, that ecological context is phage ecology (Chap. 4).

3.1 Introduction to Evolution

The modern concept of organic evolution is one of change in nucleotide sequences across organism lineages through time. Because of the ubiquity of modern sequencing technologies, such evolution or its consequences can be relatively easy to observe. Sequencing, however, tends to create biases toward recognition of evolution predominantly from genotypic rather than phenotypic perspectives. Historically, instead, changes to genotype were recognizable only from changes in phenotype, i.e., as was the case during the nineteenth and much of the twentieth centuries. Furthermore, phenotype plays a central role in Darwinian evolution, that is, in the processes of natural selection. Lastly, and somewhat equivalently (and as noted above), evolution takes place within an ecological context (Fig. 3.1), which is in terms of organisms interacting with their environments, and ecological interactions are inherently phenotypic. The study of evolutionary biology thus cannot be solely a consideration of nucleotide sequences, though nor can it be studied solely from characterization of phenotype.

We can distinguish evolution further into three categories: microevolution, macroevolution, and evolutionary ecology. Microevolution is what generally is taught as the processes of evolution, and indeed is what I have presented in Fig. 3.1. These are the noted mutation, genetic drift, migration, and natural selection, which are evolutionary forces that act below the level of species. Microevolution, that is, acts on individual populations, and all of these processes are included because they are able to modify allelic frequencies within populations, either chemically (mutation), stochastically (drift), sexually (migration), or deterministically (selection).

Macroevolution is the study of changes in lineages as those changes occur above the level of species, e.g., rates of speciation, rates of species extinctions, and the degree to which species otherwise diverge over time, the latter, i.e., in terms of genotype. Species divergence often is recognized in terms of phenotype instead, however, such as in the study of comparative fossil morphologies. As considered here, macroevolution is relevant mostly from a perspective of heterospecific horizontal gene transfer. That is, genetic migration as movement of genes between otherwise divergent lineages (different species). Macroevolution, i.e., differences

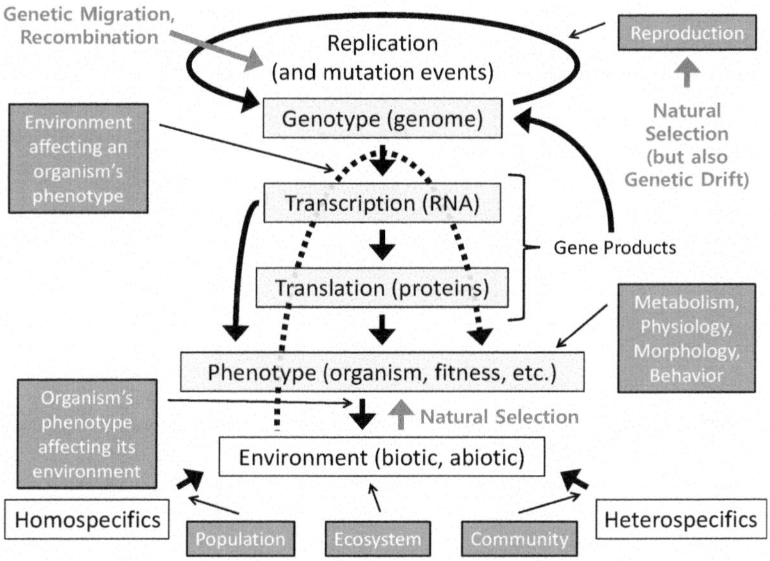

Fig. 3.1 Interface of organisms (lighter gray fill), ecology (white fill), and evolutionary processes (gray text and arrows). The central dogma of molecular genetics (replication, transcription, and translation) conspires to create phenotype (metabolism, physiology, morphology, and behavior). Phenotypes interact with environments (organismal ecology; dashed arrow), affect environments (lowest downward-facing arrow), and are affected by environments, the latter driving both phenotypic change and natural selection (lowest upward-facing arrow). Natural selection also impacts reproductive strategies (upper-right), e.g., how many offspring are produced and how often. Environments can be distinguished into abiotic and biotic aspects (together making up ecosystems). Biotic aspects can be differentiated into conspecifics ('homospecifics' in the figure) and non-conspecifics ('heterospecifics' in the figure)

between bacterial species, also plays important roles in delimiting what bacterial types a given phage type will be able to infect.

Evolutionary ecology plays a dominant role in my personal evolutionary thinking and thereby underlies a great deal of what has gone into the writing of this monograph. In simplest terms, evolutionary ecology is the study of evolutionary adaptation, that is, study of the products of natural selection. Thus why, in terms of adaptation utility, do organisms display the phenotypes that they do? See Fig. 3.2 for a look at evolutionary ecology's place in comparison to physiology, ecology, and macroevolution along with microevolution (evolutionary biology).

An evolutionary ecologist thus would ask, why as an adaptation do giraffes have such long necks? A macroevolutionist would determine what animals are most closely related to giraffes and might make an effort to document changes in neck length over time, especially from a morphological rather than ecological perspective. Both perspectives tend to involve explorations of phenotype, but evolutionary ecology in a sense is interested in phenotype as constrained by genotype, i.e., what is possible? Macroevolution, by contrast, can be an exploration of genotype often as constrained by phenotype, i.e., what has actually occurred? A microevolutionist

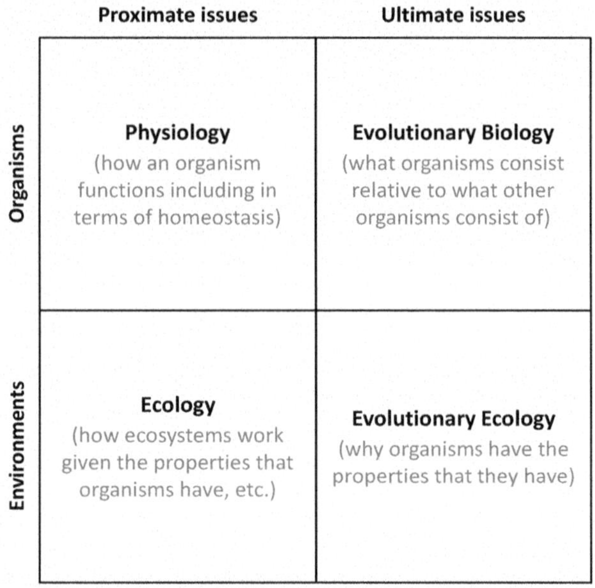

Fig. 3.2 Finding evolutionary ecology's place among related disciplines. The term 'evolutionary biology' in the figure encompasses both microevolution and macroevolution

might instead look for polymorphisms impacting giraffe neck length as exist within an existing giraffe population. They also may ask whether, e.g., the frequencies of the giraffe neck length-underlying alleles are changing over time and how. Of course, in all of these examples, genotype certainly is not irrelevant, though nor is phenotype irrelevant.

Although this chapter serves as a brief introduction to evolution generally, what is particularly important is to not move on to the next chapters before gaining an appreciation of just what microevolution consists of, as if nothing else I have organized the text around its underlying bases. Thus, there is mutation, sampling error (i.e., genetic drift), migration (i.e., genetic migration), and natural selection. In addition, there is what is known as non-random mating. All of these evolutionary processes can be influenced by phenotype, but only natural selection inherently has phenotype as its basis. Furthermore, all five phenomena represent, to a population geneticist, what are known as failures to meet the assumptions underlying something called Hardy-Weinberg equilibrium.

3.2 The No-Evolution Default Assumption

The Hardy-Weinberg equation, mathematically describing Hardy-Weinberg equilibrium, is one of the instances in which one is exposed to mathematics during introductory biology classes. As such, it can make students cringe. That equation, though actually rather simple, is not crucial to understanding this equilibrium, however. Indeed, though the equation is historically very relevant, it also can be viewed as a bit of distraction from more pertinent issues. Furthermore, the Hardy-Weinberg equation is rather specific to describing the population genetics of diploid, obligately sexually reproducing organisms. As such, it actually has little utility here, in this monograph. The underlying concept of genetic equilibrium, however, is quite important.

A system that is at equilibrium is one that does not change over time. As Hardy-Weinberg equilibrium deals with the frequency of alleles within populations—the fraction of individuals that carry specific gene variants at specific genetic loci—what Hardy-Weinberg equilibrium implies as its basis is that allele frequencies are not changing over time. The concept of evolution to a population geneticist, by contrast, is defined as *change* in allele frequencies over time. The math of the Hardy-Weinberg equation is there just to show how a lack of change in allele frequencies can be manifest despite the mixing of alleles over the course of sexual reproduction. Indeed, historically the math is important—besides subsequently serving as the basis of population genetic models of organism evolution—because it showed that changes in allele frequencies are not expected to occur, deterministically, as a consequence of the math alone. The Hardy-Weinberg equation thus is the *right* way to model the impact of sexual reproduction, particularly random mating, on allele frequencies within populations over time. Specifically, these processes don't have an impact on those allele frequencies, at least not directly.

To achieve Hardy-Weinberg equilibrium, and thus a lack of evolution or more specifically a lack of microevolution (Fig. 3.3), then four assumptions must hold, plus there is the noted random mating. It is exceptions to these assumptions that define microevolutionary forces. Thus, to understand microevolution it can be useful to start with what is *not* microevolution, which are the assumptions required to achieve Hardy-Weinberg equilibrium. These are (1) a permanence to existing nucleotide sequence, (2) infinite population sizes, (3) populations that are inviolate in terms of acquisition of nucleotide sequences from other populations, and (4) a lack of biases in the transmission of existing nucleotide sequences from one generation to the next. The additional assumption, though one that does not directly impact allele frequencies, is (5) a lack of structure within a population that could have the effect of biasing which individuals come to mate with each other. That is, as should by now be viewed as a familiar if inverted refrain, these are (1) a lack of mutation, (2) a lack of genetic drift (i.e., lack sampling error), (3) a lack genetic migration, (4) a lack of natural selection, and (5) a lack of non-random mating, respectively. Evolution thus occurs (1) if there *is* mutation, (2) if there *is* sampling error, (3) if there *is* genetic migration, or (4) if there *is* natural selection, plus (5) if the genetic structure of populations (i.e., genotype frequencies as distinct from allele frequencies) is modified through non-random mating. What is presented throughout much of this monograph are elaborations on those five phenomena, the first four of which I elaborate upon in more general terms below.

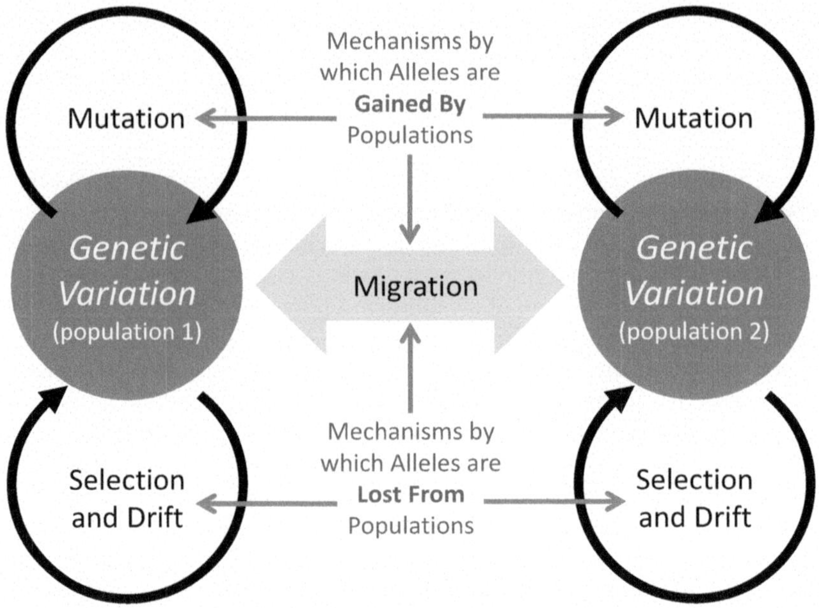

Fig. 3.3 Mechanisms of microevolution. Genetic variation can be defined in terms of allele frequencies and microevolution can be defined as changes in allele frequencies over time. Migration is movement of genetic variation (alleles) between otherwise distinct populations, whether homospecific or instead heterospecific

3.2.1 Mutational Change as a Violation of Hardy-Weinberg Equilibrium

Mutations are changes to existing nucleotide sequence (Fig. 3.4), i.e., they represent a lack of permanence to these sequences. There are a number of different ways that mutations can occur and these include (1) chemical changes to nitrogenous bases, (2) insertions or duplications of a nucleotide during replication, (3) inadvertent insertion (or deletion) of a nucleotide (again as during replication), (4) loss of longer stretches of nucleotides (deletion mutations), and (5) insertions of longer stretches of nucleotides. Recombination processes also can move around sections of chromosomes and even insert new DNA as sourced from a different organism. The latter more likely would be described as consequences of recombination or even a result of genetic migration rather than mutation. Nevertheless, if the result is a change in nucleotide sequence, then technically it is mutational no matter the cause.

As is often stated, mutation is the ultimate source of genetic variation within populations. That is, you cannot have evolutionary change, nor even polymorphisms, unless you first have mutational change. Indeed, mutation is evolutionary change and the rest of evolution has to do with what happens to those mutations.

Note that there is some evidence that phages can select for the evolution of bacterial mutator strains, that is, which display higher than wild-type mutation

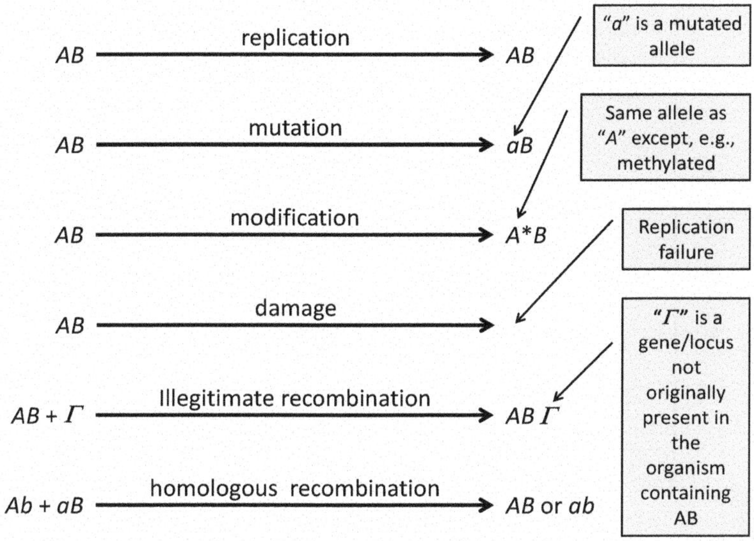

Fig. 3.4 Various types of DNA alteration. From top to bottom: unaltered, hereditarily altered (mutation), post-replicationally altered (including epigenetically), non-replicable alteration (i.e., a damaging lesion), acquisition of new gene or genes, and unlinking of adjacent alleles

rates. Presumably this could be beneficial to carrying bacteria toward increasing their potential to more rapidly or more effectively evolve phage resistance. It is questionable, though, whether these benefits are sufficient to extend to more complex environments as found outside of the laboratory such as soils (Gómez and Buckling 2013), and the issue of evolution of bacterial mutator strains otherwise will not be covered further in this volume. The role of strictly lytic phages in driving mutation accumulation by bacterial populations more generally, and indeed the evolutionary diversification of these populations, has been reviewed by Dennehy (2012).

3.2.2 Sampling Error as a Violation of Hardy-Weinberg Equilibrium: Genetic Drift

Genetic drift occurs to greater extents within smaller populations than within larger populations, and this has to do solely with the statistics of small numbers. Specifically, you are less likely to sample a population without error the larger the size of the sample. At one extreme is a sample of only one individual, which inherently will lose all genetic variation that is found within a population except that which is found in that one sampled individual. Genetic bottlenecking, for example, can give rise to genetic drift, which in turn results in the extinction of alleles and/or of genotypes. Bottlenecking, that is, is the reduction of the size of a population, particularly to a point where sampling error is high. When sampling error is high, then randomly

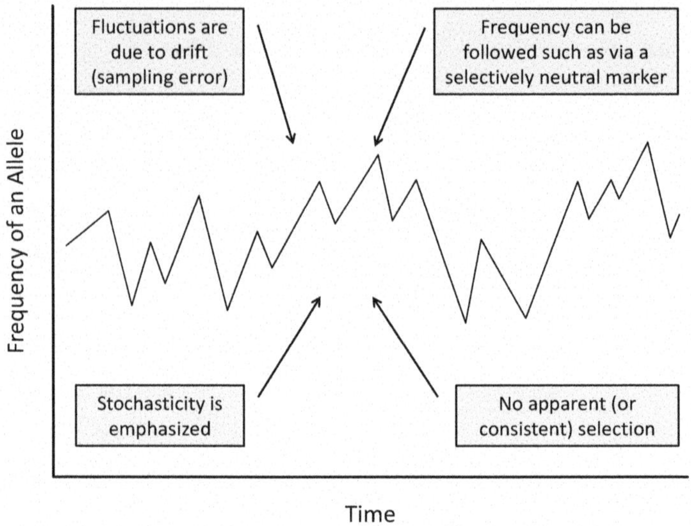

Fig. 3.5 Illustration of the impact of genetic drift as a manifestation of sampling error on allele frequencies. Shown is variation in allele frequencies as measured over time. The allele in question is assumed to be selectively neutral and variation in its frequency is assumed to be a consequence of sampling error, i.e., as resulting in genetic drift, which literally is a random drift over time in the frequency of alleles

occurring mistakes are larger in magnitude. Such mistakes can include a failure to pass even one example of a specific allele from one generation to the next, hence resulting in the extinction of that allele, at least within the population being considered. A small population thus will tend to lose genetic variation and frequencies of remaining alleles will tend to change as well. These changes with genetic drift occur solely due to random, that is, statistical processes, i.e., essentially as due to statistical noise.

We call these processes stochastic, meaning random; see Fig. 3.5 for illustration. This contrasts with the deterministic changes that one sees with natural selection, i.e., as discussed further below. Though occasionally, inadvertently, what ultimately are beneficial changes in allele frequencies might still occur as a consequence of genetic drift, more often than not genetic drift instead is a bad thing for populations.

3.2.3 Genetic Migration as a Violation of Hardy-Weinberg Equilibrium

Genetic migration simply is the movement of alleles from one population to another. As so defined, this implies that the two populations are not otherwise exchanging alleles to a substantial extent. That is, there will exist two distinct populations which display only some small degree of 'leakage' of alleles between them, but not so

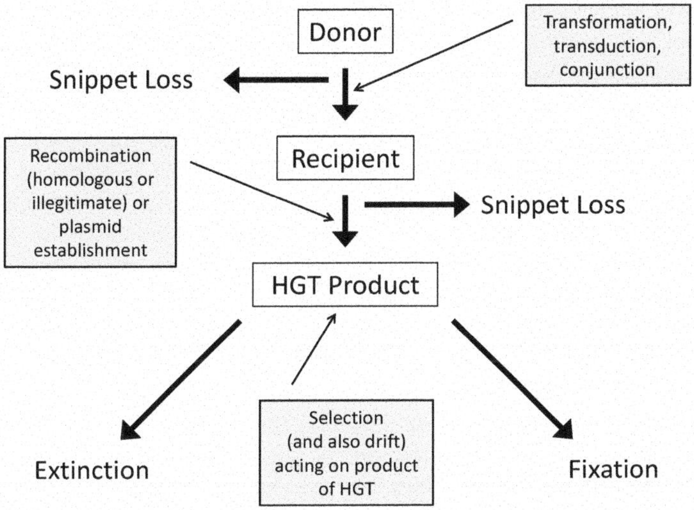

Fig. 3.6 General scheme for horizontal gene transfer (HGT) in bacteria. Note that in addition to extinction or fixation there also is establishment of stable polymorphisms, though that is not shown in the figure

much of this gene exchange that the two populations can come to be viewed as a single population. There especially is a lack of panmixis, that is, completely random mating between the two instead distinct populations. For bacteria, genetic migration can be viewed as something that occurs between strains of the same species or instead that occurs between different species. A schematic of the general process is presented in Fig. 3.6.

Note the connection between the concept of genetic migration and that of non-random mating. In a sense, that is, genetic migration occurs because mating among two (or more) populations is not randomly distributed across multiple populations, but instead is biased toward matings within rather versus between these populations. Genetic migration thus is what occurs when *barriers* to random mating to a degree break down, but don't break down so much that, in terms of mating events, two populations become one.

3.2.4 Natural Selection as a Violation of Hardy-Weinberg Equilibrium

I consider natural selection last in this introduction to Hardy-Weinberg exceptions because it is special. This is because natural selection is a deterministic rather than random (stochastic) process. With natural selection, specific alleles thus will increase in frequency and other alleles will decrease in frequency, with these changes occurring to well defined degrees. Or, in the case of stabilizing selection, frequencies of alleles will be neither increasing nor decreasing, but again this will be occurring in a quantitatively well-defined manner. This biased transmission of genetic variation also occurs due to genetically determined phenotypic interactions with environments. Thus, (1) individual genetic variants (individual alleles or whole genotypes) can differ in terms of their impact on phenotypes (though not always), (2) different phenotypes can (but do not necessarily) differ in terms of how they interact with environments, and (3) some of that variation in interactions with environments can (but does not necessarily) impact the potential for individual organisms to survive and reproduce. It is the connection between alleles and, ultimately, their potential to differentially impact reproductive success that powers natural selection (Fig. 3.7).

One can view all of the other microevolutionary forces as providing or at least manipulating the underlying genetic variation within populations upon which natural selection can act, though natural selection that has acted in the past also has shaped that underlying genetic variation. A population thus will have a genetic structure that consists of what alleles are present, the frequency of those alleles, and how those alleles have been combined together into genotypes. Mutation, genetic drift, migration, and the randomness of mating, as well as natural selection, all as have occurred in the past, will have shaped that genetic structure, modifying what alleles are present, their frequencies, and how those alleles are combined into individuals. Natural selection then comes along and further shapes that genetic structure, but does so in a unique manner, one that assures that a population

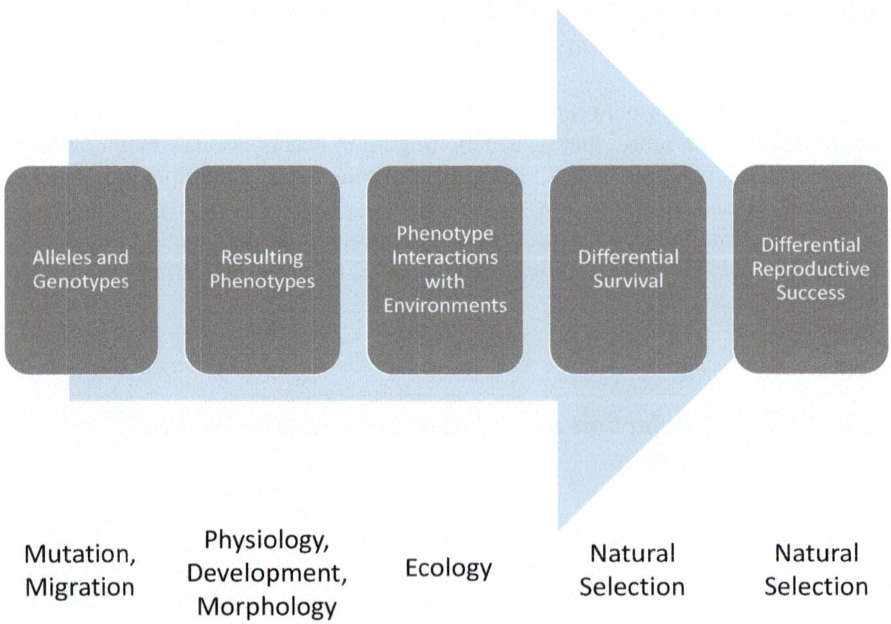

Fig. 3.7 The processes leading up to and including natural selection

following natural selection has a greater potential to survive and reproduce, or at least some individuals making up a population are better able to compete with other individuals making up the same population.

We can measure the strength of natural selection in terms of specific numbers of offspring that are produced per generation by a given genotype, where a specific genotype can be viewed as a collection of specific alleles. Without natural selection, then all genotypes should succeed in producing the same numbers of offspring with equal probability. With natural selection, by contrast, different genotypes will produce different numbers of offspring, and those numbers, so long as circumstances are invariant, and other evolutionary forces can be ignored, will also be invariant. Thus, and as noted, natural selection is deterministic.

3.3 Natural Selection and Darwinian Fitness: Relative and Absolute

In terms of our introduction to microevolution, and really serving simply as a measure of the strength of natural selection, it is important to gain an appreciation of the concept of evolutionary fitness, a.k.a., Darwinian fitness or simply fitness. Fitness is the measure that was considered at the end of the previous section (Sect. 3.2). It is the number of offspring produced by a given genotype and particularly in terms of the production of offspring which themselves go on to produce offspring. A genotype that is unable to produce offspring thus has a fitness of zero, at least when considered on its own (i.e., ignoring a concept known as inclusive fitness). A genotype that is able to produce offspring, but all of those offspring are unable to reproduce, similarly has a fitness of zero. A genotype that is able to produce offspring which themselves are not sterile has a fitness that instead is greater than zero. For example, if on average a genotype produced two such offspring, then we could say that the genotype has a fitness of 2. If another genotype on average produced three such offspring, then we say that this second genotype had a fitness of 3. More specifically, these are measures of *absolute* fitness.

Note that 3 of course is larger than 2, implying that the fitness of the first genotype (2) is lower than that of second genotype (3). We can restate that point in terms of what is known as relative fitness, i.e., the first genotype has a fitness of 0.67 (= 2/3) relative to a fitness of 1.0 (= 3/3) for the second genotype. Often relative fitness is easier to measure as well as to appreciate than absolute fitness, but when one states that one genotype (or organism) has a greater fitness than another, what that means is that we expect that the more-fit genotype will produce more non-sterile offspring than the less-fit genotype. Also, when I say that something displays a greater fitness than something else, usually implied is that I am comparing different genotypes within the same environment, or instead the same genotype but in different environments. In either case, with greater fitness comes a greater capacity to survive to produce non-sterile offspring.

Lastly, note that natural selection is a subtractive force. Conceptually, that is, natural selection can be viewed as representing a process of *removing* less-fit individuals from populations rather than one of *adding* more-fit individuals. This is because natural selection represents the constraints acting against organism survival and reproduction, where in the absence of such constraints, that is, given weaker natural selection, then one would expect greater levels of reproductive success. Thus, the stronger that natural selection is acting *against* a given genotype, then the lower that genotype's fitness. The most successful individuals—at least in terms of deterministic, a.k.a., Darwinian evolution—are those that are able to survive and then reproduce most effectively in the face of natural selection. That is, these are the genotypes that give rise to phenotypes which best 'fit' their environments. See again Fig. 3.3 for a summary of microevolutionary forces and how they interrelate.

References

Dennehy JJ (2012) What can phages tell us about host-pathogen coevolution? Int J Evol Biol 2012:396165

Gómez P, Buckling A (2013) Coevolution with phages does not influence the evolution of bacterial mutation rates in soil. ISME J 7:2242–2244

Hendrix RW (2008) Phage evolution. In: Abedon ST (ed) Bacteriophage ecology. Cambridge University Press, Cambridge, UK, pp 177–194

Chapter 4
Brief Introduction to Phage Ecology

Contents

Abstract Ecology is the study of the interaction of organisms with the abiotic and biotic components of their environments. The importance of understanding phage ecology, toward appreciating the role that phages play as drivers of evolution, is that all evolution occurs within an ecological context, which is to say in terms of the interaction of organisms with their environments. Emphasized are discussions of the interaction of phages with abiotic components of their environments (described as organismal ecology but which also can be considered to be a form of ecophysiology), the interactions of phages with other phages of the same types (i.e., phage conspecifics or homospecifics, as studied under the heading of population ecology), and phage interactions with other species, as can be denoted as heterospecifics. The latter includes other types of phages, bacterial hosts, and also other organisms besides other phages and the specific host bacteria of individual phages. These interactions are studied under the heading of community ecology, where a community is a collection of species existing together within the same environment. Some additional emphasis is placed on the ecology of lysogeny, and introduced also is the potential for various sorts of mobile genetic elements to parasitize phages.

> Phage ecology may be defined as the interaction of phage with the environment and with their host cells and the effects of such interactions on the composition and productivity of particular habitats.—Sagar M. Goyal (1987, p. 268)

Evolution takes place within ecological contexts and this of course is also true when phages are driving that evolution. Consequently, there is some utility in considering bacteriophages, as drivers of evolution, to dwell at least a bit on issues of phage

S. T. Abedon, *Bacteriophages as Drivers of Evolution*,
https://doi.org/10.1007/978-3-030-94309-7_4

Organismal Ecology: Environment-to-Phage Interactions

- Phage Adaptation to Environmental Constraints
- Constraints on Phage Movement
- Constrains of Phage Attachment to Bacteria
- Constrains on Phage-Infection "Vigor"

Population Ecology: Phage-Phage (Conspecifics) Interactions

- Coinfection versus Coadsorption
- Recombination (especially swapping of similar alleles)
- Phage-Phage Communication
- Competition between Different Phage Genotypes (Natural Selection)

Community Ecology: Phage-to-Other Species Interactions

- Phage-Phage Mosaic Evolution
- Phage Antagonistic Impact on Host Bacteria
- Phage Mutualistic Interactions with Host Bacteria
- Impacts of Phage Gene Products on non-Bacterial Host Organisms

Fig. 4.1 Some concepts of phage ecology

ecology. Specifically, these are phage interactions with their environments, with emphasis here on organismal, population, and community aspects. Phages as drivers of biogeochemical cycling, a key aspect of phage ecosystem ecology, is on the other hand of less interest here, nor also various aspects of phage biogeography. See Fig. 4.1 for a summary of the various roles that phage ecology can play in terms of phages or their interactions with environments serving as drivers of evolution. Recently I also have authored (Abedon 2019a) and co-authored (Burmeister et al. 2019; Dennehy and Abedon 2021b) brief reviews on the ecology of phages, where more information can be obtained. See also my 2008 monograph on the subject (Abedon 2008).

A complication on phage ecology is that phages exist over the course of their life cycles as two distinct entities. These are free phages or what more precisely might be described as free virions (Dennehy and Abedon 2021c) and what has been dubbed relatively recently as virocells (Forterre 2013), though of course these two states have been known to exist throughout most of the history of phage biology, e.g., (Delbrück 1946). See Fig. 4.2 for summary of the phage life cycle as involving this alternating free virions and virocells. The phage life cycle, as involving free phages, versus as involving just latent infections, thus consists of (1) acquisition of bacteria (i.e. adsorption), (2) infection of bacteria (i.e., the virocell stage), (3) release of virions from bacteria to the extracellular environment, and (4) the free phage stage. Recently I have authored and co-authored reviews on the biology the movement of phage virions and their subsequent adsorption of bacteria (Dennehy and Abedon 2021a), and another on infection and virion release (Dennehy and Abedon 2021c). In terms of phage ecology, it is important to recognize that environmental factors affect

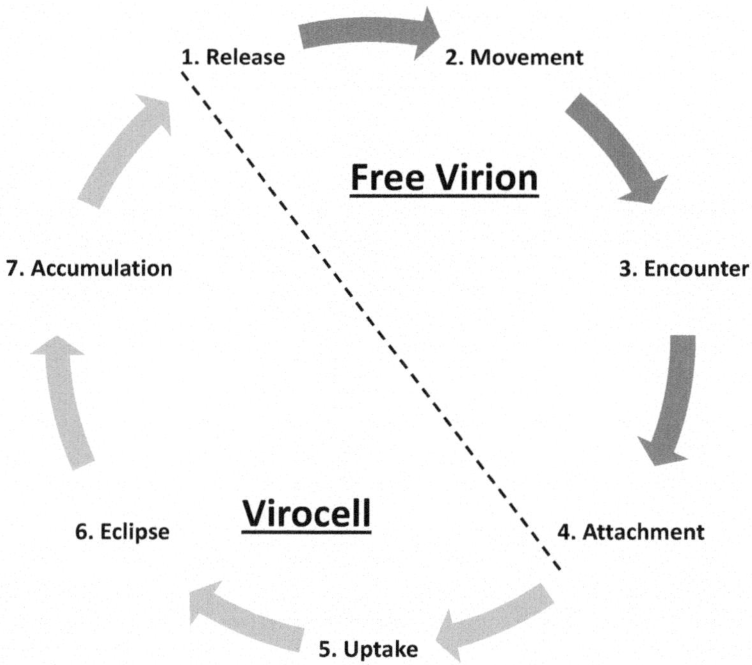

Fig. 4.2 A generalized phage life cycle, alternating between free virions and virocells. 1. Release is a phage's transition from an intracellular to extracellular state. Virions 2. move about randomly, 3. encounter bacteria, and then 4. attach to those bacteria. Successful attachment is followed by 5. genome uptake, 6. a period of intracellular gearing up for production of phage progeny (eclipse), 7. a post-eclipse period of intracellular virion accumulation (especially for lytic phages), and 1. again release. Not elaborated upon in the figure are lysogenic cycles (virocells in which phage genomes replicate but progeny virions do not accumulate) nor chronic infections (in which virocells persist despite ongoing virion release)

phages or phages affect environmental factors in different ways depending on whether phages are currently existing as virocells versus as free virions.

4.1 Phage Organismal Ecology

Phages are organisms and as organisms they interact with their environments. Aspects of environments can be phages of the same type (as studied especially in population ecology) or organisms other than of the same phage type (as studied especially in community ecology; e.g., 'b' in Fig. 4.3). There also are the abiotic (non-living) aspects of environments. Ecosystem ecology considers the impact of organisms on abiotic aspects of environments, again, e.g., as having to do with biogeochemical cycling. On the other hand is the impact of abiotic aspects of

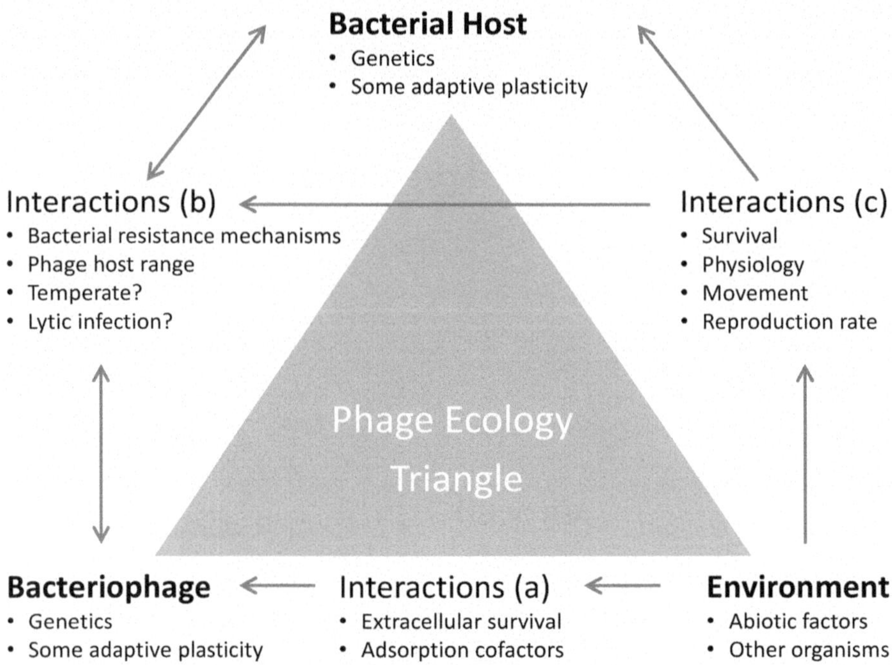

Fig. 4.3 Phage-bacterium-environment interactions especially from the perspective of a specific phage type and its bacterial host. (a) Interactions between virions and their non-host environment (organismal ecology for abiotic environmental factors and community ecology for biotic abiotic factors). (b) Interactions between phages and host bacteria (community ecology). (c) Interaction between host bacteria and their environment, other than in terms of the focus phage population (organismal ecology for abiotic environmental factors and community ecology for biotic factors). Environment-host bacterium interactions also can impact phage-host bacterium interactions (upper horizontal arrow)

environments on phages (part of 'a' in Fig. 4.3), where here I am defining abiotic as something that is neither alive nor a relatively intact organism that was once alive. Regardless of how one defines it, the consequence of abiotic impact is what I am describing here as being under the purview of phage organismal ecology, though one also can describe this as phage ecophysiology. An issue with that latter designation, however, is that phages can be impacted by the environment in ways that don't actually impact phage physiology. The latter point is particularly relevant to the extent that aspects of virion movement within environments are affected by environmental parameters (Dennehy and Abedon 2021a).

For example, once released, movement of virions can be by diffusion which occurs at a rate that can be impacted by the presence of colloidal materials, can be mediated by fluid flow (including turbulent fluid flow), can occur relative to surfaces, or can occur given phage association with other moving entities including dust and marine snow, though also animals (Dennehy and Abedon 2021a). Though not free virion movement, phages also can move from one abiotic location to another while infecting motile bacteria. One environmental aspect that can impact rates of diffusion by at least some virions is as abiotic conditions impact the extension (slower diffusion) and retraction (faster diffusion) of phage tail fibers, such as those of phage T4. Indeed, in the case of phage T4 these two states likely are determined by chemical and physical differences between colonic and extracolonic environments for this phage.

The likelihood of virion encounter with a bacterium is in part a function of rates of relative movement between virion and bacterium. Thus, both diffusion and likely even more so turbulent fluid flow will cause virions to collide with bacteria with greater likelihood than if diffusion is slowed or environments are otherwise not well mixed. Virions also potentially can move relative to bacteria as fluids flow; for example, the flow of fluids over and through bacterial biofilms, though also relative to planktonic bacteria given turbulent fluid flow. The likelihood of virion attachment to a bacterium also can in part be a function of chemical and physical aspects of environments. For instance is the requirement for monovalent or divalent cations by many phages to successfully attach to bacteria.

Once a phage has infected a bacterium, then there are a number of extracellular abiotic factors that can impact phage latent period lengths and burst sizes. This is other than intracellular factors as would be considered instead under the heading of community ecology. Most notably would be the availability of nutrients and energy, with an expectation that lower amounts of either could lead to lengthened phage latent periods or smaller phage burst sizes (Hadas et al. 1997). Physical conditions such a temperature extremes or ionizing radiation that have harsh effects on bacterial physiology also can be expected to lengthen latent periods or reduce burst sizes. So too can antibiotics and other antagonists of bacterial physiologies such as KCN have this effect (Adams 1959; Abedon 2019c; Danis-Wlodarczyk et al. 2020). We also can consider abiotic factors impacting the evolution of optimal lengths of phage infections, such as rates of virion diffusion (Abedon 2022). The length of phage infection periods can also vary as a consequence of phage interaction with other phages (Abedon 2019b), though that technically would be considered under a heading of phage population ecology or community ecology, depending on whether the other phages are homospecifics versus heterospecifics.

4.2 Phage Population Ecology

Population ecology is the study of interactions between conspecifics, a.k.a., homospecifics, with these conspecific individuals thereby considered to be an aspect of an organism's environment. Interactions between phage conspecifics—notwithstanding how it is that we might define a phage "species" (Ackermann et al. 1984, Jarvis 1984, Jarvis et al. 1991, Adriaenssens and Brister 2017, Bobay and Ochman 2018, International Committee on Taxonomy of Viruses 2020) or instead phage "quasi-species" (Casjens et al. 1992)—generally happen during the interaction of two or more phages with or within the same bacterial cell. These interactions we can differentiate into coadsorptions and coinfections, with coadsorptions not necessarily always leading to coinfections (Abedon 2015). Coinfections nevertheless probably are somewhat common, though the most common of coinfections likely are between a recently infecting phage and a phage that is already lysogenically infecting (Campbell 1996). Note, though, that just because a coinfection is occurring doesn't mean that it is a coinfection between conspecific phages and thus could be considered instead under a heading of phage community ecology (interactions between heterospecific phages) rather than phage population ecology (interactions between homospecific phages). The latter likely is particularly true for coinfections between newly adsorbing phages and prophages.

We can also differentiate interactions that occur between phages into ones that are antagonistic versus those that are potentially beneficial to the interacting phages. Antagonistic interactions often are seen when the first adsorbing phage (the primary phage) in some manner interferes with infection by a second phage that adsorbs to the same bacterium (a secondary phage, as discussed further in Chap. 16). Beneficial interactions can occur following resulting coinfection of a single bacterium by two phages, but those benefits can be lopsided, with one of the participating phages benefiting and the other less so. Such lopsided interactions can be particularly the case as helper phages are parasitized by other phages (as considered in Sect. 4.5, and also in Chaps. 13 and 20). Benefits, at least evolutionarily beneficial, also can occur as a consequence of recombination between infecting phage genomes, whether coinfecting as productive infections or instead with one phage infecting an already established lysogen; for some thoughts on the latter, see Khan and Wahl (2020). In this case, one phage usually acquires genes from another phage, rather than the two phages equally exchanging genetic material. Note again that we are considering explicitly homospecific interactions with this example, and in this section, though in principle recombination can occur between heterospecific phages as well. Benefits to phages also can include where the presence of the second phage signals the first phage to display a modified infection cycle, such as an extended phage latent period or display of lysogeny rather than a productive cycle (Abedon 2022). Together, these latter phenomena represent a form of homospecific phage-phage communication (Abedon 2019b, 2020; Igler and Abedon 2019).

Also included under phage-phage interactions is what is known as exploitative competition. Here, two phage types—conspecifics in the case of considerations of population ecology—are competing over the same resources but are not necessarily otherwise interacting, e.g., such as competing over access to the same population of bacterial hosts though not necessarily competing over the exact same bacterium at the same time. As also is the case for recombination, exploitative competition can be between different phage species instead of just among conspecifics and therefore can be a theme also of phage community ecology.

4.3 Phage Community Ecology

Communities consist of more than one species found within a single environment. In terms of phages and phage ecology, this can be more than one species of phages, one phage species in combination with one potential host bacterial species, or phages in combination with all other possible organisms. That is, population ecology in combination with community ecology considers an organism's interactions with the biotic components of environments, homospecifics versus heterospecifics, respectively.

Clearly the noted recombination between different phage types (heterospecifics) is possible and is an important aspect of phages as drivers of evolution, i.e., as leading to phage mosaic evolution, e.g., (Aziz et al. 2005; Hatfull 2015). A lot of this phage-genome mosaicism is found among temperate phages (Mavrich and Hatfull 2017) and may be driven in part by low-fidelity recombinases expressed by these phages, that is, rather than resulting predominantly or exclusively from non-homologous, i.e., illegitimate recombination (Campbell and Botstein 1983; Morris et al. 2008; De et al. 2014) (Chap. 13, Sect. 13.2). Also of interest in terms of phages as drivers of *evolution* within a community ecological setting is, of course, their interactions with host bacteria, as can lead to recombination (Chaps. 11 and 13), natural selection for certain phage traits (such as seen with antagonistic evolution; Chap. 25), in some cases bacterial mutation (Chap. 5), and even a potential for sampling error (genetic drift) to the extent that phages can reduce bacterial populations sufficiently in size (Chap. 7). It is especially phage evolutionary impacts on their bacterial hosts that is the emphasis here, however, rather than highlighting the phage impact on phage genome evolution. Phages also can interact with eukaryotes, perhaps most notably in terms of phage carriage of bacterial exotoxin genes and other virulence-factor genes (see Chap. 15). In addition to phage carriage of genes that are antagonistic to eukaryotic organisms, i.e., as following the expression of these genes by host bacteria, genes also could be carried by phages that are helpful to eukaryotes, e.g., such as helpful by enhancing the strength of bacterium-eukaryote mutualistic interactions. Phages also can enter into eukaryotic cells via various not specifically phage-driven processes, resulting in enrichment in association with certain organelles within those cells and potentially interacting with receptors associated with innate immune responses (Bodner et al. 2021). To the extent that degraded virions might enter the nucleus, then there exists the possibility of recombination of phage DNA into the genomes of these cells (Doolittle 1998), a process that could even be aided by phage enzymes if those should be present along with phage DNA within the nucleus (Bodner et al. 2021).

4.4 Prophages Are at the Interface of Phage Population and Community Ecologies

A prophage is a phage genome as it exists during lysogenic cycles. These can be found integrated into host DNA, i.e., as seen with the canonical temperate phage, phage λ. Alternatively, prophages can exist as plasmids. An example of the latter is phage P1, which like phage λ is also a coliphage (Fig. 1.5). Even stranger are linear plasmid prophages, as typified by the also coliphage, N15 (Ravin 2006). What all of these temperate phages as prophages have in common is that upon forming lysogens they exist as both phages and as long-term molecular components of bacteria. Indeed, prophages can contribute to bacterial phenotypes in positive ways, typified by display of the noted lysogenic conversion (Chap. 15).

To retain the prophage state, prophages express repressor genes, the products of which prevent the expression of genes that contribute to virion-productive life cycles, such as lytic infections. The resulting repressor proteins can interact with coinfecting phages of similar types, such as other phages serving as conspecifics. Thus, prophage repressors play a role in the population ecology of temperate phages. Specifically, repressor proteins interfere with the ability of newly infecting phages to display successful infections. As some of these infections might otherwise be lytic, a result not only is prophage protection from being destroyed by a newly infecting phage, but so too is the bacterial lysogen protected from these other phages (Chap. 16). Thus, the action of prophage repressor proteins not only can result in phage-phage interactions but also can result in phage-bacterium interactions, the latter at least twice: first involving a bacterial symbiotic relationship with the prophage and second involving the blocking of successful infection of the bacterium by a subsequent phage.

Prophage expression of repressor genes can result in a community ecological interaction in an additional way, not directly involving the bacterial host. That is, newly infecting phages whose infections are blocked by repressor proteins are said to be of the same immunity type as the repressor gene-expressing prophage. These phages consequently are said to be homoimmune. This is not to be confused with 'homospecific', though mostly homospecific temperate phages can be thought of as also being homoimmune to each other. Homoimmune phages nevertheless are not necessarily always highly similar, that is, not necessarily conspecifics. Instead, just the DNA sequence that the repressor protein binds to needs to be similar. Thus, the interaction between prophages and newly infecting phages, depending on their mutual characteristics, may be studied ecologically under a banner of either population or community ecologies, depending on the degree of similarity of the two phages, with those two phages potentially heterospecific even if they are homoimmune. An additional set of temperate phages may be described as heteroimmune. In this case, the prophage's repressor protein by definition fails to block the infection of the incoming phage, and this is even if the two phage types are otherwise similar (thus, at least in principle two phages can be both homospecific and heteroimmune). Typically, though, heteroimmunity we would expect to be an interaction between heterospecific phages, and therefore something that would be studied as an aspect of community ecology rather than population ecology.

As a generalization, the less phages are able to specifically interact, e.g., such as in terms of homologous recombination or homoimmunity, then the less likely that the interaction is occurring within a population, i.e., between homospecifics, versus occurring between heterospecifics (the latter thus within a community). Exceptions to this statement are seen, however, when phages encode anti-phage defenses (Chap. 16). Also, as Javan et al. (2019) note (p. 1)—as considered more in terms of the hosts of prophages rather than the prophages themselves—at least for streptococci, "Cross-species transmission of prophages is not uncommon". This however appears to have been over evolutionary time ("same phylogenetic cluster", p. 2) rather than necessarily implying recent transmissions. Another exception, also seen with temperate phages, is when one species of phage, or mobile genetic element, specifically parasitizes another species of phage.

4.5 Mobile Genetic Elements Parasitizing Phages

In the past decade there was much ado over the discovery of so-called virophages, which are parasitic viruses of viruses but which, at least as originally defined, were not bacteriophages. Instead, these were relatively small eukaryotic viruses which are only able to successfully infect cells that were already infected by so-called giant eukaryotic viruses (Fischer 2018). Though certainly an interesting concept, in fact it was not a purely original one as in fact true phages have been known for decades which can productively infect only given the presence of a second, coinfecting phage, the latter known as a helper phage (Domingo-Calap et al. 2020; Mitarai 2020). These can be described as satellite phages, and satellite phages especially are phages that parasitize other phages, ones that are heterospecific, hence making this an intra-community rather than intra-population interaction. Also possibly serving as phage parasites, though within phage genomes themselves, are mobile introns, e.g., (Belfort 1989; Landthaler and Shub 1999), as well as other transposable elements (below).

Prominent among satellite phages is coliphage P4, which parasitizes the helper temperate coliphage P2 as well as P2-related phages (Dehò and Ghisotti 2006). Especially, phage P4 parasitizes phage P2 lysogens. Thus, the phage's host for productive infections is a specific combination of bacterial and prophage types. As a further complication, however, phage P4 itself is a temperate phage, and one that is capable of existing either as an integrated prophage or instead as a plasmid prophage. In either of these latter states, superinfection by phage P2 also gives rise to a lytic phage P4 infection. Thus, phage P4 is able to detect the presence of phage P2 under a variety of circumstances, with the result being phage P4 transitioning to a productive infection in which it parasitizes phage P2. Different circumstances of productive infections yield different levels of phage P4 productivity (burst size) and it is particularly phage P4 infection of phage P2 lysogens that seems to result in the greatest potential for phage P4 replication. Dehò and Ghisotti summarize by noting that (p. 391), "Because of its diverse modes of propagation, P4 may be considered as an integrative plasmid that has evolved the potential for horizontal transfer by a very specialized phage-mediated transduction mechanism."

Also described as satellite phages, or phage satellites, are phage-inducible chromosomal islands (Penadés and Christie 2015). These are covered further in Chap. 13, Sect. 13.3. Altogether, a number of satellite phages of various types have been identified in the genomes of *S. aureus* (Penadés and Christie 2015), *Streptococcus* spp. (Javan et al. 2019), and *V. cholerae* (Faruque and Mekalanos 2012) as well as the above-note phage P4 of *E. coli*.

Another type of parasitizing virus is known as defective interfering particle. Unlike satellite phages, with defective interfering particles the parasitized phage is homospecific to the parasitizing phages, i.e., serving as within-species cheaters or exploiters (Horiuchi 1983; Turner and Duffy 2008; Secor and Dandekar 2020). Defective interfering particles evolve in response to phage propagation under high multiplicity of infection conditions. Specifically, if wild-type phages are likely to be present within homospecifically coinfected bacteria, then phage mutants that display within-bacterium replication advantages at the expense of coinfecting wild-type phages can be selected for. At an extreme, these phage mutants lose an ability to productively infect without coinfecting with their wild-type helper phages. Hence, the mutant phages can be 'defective', and as their replication can come at the expense of wild-type helper phages, they are 'interfering' as well. Lastly, successful replication of these defective interfering viruses, as occurs especially in the presence of helper phages, results in the production and release of virion 'particles' carrying the mutant phage DNA. Hence, 'defective interfering particles'.

References

Abedon ST (2008) Bacteriophage ecology: population growth, evolution, and impact of bacterial viruses. Cambridge University Press, Cambridge, UK

Abedon ST (2015) Bacteriophage secondary infection. Virol Sin 30:3–10

Abedon ST (2019a) General ecology of bacteriophages. In: Mahy BWJ, Van Regenmortel MHV (eds) Encyclopedia of virology, 4th edn. Elsevier, Oxford

Abedon ST (2019b) Look who's talking: T-even phage lysis inhibition, the granddaddy of virus-virus intercellular communication research. Viruses 11:951

Abedon ST (2019c) Phage-antibiotic combination treatments: antagonistic impacts of antibiotics on the pharmacodynamics of phage therapy? Antibiotics 8:182

Abedon ST (2020) Phage-phage, phage-bacteria, and phage-environment communication. In: Witzany G (ed) Biocommunication of phages. Springer, New York, pp 23–70

Abedon ST (2022) Evolution of bacteriophage latent period length. In: Dickins TE, Dickens BJA (eds) Evolutionary biology: new perspectives on its development. Springer

Ackermann H-W, Cantor ED, Jarvis AW, Lembke J, Mayo JA (1984) New species definitions in phages of gram-positive cocci. Intervirology 22:181–190

Adams MH (1959) Bacteriophages. InterScience, New York

Adriaenssens E, Brister JR (2017) How to name and classify your phage: an informal guide. Viruses 9:70

Aziz RK, Edwards RA, Taylor WW, Low DE, McGeer A, Kotb M (2005) Mosaic prophages with horizontally acquired genes account for the emergence and diversification of the globally disseminated M1T1 clone of *Streptococcus pyogenes*. J Bacteriol 187:3311–3318

Belfort M (1989) Bacteriophage introns: parasites within parasites? Trends Genet 5:209–213

Bobay LM, Ochman H (2018) Biological species in the viral world. Proc Natl Acad Sci U S A 115:6040–6045

Bodner K, Melkonian AL, Covert MW (2021) The enemy of my enemy: new insights regarding bacteriophage-mammalian cell interactions. Trends Microbiol 29:528–541

Burmeister AR, Abedon ST, Turner PE (2019) Bacteriophage ecology. Encyclopedia of microbiology. Elsevier

Campbell AM (1996) Cryptic prophages. In: Neidhardt FC (ed) *Escherichia coli* and *Salmonella* cellular and molecular biology. ASM Press, Washington, D.C., pp 2041–2046

Campbell A, Botstein D (1983) Evolution of the lambdoid phages. In: Hendrix RW, Roberts JW, Stahl FW, Weisberg RA (eds) Lambda II. Cold Spring Harbor Laboratory, Cold Spring Harbor, N.Y, pp 365–380

Casjens S, Hatfull GF, Hendrix RW (1992) Evolution of dsDNA tailed-bacteriophage genomes. Semin Virol 3:383–397

Danis-Wlodarczyk KM, Cai A, Chen A, Gittrich M, Sullivan MB, Wozniak DJ, Abedon ST (2020) Friends or foes? Rapid determination of dissimilar colistin and ciprofloxacin antagonism of *Pseudomonas aeruginosa* phages. Pharmaceuticals (Basel) 14:11

De PM, Hutinet G, Son O, Amarir-Bouhram J, Schbath S, Petit MA (2014) Temperate phages acquire DNA from defective prophages by relaxed homologous recombination: the role of Rad52-like recombinases. PLoS Genet 10:e1004181

Dehò G, Ghisotti D (2006) The satellite phage P4. In: Calendar R, Abedon ST (eds) The bacteriophages. Oxford University Press, Oxford, pp 391–408

Delbrück M (1946) Bacterial viruses or bacteriophages. Biol Rev 21:30–40

Dennehy JJ, Abedon ST (2021a) Adsorption: phage acquisition of bacteria. In: Harper D, Abedon ST, Burrowes BH, McConville M (eds) Bacteriophages: biology, technology, therapy. Springer, New York, pp 93–117

Dennehy JJ, Abedon ST (2021b) Bacteriophage ecology. In: Harper D, Abedon ST, Burrowes BH, McConville M (eds) Bacteriophages: biology, technology, therapy. Springer, New York, pp 253–294

Dennehy JJ, Abedon ST (2021c) Phage infection and lysis. In: Harper D, Abedon ST, Burrowes BH, McConville M (eds) Bacteriophages: biology, technology, therapy. Springer, New York, pp 341–383

Domingo-Calap P, Mora-Quilis L, Sanjuan R (2020) Social bacteriophages. Microorganisms 8:533

Doolittle WF (1998) You are what you eat: a gene transfer ratchet could account for bacterial genes in eukaryotic nuclear genomes. Trends Genet 14:307–311

Faruque SM, Mekalanos JJ (2012) Phage-bacterial interactions in the evolution of toxigenic Vibrio cholerae. Virulence 3:556–565

Fischer MG (2018) The virophage family *Lavidaviridae*. In: Hyman P, Abedon ST (eds) Viruses of microorganisms. Caister Academic Press, Norwich, pp 271–294

Forterre P (2013) The virocell concept and environmental microbiology. ISME J 7:233–236

Goyal SM (1987) Methods in phage ecology. In: Goyal SM, Gerba GP, Bitton G (eds) Phage ecology. Wiley, New York, pp 267–287

Hadas H, Einav M, Fishov I, Zaritsky A (1997) Bacteriophage T4 development depends on the physiology of its host *Escherichia coli*. Microbiology 143:179–185

Hatfull GF (2015) Dark matter of the biosphere: the amazing world of bacteriophage diversity. J Virol 89:8107–8110

Horiuchi K (1983) Co-evolution of a filamentous bacteriophage and its defective interfering particles. J Mol Biol 169:389–407

Igler C, Abedon ST (2019) Commentary: a host-produced quorum-sensing autoinducer controls a phage lysis-lysogeny decision. Front Microbiol 10:1171

International Committee on Taxonomy of Viruses (2020) ICTV master species lists, https://talk.ictvonline.org/files/master-species-lists/
Jarvis AW (1984) Differentiation of lactic strococcal phages into phage species by DNA-DNA homology. Appl Environ Microbiol 47:343–349
Jarvis AW, Fitzgerald GF, Mata M, Mercenier A, Neve H, Powell IB, Ronda C, Saxelin M, Teuber M (1991) Species and type phages of lactococcal bacteriophages. Intervirology 32:2–9
Javan RR, Ramos-Sevillano E, Akter A, Brown J, Brueggemann AB (2019) Prophages and satellite prophages are widespread in *Streptococcus* and may play a role in pneumococcal pathogenesis. Nat Commun 10:4852
Khan A, Wahl LM (2020) Quantifying the forces that maintain prophages in bacterial genomes. Theor Popul Biol 133:168–179
Landthaler M, Shub DA (1999) Unexpected abundance of self-splicing introns in the genome of bacteriophage Twort: introns in multiple genes, a single gene with three introns, and exon skipping by group I ribozymes. Proc Nat Acad Sci USA 96:7005–7010
Mavrich TN, Hatfull GF (2017) Bacteriophage evolution differs by host, lifestyle and genome. Nat Microbiol 2:17112
Mitarai N (2020) How pirate phage interferes with helper phage: comparison of the two distinct strategies. J Theor Biol 486:110096
Morris P, Marinelli LJ, Jacobs-Sera D, Hendrix RW, Hatfull GF (2008) Genomic characterization of mycobacteriophage Giles: evidence for phage acquisition of host DNA by illegitimate recombination. J Bacteriol 190:2172–2182
Penadés JR, Christie GE (2015) The phage-inducible chromosomal islands: a family of highly evolved molecular parasites. Annu Rev Virol 2:181–201
Ravin NV (2006) N15: the linaer plasmid prophage. In: Calendar R, Abedon ST (eds) The bacteriophages. Oxford University Press, Oxford, pp 448–456
Secor PR, Dandekar AA (2020) More than simple parasites: the sociobiology of bacteriophages and their bacterial hosts. MBio 11:e00041-20
Turner PE, Duffy S (2008) Evolutionary ecology of multi-phage infections. In: Abedon ST (ed) Bacteriophage ecology. Cambridge University Press, Cambridge, UK, pp 195–216

Part II
Mutation, Genetic Drift, and Non-Random Mating

Chapter 5
Bacterial Mutation Associated with Phages: Insertions

Contents

Abstract One means by which phages can drive the evolution of bacteria is by the disruption of bacterial genes, as can occur in the course of prophage integration. The results of this are insertion mutations. In many cases, prophages employ mechanisms to ensure that the resulting mutations are not detrimental to either their host bacteria or themselves. Sometimes, even for those temperate phages, however, detrimental mutations can still occur, such as if integration occurs at non-primary but instead secondary integration sites. For certain types of temperate phages, though, there are no preferred insertion sites and therefore relatively little effort is put forth by their prophages to avoid integrating detrimentally into their host's chromosomes. These latter phages are typified by the *E. coli* phage Mu, which in fact was named for the ease with which it can disrupt bacterial genes during its lysogenic cycles. Introduced as well in this chapter are CRISPR-Cas systems, which also involve DNA insertions into bacterial chromosomes, though insertions that are beneficial to the carrying bacteria rather than detrimental, as they are used to defend against foreign DNA such as that associated with infecting bacteriophages.

> Mutations are most precisely described as changes in the sequence of the four bases that provide the informational content of DNA.—John W. Drake et al. (1983, p. 621)

Darwinian evolution is that which is mediated by natural selection. Non-Darwinian evolution consequently is that which is associated with evolutionary phenomena other than natural selection. These non-Darwinian processes, in the order considered

S. T. Abedon, *Bacteriophages as Drivers of Evolution*,
https://doi.org/10.1007/978-3-030-94309-7_5

here, are mutation (Chaps. 5 and 6), genetic drift (Chap. 7 but also Chap. 8), non-random mating (Chap. 9), and genetic migration (Chaps. 10 through 17). Natural selection along with its consequences, all as driven by phages, is then covered in various ways over the course the remainder of this volume (Chaps. 18 through 25). See Chap. 3 for an introduction to these evolutionary forces. The current chapter considers especially the role that phages can directly play in the acquisition of changes in genetic sequence by bacteria, particularly in terms of insertional mutagenesis. This emphasis differs from any natural selection acting on such changes once they have occurred, but rather emphasis is on the means of formation of the changes themselves. Nevertheless, note that in terms of evolution, the key question regarding mutations is often whether they have negative (detrimental), neutral, or positive (beneficial) effects on the Darwinian fitness of the mutant.

5.1 Mutations, Mutants, Mutagenesis

A mutation is a change in a nucleotide sequence, especially a change that persists and where the nucleic acid sequence consists of the four standard bases. Mutations thus differ from epigenetic changes in genomes, i.e., as are associated with methylation of nitrogenous bases rather than changes in base sequence. Among mutations, at least as considered here, are also changes in nucleotide sequence resulting from recombination, though clearly products of especially homologous recombination represent a special kind of 'mutation', i.e., the acquisition of a mutation that originated in a different location, including in a different organism.

A mutant is an organism that carries a mutation. Thus, a bacterium that carries a mutation conferring phage resistance, for example, would be described as a phage-resistant mutant (Chap. 21). Mutants differ from parental genotypes and often though not quite always they differ from their parent organism's genotype in relatively few ways. An *isogenic* 'mutant', for instance, differs from the parental genotype in only, for example, a single known way. Isogenic mutants are extremely useful toward understanding the phenotypic impact of mutations, and though such isogenic 'strains' can be created via recombination processes, it is worth keeping in mind that the products of recombination themselves are not often described using the term, 'mutant'. Indeed, generally the concepts of 'mutant' and 'mutation' are reserved to describe acquired changes in genotype that are not associated especially with homologous recombination, nor for that matter with plasmid acquisition. Again, here I am lumping together all changes in the sequence of existing nucleotides as at least mutation-like in terms of their consequences to the harboring organism, keeping in mind though that sequence that has been newly acquired by homologous recombination generally—and unlike newly occurring mutations, *sensu stricto*—has previously been subject to natural selection.

Mutagenesis is the formation of mutations. Mutagenesis can be a spontaneous process. In particular, mistakes happen in the course of nucleic acid replication. Alternatively, chemical structures can decay over time in ways that alter perceived nucleotide sequences. Mutations also can be motivated by physical, chemical, or biological agents and processes. The latter would represent non-spontaneous or, rather, induced mutations. Whether spontaneous or non-spontaneous, mutations also can occur purely randomly or instead can occur in biased manners in terms of what sequences are affected and how. Regarding point mutations, purely random, especially without bias, would mean that every nucleotide found within a genome may mutate to any other nucleotide with equal probability. In terms of insertion mutations, purely random would mean insertion between any two nucleotides could occur with equal probability. In terms of deletion mutations, this would mean that any given stretch of nucleotide sequence could be lost with equal probability.

The changes in base sequences associated with mutations typically are difficult to predict *a priori*, e.g., which bases will change and how. More generally, however, the concept of mutation can refer to more predictable changes such as changes in chromosome base sequences that can occur due to various forms of recombination, e.g., as products of transduction events. Here too, though, where recombination events will begin and where they will end can occur purely randomly, without bias. In either case—chemical alterations in nucleotide bases or due to recombination events—the result of mutation is the introduction typically of new genetic variation into a population; see, though, Table II of Davey and Reanney (1980) for a far more comprehensive list of ways that new genetic variation in bacteria can be acquired. This newly acquired genetic variation, however it has been acquired, may then be subject to natural selection.

Phages can give rise to changes in the base sequences of bacterial chromosomes via random mutagenesis (typically insertion mutations; this chapter), by facilitating recombination (Chaps. 10 through 13), due simply to prophage acquisition (as covered in various ways in Chaps. 14 through 17), and by motivating the acquisition of anti-phage DNA sequences by bacteria, as seen with CRISPR-Cas systems (Sect. 5.4 and Chaps. 18 through 20). It is especially insertional mutagenesis associated with prophage acquisition that is considered in this chapter, some examples of which have been described by Fortier and Sekulovic (2013). Briefly discussed as well is the noted CRISPR-Cas (Sect. 5.4).

5.2 Secondary Insertion Sites

As noted in Chaps. 1 and 2, we can differentiate among bacteriophages into those that are temperate versus those that are not. Temperate phages upon infection are able to display either lysogenic cycles or instead productive cycles, whereas non-temperate phages are only able to display productive cycles. Lysogenic cycles, in turn, can be differentiated into those that involve prophage insertion into the bacterial chromosome and those that do not involve such insertion. In the latter case, prophages exist as replicating plasmids. See Fig. 5.1 for a summary plus additional issues also as discussed in this section. Our primary interest in this section is that prophage insertion can result in insertional mutagenesis. Though mostly not an

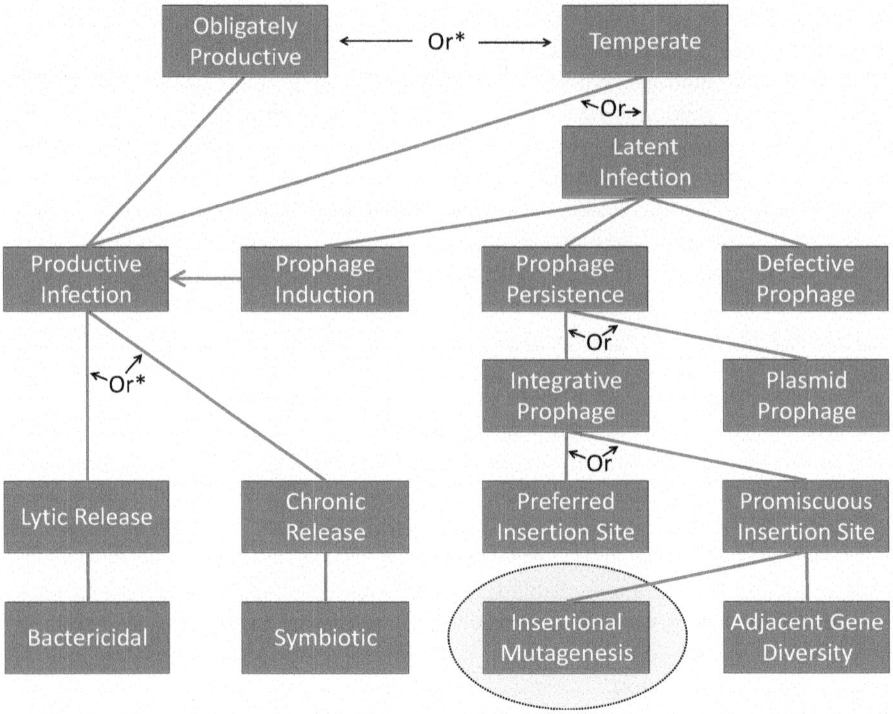

Fig. 5.1 Diversity of phage types with emphasis on the diversity of latent infection outcomes and types. Note that the indicated prophage persistence can be followed at any time with prophage induction or instead with loss of functionality (creating a defective prophage). Generally flow otherwise is from top to bottom while clarification with an arrow is provided in one instance where flow instead is from right to left. "Or" denotes, "Depending on the type of phage or circumstances." "Or*" implies "Depending only on the type of phage"

emphasis of this section, note that plasmids also can insert into bacterial chromosomes. Like inserting prophages, inserting plasmids also can employ site-specific recombinase enzymes such as integrases, or *int* (Campbell 1992).

5.2.1 Prophage Integration and Insertional Mutagenesis

Insertion of a prophage into a bacterial chromosome inherently involves disruption of bacterial chromosome sequence at the point of insertion (Campbell 2006). The result is a bacterial mutation since it represents a change in the sequence of a bacterial chromosome. This to a degree is obvious because now there is a prophage present (new sequence!) where previously there was none. But, evolutionarily, of particular importance is whether the prophage insertion has disrupted a bacterial gene. For many temperate phages, this point of prophage insertion nevertheless is

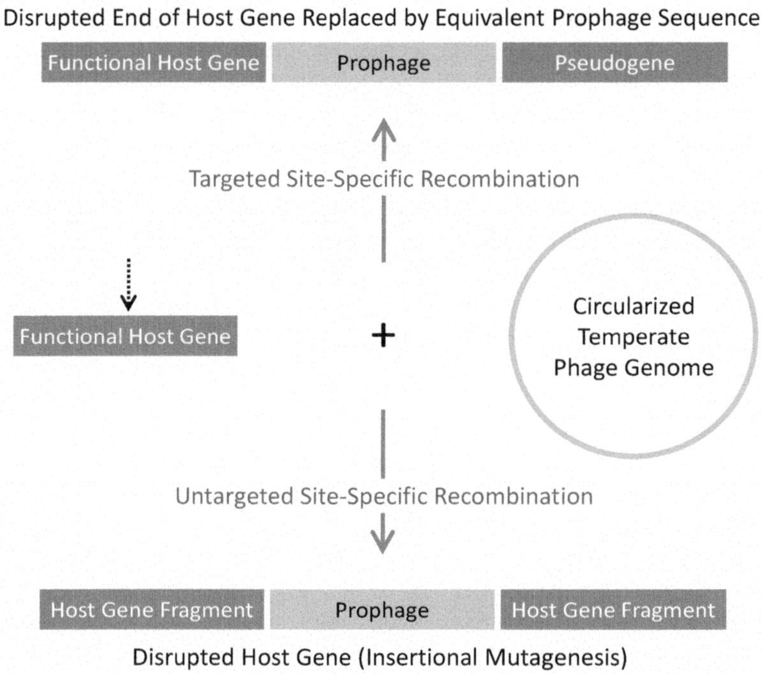

Fig. 5.2 Consequences of prophage insertion into a bacterial genome. If a preferred insertion site is used, then generally host-gene functionality is retained such as by partial replacement of the host gene on one side of the prophage genome and generation of a pseudogene on the other side of the prophage. Alternatively, if prophage insertion is not targeted, then host gene disruption by insertional mutagenesis instead occurs. A hypothetical position of prophage insertion is indicated with the dotted, downward-facing arrow

into a bacterial gene, though the insertion point can be well defined in terms of what gene and what location within that gene. The resulting prophage integration-associated insertion mutations thus are often somewhat limited in their negative impact on lysogen fitness, and this is true because prophages typically will compensate by supplying sequences at prophage insertion sites that complete the now-disrupted genes (Fig. 5.2), thereby minimizing the phenotypic impact.

Prophage insertion often occurs into tRNA genes (Campbell 2003). Such genes are only RNA encoding and therefore lack at least reading frames to disrupt. Furthermore, tRNA genes tend to be more evolutionarily stable than protein-encoding genes in terms of their nucleotide sequences, plus their small size makes it easier for phages to carry DNA sequences necessary to replace gene sequence upon prophage integration (Williams 2002); the mechanism underlying the latter is briefly discussed at the end of this paragraph. In addition, tRNA genes are also both essential and ubiquitous (Campbell 2003), so not fully deletable by bacteria to prevent prophage insertions, and presumably these genes therefore are useful as a commonly available targeted insertion site that can allow temperate phages to expand their lysogenizing host ranges. Furthermore, prophages themselves can encode tRNAs (Ventura et al. 2003; Fortier and Sekulovic 2013). Preferred prophage insertion into intergenic as well as protein-encoding genes, however, occurs as well (Campbell 2003). Insertion into genes, including into tRNA genes, generally results in completion of the gene using prophage sequence, though with exceptions (Ventura et al. 2003), but an incomplete pseudogene on the opposite end of the prophage generally is created (Campbell 1992, 2003). Thus, bacterial functional-gene copy number is not modified, though prophage genes certainly are otherwise added to the bacterium's gene repertoire.

There are a few examples of prophage insertion into bacterial genes where the result is gene inactivation (Fortier and Sekulovic 2013). For example, insertion by the phage ϕ13 and phage ϕ42 prophages of *S. aureus* results in inactivation of the β-hemolysin gene, *hlb* (Coleman et al. 1991; Dempsey et al. 2005). In addition, there is what has been termed, "Active lysogeny" (Feiner et al. 2015). Here, integrating prophages also have inserted into bacterial genes without providing genetic compensation for the genes they've disrupted. Those now-disrupted bacterial genes do not remain inactivated indefinitely, however, but instead only until the occurrence of prophage excision, whereupon expression of the previously disrupted genes is again possible. This prophage excision can be reversible or non-reversible, where for the former the prophage remains active whether integrated or not, i.e., the prophage is truly episomal, while the gene it integrates into is active only while the prophage is not integrated. This thereby allows for a prophage to reversibly regulate, both insertionally and excisionally, bacterial gene expression. Non-reversible excision, by contrast, occurs in the course of terminal differentiation of certain bacterial cells, specifically mother cells during sporulation, which do not survive the sporulation process, and cyanobacterial nitrogen-fixing heterocysts, which do not replicate. As the authors note, in neither of these cases is it likely that the original prophage insertion was evolutionarily recent. Other, less well understood mechanisms are also discussed in the same publication (Feiner et al. 2015).

5.2.2 Secondary Insertion Sites

Besides the normal sites for prophage insertion, which for most integrating prophages seem to be well defined, there is a potential also for prophages to integrate into non-primary insertion sites, and particularly so when the normally used (primary) insertion site is not available (Rao and Smith 1968; Shimada et al. 1972; Weisberg and Landy 1983). This can occur because the normal insertion site is already occupied by a prophage, including by defective prophages (Campbell 1996). Resulting insertions into secondary integration sites can give rise to gene inactivation via insertional mutagenesis (Shimada et al. 1972). This can give rise to negative fitness consequences for the resulting lysogen to the degree that the prophage is less able to compensate in terms of supplying gene-completing sequence. As pointed out by Brüssow et al. (2004), gene disruptions can occur even if the second site is also a tRNA. They then go on to describe the resulting reduced bacterial fitness as a (p. 577), "negative lysogenic conversion phenotype", though they do note as well that not all gene inactivations are costly to the carrying organism, an issue returned to below in terms of the evolution of phage resistance (e.g., Chaps. 22 and 25). The phage itself can also be directly impacted, as excision from the host chromosome in the course of induction can be negatively affected given insertion into non-primary locations (phage λ; Shimada et al. 1972), though that is not true for all phage types (phage P22; Rao and Smith 1968).

It follows that there should be selective advantages to prophages displaying better compensating end sequences to the genes they integrate into. Also, these sequences must be better adapted to inserting into some hosts than they are into others, i.e., given that genetic sequences between bacterial strains and species vary. Thus, natural selection should favor those temperate phages—in terms of lysogen formation, lysogen competitive ability, and to some degree phage functionality—which both succeed in inserting at their preferred insertion site and which otherwise provide more effective insertion-compensating sequence at their sites of insertion (keeping in mind, though, that not all prophages are insertional, i.e., versus existing as plasmids; Fig. 1.5).

For phage λ, secondary insertions occur at about 0.5% of the efficiency of insertion into its primary site if the latter is missing (Shimada et al. 1975). Though secondary insertions can result in less fit lysogens than when insertion occurs at primary sites, i.e., as due to insertion mutagenesis, not all bacterial genes may be equivalently susceptible to such secondary insertion (Shimada et al. 1972, 1973). Interestingly, and perhaps supportive of the idea of lower host fitness, the burst sizes of induced λ lysogens mostly were lower given use of a secondary insertion site (9 of 11 "Groups" of secondary insertion sites) versus primary insertion, though in two cases instead burst sizes actually were higher, i.e., 160 or 200 rather than 110 phages/bacterium (Shimada et al. 1972), and can be higher as well for phage P22 secondary site insertions (Rao and Smith 1968). It is also possible for prophages to 'insert' into non-preferred sites by recombining with existing prophages, including with defective prophages (Campbell 1996), with fitness consequences that presumably are situation dependent.

Likely balanced to a degree by the costs of such insertional mutagenesis, there also appears to exist diversifying selection for primary insertion sites (Campbell 2003; Nilsson and Ljungquist 2006). This presumably would result in reduced competition for insertion sites among temperate phages as they come to seek out different locations on bacterial chromosomes within which to integrate, i.e., as a form of stabilizing frequency-dependent selection (Chap. 24, Sect. 24.6). In addition, some temperate phages possess better defined secondary insertion sites, e.g., phage P2 in *E. coli* K12 can access multiple secondary insertion sites, with (p. 375) "up to 37% mismatches within the core sequence" (Nilsson and Ljungquist 2006). Presumably insertional mutagenesis into these similar but not identical insertion sites is less severe than were secondary insertion sites otherwise unrelated to primary sites.

5.2.3 Insertions by Non-phage Mobile Genetic Elements

Some integrative and conjugative elements (ICEs, a.k.a., conjugative transposons)—which could be viewed as the plasmid equivalent of integrative prophages—also integrate into bacterial chromosomes at preferred locations. Other ICEs are less particular about what locations they insert into, as also can be seen among certain integrative prophages (see phage Mu, below). Paralleling prophages, conjugative elements thus may be differentiated into those that do not integrate into the bacterial chromosomes, those that integrate at specific locations, and those that are less choosy in where they integrate. Furthermore, in at least in one case, ICE insertion into non-preferred sites—as occurs spontaneously for about 10^{-4} or more insertions as measured for the *B. subtilis* ICE, ICE*Bs1*—has been shown to reduce the fitness of both ICE and host (Menard and Grossman 2013). As noted by those authors (p. 1),

> Each strategy for integration has its benefits. The more promiscuous elements can acquire a wider range of genes adjacent to the integration sites, and their spread is not limited to organisms with a specific attachment site. On the other hand, site-specific elements are much less likely to disrupt important genes.

So too for integrative phages: promiscuous insertions can result in "novel types of specialized transducing phages" (p. 485), though insertion efficiency is reduced, for phage λ, by about 200 fold (Shimada et al. 1972). Plasmid prophages, and indeed plasmids generally, would meanwhile seem to sidestep all of these various issues (de Jonge et al. 2019), though at a cost of requiring alternative means of assuring segregation to daughter bacteria.

Another form of insertion involving prophages is their own acquisition of insertion elements, which are forms of transposable elements, such as IS2 in *E. coli* (Sengstag and Arber 1983). While potentially leading to formation of defective prophages (Sengstag and Arber 1983), if the phage remains capable of producing infective virions, then this can serve as one means of infectious acquisition of insertion elements by bacteria, particularly in addition to via the transfer of insertion element-carrying plasmids. Larger transposable elements, though, should be less efficiently transferred to bacteria by temperate phages to the extent that transposable element size would interfere with encapsidation of sufficient phage DNA to allow for subsequent establishment of a lysogenic cycle.

5.3 For Phage Mu, Insertional Mutagenesis Is a Key Aspect of Its Infection Process

Notwithstanding advantages to temperate phages of utilizing preferred insertion sites, some temperate phages have dispensed with primary or even secondary insertion sites altogether. The best studied of such phages is the coliphage Mu. Here 'Mu' literally stands for 'Mutation' and these mutations are insertional (Paolozzi and Ghelardini 2006; Harshey 2014). A number of other such phages are

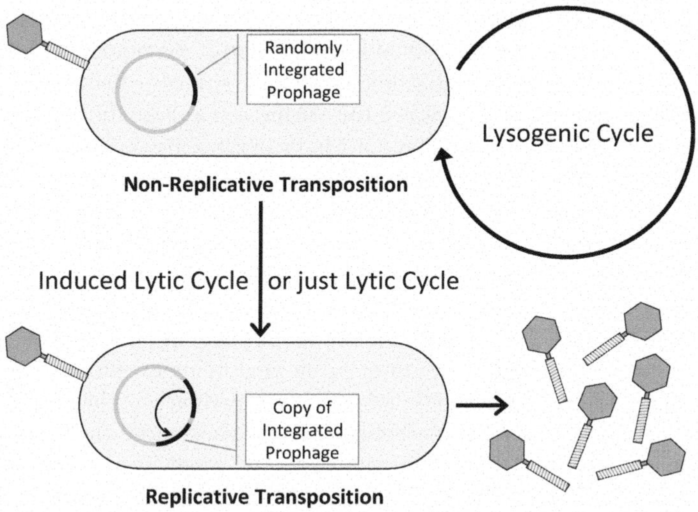

Fig. 5.3 Infection cycle of temperate phage Mu. See the main text for additional discussion

known, especially as prophages, which have Mu-like characteristics (Toussaint and Rice 2017). As the basis of its insertional mutagenesis, phage Mu serves not just as a temperate bacteriophage but also as a transposon.

An outline of phage Mu's lifecycle is shown in Fig. 5.3 and described as follows: Upon genome uptake, the phage Mu chromosome is circularized and then undergoes what is known as a non-replicative or conservative transposition. That is, Mu DNA is not subject to replication prior to this process and thus non-replicative transposition involves only that DNA that had been carried into the bacterium from the adsorbing virion. The process then involves insertion of the Mu chromosome into the bacterial chromosome at random locations. As these locations vary with lysogen initiation, each new lysogen in effect possesses an insertion mutation in a new location. Though seemingly equivalent to the generation of a chromosomally inserted prophage, in fact in no more than about 10% of infections (1–10%) does Mu insertion result in establishment of a lysogenic cycle, despite all infections first going through non-replicative transposition and therefore insertion of the Mu genome into the host chromosome. For the other 90 to 99% of Mu infections, a lytic cycle is established, resulting of course in the death of the infected bacterium and therefore no establishment of a Mu-mutagenized bacterial lysogen. It is during the lytic cycles that Mu subsequently displays what is known as replicative transposition, though as already noted, successful lytic cycles do not have an evolutionarily relevant mutational impact on bacteria as the host bacteria are killed.

Mu lysogens, as a consequence of Mu-mediated insertional mutagenesis, can contain detrimental mutations. In terms of sites of insertion, in fact Harshey (2014, p. 680), states that, "Mu is the most promiscuous of known mobile elements." For instance, 1–3% of Mu lysogens are auxotrophic mutants (Paolozzi and Ghelardini 2006). Mu, though, does seem to have preferred insertion areas, and also may be biased in its integration against actively transcribed regions of the bacterial chromosome. Thus, perhaps phage Mu, though molecularly committed to harming its bacterial host upon lysogen establishment, at least strives to avoid harming that host excessively, at least in the short term. At the same time, though of unknown utility so far as I am aware, acquisition of Mu prophages also is associated with the release of short pieces of DNA, 100–150 base pairs in length, that had flanked the inserted Mu genome prior to its packaging. That DNA comes from the previously infected bacterium and thus Mu lysogens not only randomly mutagenize their hosts obligatorily but also are randomly transducing phages. In a more traditional sense, however, Mu is also a genuine generalized transducing phage (Howe 1973).

Golais et al. (2013) argue that (pp. 183–184), "It seems as if the persistence of Mu prophage in bacterial genome was an ultimate goal of its evolution, and the lytic development associated with the production of virus particles might be considered as an evolutionary conserved relict mechanism." This speculation was based on a lack of known mechanisms of induction of phage Mu prophages. That perspective, however, is clearly biased toward thinking about phage Mu from the perspective of its lysogenic cycles rather than lytic cycles. Thus, an alternative interpretation is that phage Mu is primarily a lytically infecting phage which only occasionally takes on a lysogenic life style, a perspective indeed that may be the case for most temperate phages. This is because all Mu infections start with conservative transposition, with only a fraction of those insertions just happening to progress into prophages. When studying exclusively bacteria and their properties, rather than as hosts for lytic phage infections, it of course is only the prophages that one is aware of, presumably thus potentially resulting in such biased thinking.

The big question I would then ask is whether phage Mu's propensity to mutagenize its host is in some manner adaptive (a feature!) or instead simply something that the phage must put up with given its transpositional infection strategy (a bug. . .). I'm unsure, however, how often insertion mutations might be beneficial to the affected bacterium and therefore to the resulting lysogen, though see Davies et al. (2016) for experimental evidence of the occurrence at least temporarily beneficial insertions in *Pseudomonas aeruginosa* by the randomly inserting phage ϕ4. Notwithstanding the Davies et al. example, and see also (O'Brien et al. 2019; Marshall et al. 2021), I would default in assuming that phage Mu's eponymous feature tends to be a cost of its infection strategy rather than existing as a means of aiding its lysogenic cycles; from Hendrix and Casjens (2008, p. 92), "Presumably, the rare occasions when such phages integrate into essential genes are not detrimental enough to the phage to force a lifestyle change." Alternatively, as it is only those lysogens that survive to be discovered which are known and studied, it would not be at all surprising were there a discovery bias in terms of insertions of Mu-like prophages into sites in bacterial chromosomes that are at least less costly to the resulting lysogen if not necessarily also beneficial.

5.4 CRISPR Spacer Sequences Are and Are Not Insertion Mutations

A very specialized, bacteria-mediated, but nevertheless often phage-driven form of insertional mutagenesis is seen with what are known as CRISPR-Cas systems. Use of the term ‘mutagenesis’ here might be viewed as misleading, but I will suggest that it is not. Use of the qualifier ‘insertional’, however, is a bit inaccurate. If we define mutation broadly as a change in nucleotide sequence, then any addition of nucleotides is a mutation (though perhaps not quite *any*, as one could argue effectively against the impact of telomerase enzymes as mutational). An *insertion* mutation, on the other hand, generally is disruptive of nucleic acid-encoded information, or at least is *potentially* disruptive, and that is certainly *not* the case with CRISPR-Cas systems. Indeed, CRISPR-Cas systems are set up so that these insertions of new DNA explicitly do *not* result in disruption of otherwise useful surrounding bacterial DNA.

So what then are CRISPR-Cas systems? Basically, they represent an aspect of bacterial immune systems against phage DNA sequence, as well as against other ‘dangerous’ forms of especially infectious DNA. Unlike deletion mutagenesis, which is posited to operate after acquisition of foreign DNA, CRISPR-Cas systems operate prior to foreign DNA establishment. Furthermore, and truly unique to CRISPR-Cas systems, at least so far as we know, CRISPR-Cas systems represent a form of bacterial (and archaeal) adaptive immunity (Marraffini 2015), whereas all other forms of bacterial immunity against, e.g., phages, can be viewed instead as innate (Abedon 2012). All of these various, often not-chromosomally encoded anti-phage systems (Chaps. 16 and 18 through 20), whether innate or adaptive, and unlike deletion mutagenesis (Chap. 6), act especially by first allowing bacteria to distinguish self (bacterial) from non-self (e.g., phage), often in terms of DNA properties (Goldberg and Marraffini 2015; Gurney et al. 2019).

CRISPR-Cas itself stands for Clustered Regularly Interspaced Short Palindromic Repeats-CRISPR-associated (proteins). The nucleic acid-sequence repeats represent the DNA framework within which CRISPR spacer sequences are placed. Spacer sequences in turn represent the novel changes to the bacterial nucleic acid (i.e., the ‘mutations’/new alleles) that are associated with the adaptive aspect of CRISPR-Cas systems (Fig. 5.4). These spacer sequences are short segments of DNA, tens of bases long, that have been acquired from foreign DNA that has found its way into the cytoplasm of a cell. This foreign DNA carries ‘proto-spacer’ sequences which are acquired as spacer sequences, while spacer sequences are used to recognize future foreign DNA and thereby protect a cell from exposure to entities containing this same sequence (Marraffini 2015). The spacers are added sequentially so that older spacers are found at one end of spacer arrays and newer spacers are found at the other end. In principle, it is possible to match spacer sequences to presumptive proto-spacer sequences and gain an appreciation both of past exposure of a lineage to a given virus, but also to appreciate the relative order in which the lineage has been

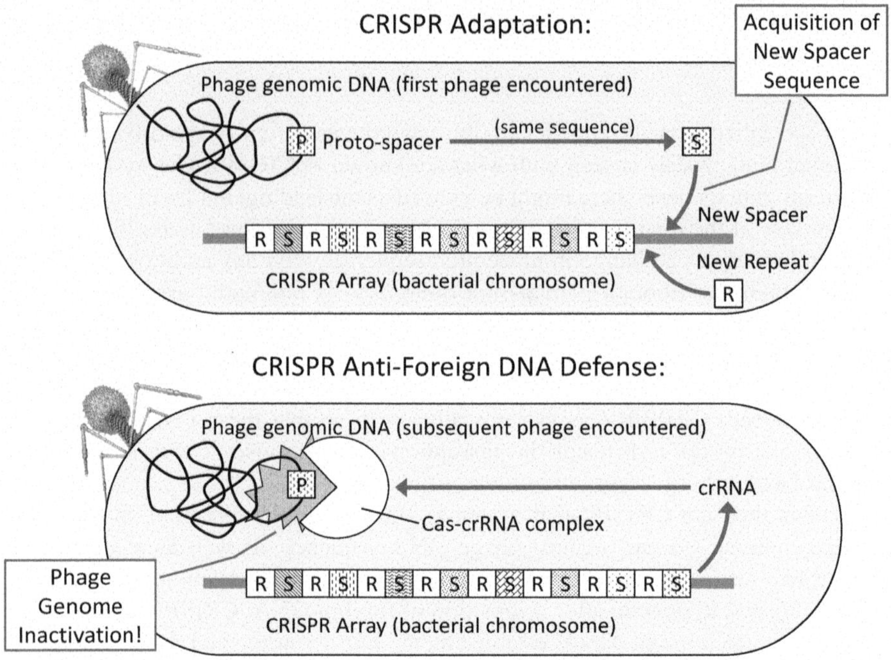

Fig. 5.4 CRISPR adaptation and use as an anti-foreign DNA defense. Not shown are the *cas* genes which are found adjacent to the shown CRISPR Array. Note the consistency of the repeat sequence ("R", indicated as same background pattern) and inconsistency of the spacer sequences ("S", indicated using different background patterns). The active RNA products of CRISPR arrays are indicated as crRNA and are used to recognize proto-spacer sequences encoded by foreign DNA

exposed to each proto-spacer sequence, i.e., older exposures to newer exposures and even latest exposure (Andersson and Banfield 2008).

Part of the process of CRISPR-Cas functioning involves not just acquisition of new spacer sequences but also loss of old spacer sequences. Similarly, part of the process of acquisition of new prophages is also the eventual mutational loss of prophages from bacterial genomes. Indeed, generally speaking, bacteria gain DNA sequences and bacteria mutationally lose DNA sequences over time, resulting in the bacterial genomes that we see today. The removal of DNA occurs via deletions, as is the topic of the following chapter (Chap. 6).

References

Abedon ST (2012) Bacterial 'immunity' against bacteriophages. Bacteriophage 2:50–54

Andersson AF, Banfield JF (2008) Virus population dynamics and acquired virus resistance in natural microbial communities. Science (New York, NY) 320:1047–1050

Brüssow H, Canchaya C, Hardt WD (2004) Phages and the evolution of bacterial pathogens: from genomic rearrangements to lysogenic conversion. Microbiol Mol Biol Rev 68:560–602

Campbell AM (1992) Chromosomal insertion sites for phages and plasmids. J Bacteriol 174:7495–7499

Campbell AM (1996) Cryptic prophages. In: Neidhardt FC (ed) *Escherichia coli* and *Salmonella* cellular and molecular biology. ASM Press, Washington, DC, pp 2041–2046

Campbell A (2003) Prophage insertion sites. Res Microbiol 154:277–282

Campbell AM (2006) General aspects of lysogeny. In: Calendar R, Abedon ST (eds) The bacteriophages. Oxford University Press, Oxford, pp 66–73

Coleman D, Knights J, Russell R, Shanley D, Birkbeck TH, Dougan G, Charles I (1991) Insertional inactivation of the *Staphylococcus aureus* β-toxin by bacteriophage ϕ13 occurs by site- and orientation-specific integration of the ϕ13 genome. Mol Microbiol 5:933–939

Davey RB, Reanney DC (1980) Extrachromosomal genetic elements and the adaptive evolution of bacteria. In: Hecht MK, Steere WC, Wallace B (eds) Evolutionary biology, vol 13. Plenus Press, New York, pp 113–147

Davies EV, James CE, Williams D, O'Brien S, Fothergill JL, Haldenby S, Paterson S, Winstanley C, Brockhurst MA (2016) Temperate phages both mediate and drive adaptive evolution in pathogen biofilms. Proc Natl Acad Sci USA 113:8266–8271

de Jonge PA, Nobrega FL, Brouns SJJ, Dutilh BE (2019) Molecular and evolutionary determinants of bacteriophage host range. Trends Microbiol 27:51–63

Dempsey RM, Carroll D, Kong H, Higgins L, Keane CT, Coleman DC (2005) *Sau*42I, a *Bcg*I-like restriction-modification system encoded by the *Staphylococcus aureus* quadruple-converting phage ϕ42. Microbiology 151:1301–1311

Drake JW, Glickman BW, Ripley LS (1983) Updating the theory of mutation. Am Scientist 71: 621–630

Feiner R, Argov T, Rabinovich L, Sigal N, Borovok I, Herskovits AA (2015) A new perspective on lysogeny: prophages as active regulatory switches of bacteria. Nat Rev Microbiol 13:641–650

Fortier LC, Sekulovic O (2013) Importance of prophages to evolution and virulence of bacterial pathogens. Virulence 4:354–365

Golais F, Holly J, Vitkovska J (2013) Coevolution of bacteria and their viruses. Folia Microbiol 58: 177–186

Goldberg GW, Marraffini LA (2015) Resistance and tolerance to foreign elements by prokaryotic immune systems - curating the genome. Nat Rev Immunol 15:717–724

Gurney J, Pleska M, Levin BR (2019) Why put up with immunity when there is resistance: an excursion into the population and evolutionary dynamics of restriction-modification and CRISPR-Cas. Philos Trans R Soc Lond Ser B Biol Sci 374:20180096

Harshey RM (2014) Transposable phage Mu. In: Mobile DNA. American Society for Microbiology, Washington, DC, pp 669–691

Hendrix RW, Casjens SR (2008) The role of bacteriophages in the generation and spread of bacterial pathogens. In: Hensel M, Schmidt H (eds) Horizontal gene transfer in the evolution of pathogenesis. Cambridge University Press, Cambridge, UK, pp 79–112

Howe MM (1973) Transduction by bacteriophage MU-1. Virology 55:103–117

Marraffini LA (2015) CRISPR-Cas immunity in prokaryotes. Nature (London) 526:55–61

Marshall CW, Gloag ES, Lim C, Wozniak DJ, Cooper VS (2021) Rampant prophage movement among transient competitors drives rapid adaptation during infection. Sci Adv 16:eabh1489

Menard KL, Grossman AD (2013) Selective pressures to maintain attachment site specificity of integrative and conjugative elements. PLoS Genet 9:e1003623

Nilsson AS, Ljungquist EH (2006) The P2-like bacteriophages. In: Calendar R, Abedon ST (eds) The bacteriophages. Oxford University Press, Oxford, pp 365–390

O'Brien S, Kummerli R, Paterson S, Winstanley C, Brockhurst MA (2019) Transposable temperate phages promote the evolution of divergent social strategies in *Pseudomonas aeruginosa* populations. Proc Biol Sci 286:20191794

Paolozzi L, Ghelardini P (2006) The bacteriophage Mu. In: Calendar R, Abedon ST (eds) The bacteriophages. Oxford University Press, Oxford, pp 469–496

Rao RN, Smith HO (1968) Phage P22 lysogens of a *Salmonella typhimurium* mutant deleted at the normal prophage attachment site. Virology 36:328–330

Sengstag C, Arber W (1983) IS2 insertion is a major cause of spontaneous mutagenesis of the bacteriophage P1: non-random distribution of target sites. EMBO J 2:67–71

Shimada K, Weisberg RA, Gottesman ME (1972) Prophage lambda at unusual chromosomal locations. I. Location of the secondary attachment sites and the properties of the lysogens. J Mol Biol 63:483–503

Shimada K, Weisberg RA, Gottesman ME (1973) Prophage lambda at unusual chromosomal locations. II. Mutations induced by bacteriophage lambda in Escherichia coli K12. J Mol Biol 80:297–314

Shimada K, Weisberg RA, Gottesman ME (1975) Prophage lambda at unusual chromosomal locations. III. The components of the secondary attachment sites. J Mol Biol 93:415–429

Toussaint A, Rice PA (2017) Transposable phages, DNA reorganization and transfer. Curr Opin Microbiol 38:88–94

Ventura M, Canchaya C, Pridmore D, Berger B, Brüssow H (2003) Integration and distribution of *Lactobacillus johnsonii* prophages. J Bacteriol 185:4603–4608

Weisberg RA, Landy A (1983) Site-specific recombination in phage lambda. In: Hendrix RW, Roberts JW, Stahl FW, Weisberg RA (eds) Lambda II. Cold Spring Harbor Laboratory, Cold Spring Harbor, NY, pp 211–250

Williams KP (2002) Integration sites for genetic elements in prokaryotic tRNA and tmRNA genes: sublocation preference of integrase subfamilies. Nucl Acids Res 30:866–875

Chapter 6
Bacterial Mutation Associated with Phages: Deletions

Contents

Abstract This chapter addresses in particular the article written by Lawrence et al. (Trends Microbiol. 9:535–540, 2001), titled, "Where are the pseudogenes in bacterial genomes?" The authors in that article suggest that bacteria can display elevated rates of deletion mutagenesis especially toward removing detrimental or potentially detrimental prophages from their genomes. While I have long had an appreciation of this hypothesis, here I suggest that it may not necessarily be prophages that are driving an existence of elevated rates of deletion mutations in bacteria. Instead, I point to a countering suggestion that it is bacterial pseudogenes that may instead or in addition be the primary targets of these deletion mutation rates, though with the caveat that the relative costs of prophages versus pseudogenes may vary depending on the bacterium. Discussed as well is the utility to bacteria of generating mutations to interfere with new phage infections, i.e., as a mechanism of phage resistance. Briefly considered also is the potential for prophages to contribute to bacterial genome rearrangements, as can give rise to bacterial resistance to phages as well.

> There appears to be a strong 'deletion' bias in bacterial genomes that prevents the accumulation of transposons, pseudogenes, and other segments of DNA that are not under positive selection for function.—Jeffrey G. Lawrence and Heather Hendrickson (2008, p. 10)

The topic of this chapter is explicitly inspired by the 2001 opinion article by Lawrence et al. (2001), titled, "Where are the pseudogenes in bacterial genomes?" Unlike insertion mutations that are directly caused by phage action, i.e., prophage integration (Sect. 5.2.1), instead the deletion mutagenesis discussed in this chapter

S. T. Abedon, *Bacteriophages as Drivers of Evolution*,
https://doi.org/10.1007/978-3-030-94309-7_6

explicitly is only 'indirectly' caused by phage action, at least as hypothesized, though still hypothesized to be 'caused' by inserted prophages. Specifically, bacterial genomes may be more prone to acquiring deletions because deletion mutations can provide the benefit of curing, fully or partially, bacterial lineages of inserted prophages, though at the same time leaving plasmids and plasmid prophages untouched. Spoiler alert: I will be arguing against at least the universal correctness of the Lawrence et al. hypothesis. I should note as well that I will not be considering as an alternative that individual deletions may be selected in bacteria solely for the sake of limiting bacterial genome sizes and this is even though ultimately smaller genome sizes may be beneficial to bacteria, i.e., a consideration as consistent with the perspective of Mira et al. (2001).

6.1 Pseudogenes

A pseudogene, as the term suggests, is a gene-like entity—as defined in terms of its nucleotide sequence—but that both does not and cannot serve any functional purpose to the organism it is found in. Pseudogenes mostly exist because of a mutational degradation of a functional gene, either (1) because the functional gene is no longer needed due to longer-term ecological or physiological changes, (2) because the gene is a duplication of an existing gene so therefore is redundant, or (3) because of fixation of otherwise deleterious, pseudogene-creating mutations within populations due to stochastic processes (see Chapters 7 and 8 for the latter). Basically, if a gene is not subject to some degree of natural selection for its retention and/or if a gene cannot produce a functional product, then over time mutations will accumulate in that gene that will serve to further degrade its potential for usefulness. Pseudogenes thus tend to represent genomic 'debris' on two levels: (1) they don't work, and (2) even if they did work (and therefore weren't pseudogenes) they probably still might not be useful to the encoding organism.

Pseudogenes tend to be physically removed from genomes solely by randomly acting deletion mutagenesis, thereby reducing the amount of DNA carried by an organism. As deletion mutations are random in terms of what DNA is deleted, however, even individual deletion mutations that include pseudogenes will not necessarily be useful to an organism, depending on what else is deleted along with the pseudogene. Thus, pseudogene removal by bacteria is certainly possible, but isn't necessarily easily accomplished, particularly without cost.

6.2 Genomic Immunity Against Prophages?

With these ideas in mind, note that Lawrence et al. (2001) suggested that many bacteria tend to possess few pseudogenes because the loss of those pseudogenes is a byproduct of a utility to trimming, over time, both inserted prophages and transposons from bacterial genomes; together, this is removal of "Dangerous" DNA. More generally, organisms that possess genomes that are consistent in size over evolutionary time can be viewed as displaying a genome-size steady state, which means that additions of genetic material—either by gene duplications, horizontal gene transfer of various means, transposons, acquisition of integrating plasmids, or acquisition of integrating prophages—must be balanced by some sort of removal of genetic material. As Lawrence et al. state (p. 535),

> We propose that the influx of dangerous genetic elements such as transposons and bacteriophages selects for the maintenance of relatively high deletion rates in most bacteria; the sheltered lifestyle of intracellular parasites removes this threat, leading to reduced deletion rates and larger pseudogene loads.

This idea analogously suggests that deletion-mutation tendencies serve lineages as genomic immune systems, allowing bacteria to remove 'foreign' DNA over time, albeit inefficiently. Such a genome 'immune system' function would only be needed, however, to the extent that organisms are exposed to foreign DNA that is potentially "dangerous" to the carrying organisms. As bacteria that exist as intracellular parasites are less exposed to foreign DNA, so too, as they argue (see the quotation immediately above), their anti-foreign DNA 'immune system' need not be as robust, though at a cost also of not efficiently removing pseudogenes.

6.3 Evidence for the Existence of Genomic Immunity?

Consistent with deletion mutations removing prophage DNA, Bobay et al. (2014) found that a substantial fraction, e.g., 37% or more of prophages they identified in *E. coli* and *Salmonella enterica* likely possessed deletions as based on their smaller than expected sizes for prophages in these hosts (less than 30 kb). That same study also suggests that potential costs associated with deleting useful genes along with prophage genes could be reduced when the size of deletions in nucleotides are smaller rather than larger, and that likely there indeed are biases toward smaller rather than larger deletions, with median deletion lengths of 500 nucleotides. Thus, an anti-prophage genomic immune system in bacteria that are commonly exposed to prophages may exist that may even be tuned, i.e., with smaller deletion sizes, toward increasing the potential to delete prophage DNA without also deleting surrounding DNA. On the other hand, all phages have genomes that are somewhat larger in size than 500 nucleotides (Abedon 2011, Hyman and Abedon 2012).

Even with that tuning toward smaller deletions, and unlike actual immune systems, such genome editing presumably would have a high likelihood of attacking 'self' DNA rather than just foreign DNA. Thus, the immune system analogy breaks down, but rather than avoiding attacking self through physiological or molecular mechanisms, natural selection instead is relied upon to purge inappropriate deletions from populations. Thus, deletion mutations that result in removal of costly, e.g., dangerous, and otherwise less useful DNA sequence would tend to increase in frequency within populations—as allele equivalents—due to natural selection, though only if the *benefits* to the carrying organisms are sufficiently high. At the same time, deleterious deletion mutations such as those attacking self would be expected to be removed from populations (decrease in frequency), though equivalently only if the *costs* of continuing to possess those mutations also are sufficiently high.

This then brings us back to pseudogenes not being rare in various especially intracellularly parasitic bacteria lineages. This could be because there is no selection in place for removal of dangerous (foreign) DNA in these bacteria, but it also could be that costs of inadvertently deleting potentially useful DNA is higher. That is, if a population were less able to recover useful genes/genotypes, such as via various forms of horizontal gene transfer (Chapters 8 and 9), or less able to avoid fixing mildly deleterious alleles due to small population sizes (Chap. 7), then displaying a typical propensity toward deletion mutations may be not just less useful but could be more costly as well. Indeed, just the fact that these lineages seem to accumulate mutations would seem to suggest that including yet additional means to generate mutations, such as more normal propensities toward deletion mutations, might be excessively costly.

6.4 Just how Dangerous Are Prophages?

A further quibble that I have with these ideas, though generally I feel that the concept of bacteria possessing genomic 'immune systems' as Lawrence et al. (2001) implicitly suggest is fairly brilliant, is that I am not convinced that individual prophages are necessarily all that costly to bacteria. Specifically, the concern expressed by Lawrence et al. is that prophages serve as,

> [1] . . .dangerous molecular time bombs that can kill the cell upon their eventual induction. More insidiously, prophage induction [2] does not kill just one host cell, but will systematically lyse many of the descendents of the original lysogen which can experience coincident lysogen induction when environmental conditions change appropriately. In addition, [3] many descendants of uninfected siblings of the initially infected cell will also perish, as they will inhabit the same or very similar niches.

There are three ideas there to unpack, as indicated by the numbers added within the brackets.

First, while I agree that induction should be of obvious cost to a bacterium, since resulting lytic infections kill the induced lysogen, I am not convinced that rates of spontaneous lysis are always high enough for this cost to be efficiently impacted by natural selection. These, for example, can range from 0.5 to 1% for *Acinetobacter calcoaceticus* up to 1 to 2% for *Staphylococcus epidermidis* but down to 0.01 to 0.1% for *E. coli* K12λ (Berenstein 1986). In addition, not only can prophages be beneficial to bacteria in the uninduced state (see lysogenic conversion as well as superinfection immunity, Chapters 15 and 16), but even occasional spontaneous inductions might also be beneficial at least to the carrying genotype (see lysogen allelopathy; Chap. 24, Sect. 24.4). In addition, bacteria have means of inhibiting prophage inductions without deleting prophages, called xenogeneic silencing, though phages in turn can evolve means of overcoming these mechanisms (Pfeifer et al. 2019). Note, though, that generally speaking, while Iranzo and Koonin (2018) might agree with this assertion that prophages can be sufficiently deleterious to be worth it to bacteria to get rid of, they do temper that idea with this caveat (p. 7): "Because of these multiple fitness effects of genetic parasites, any evolutionary meaningful estimate of their cost must consider averages over long times and multiple environmental conditions." Thus, at a minimum, I would argue against the universality of the idea that bacteria display high rates of deletion mutations specifically for the sake of deleting prophages because of the costs of continuing to carry those prophages.

The second idea references the fact that prophages can be induced by agents such that induction of a lysogen population might occur *en masse*. Obviously, this could be costly particularly to a lysogen population. The usual example of how this may occur involves host DNA-damage motivated prophage induction, such as seen upon host exposure to ultraviolet radiation or to the DNA damaging agent, mitomycin C. In this case, the prophage in effect is making a calculation, i.e., following an evolutionary algorithm that says that a phage is more likely to survive as virions than as a lysogen, e.g., such as because the target size of individual virions in terms of being DNA damaged is smaller than that of host bacteria. In addition, the number of individual virion targets will be even larger following induction, as resulting in the production of a burst size of virions, should the causing DNA damaging agent persist (Abedon 2008). That is, would you rather be one of 100 ducks (new phage virions) when someone is shooting a shotgun at you, or instead one lone turkey (an uninduced lysogen)?

Third, the idea of a bacterial population deleting a prophage to avoid killing otherwise clonally related bacteria that happen to be not carrying the same prophage sounds at best to be selectively weak. How, for example, might we expect one clonal population (the one that could be carrying the prophage) to be protective of a second, clonal population, the one that is vulnerable because it is not carrying the prophage? My guess is that we cannot expect this occur. Indeed, the time frames over which prophage deletions might occur likely are very long relative to the time frames over which two hypothetical related bacterial populations, one containing the prophage in question and the other not, may be found physically adjacent. That is, from a perspective of kin selection, it is difficult to imagine that one bacterial population will acquire a prophage by only a single member and then for that member to protectively lose that prophage deletionally before either replicating (via binary fission, and thereby providing a requirement for deletional removal of the prophage from more than one bacterial genome) or generating killer virion progeny.

In the second scenario, if indeed the prophage is responding to inducing signals because their evolutionary history indicates that their survival is more likely as a lytic cycle than as a lysogenic cycle, then that would suggest that the bacteria in these scenarios likely would die anyway, independent of lysogen induction. In my opinion, however, this could still be the strongest argument for why carrying prophages can be costly to bacteria. Nevertheless, I would want to understand a great deal more about lysogenic cycle biology and ecology before reaching any firm conclusions as to the possible validity of this second selective scenario. That is, prophages would have to tend to be induced despite reasonably high potentials for lysogens to survive. In addition, these mass inductions would have to be over relatively broad areas or volumes else, at an extreme, we would be simply approximating the first scenario of relatively small fractions of lysogen populations being affected. Thus, we would have to consider as selective the perhaps narrow set of circumstances in which large fractions of bacterial populations are exposed to sufficient quantities of prophage-inducing agents that prophages are induced but the bacteria nevertheless would have survived the DNA damage had they not been carrying prophages. In addition, the deletions protecting the bacterial population from this outcome would have had to have occurred well before the bacterial population came into existence since, as with the third scenario, while prophage deletion from one bacterium should be a fairly low likelihood event, independent deletion of the same prophage from two or more clonally related bacteria should be especially unlikely.

6.5 Just how Dangerous Are Pseudogenes?

Above I suggest that perhaps prophages are not sufficiently dangerous or at least that they are insufficiently costly to their bacterial carries to warrant the existence of mechanisms of elevated occurrence of deletion mutations to get rid of them. Kuo and Ochman (2010) interestingly suggest that pseudogenes themselves, particularly ones that encode still-expressed though defective, indeed "Toxic" proteins, could be costly to their bacterial carriers, thereby selecting for mechanisms specifically for pseudogene deletional removal. Thus, rather than natural selection favoring elevated rates of occurrence of deletion mutations toward removal of prophages, thereby resulting also in the deletion of what otherwise would be excessive numbers of pseudogenes, it could be that natural selection favors those same elevated rates of deletion mutagenesis toward removal of pseudogenes, thereby resulting also in the deletion of prophages.

An interesting twist on this latter idea is provided by Shen et al. (2018). They suggest that bacteria self-generate deletion mutations to mutate to phage resistance, in this case resistance to lytic infections rather than to eliminate prophages. One could ask, why bother mutating the gene encoding, e.g., a phage receptor by deletion when just a point mutation alone could do? The hypothesis of Kuo and Ochman (2010) from the previous paragraph, however, could provide the answer, i.e., non-deletion mutations could result in dangerous pseudogenes that are still expressed, while deletion mutations might be less likely to give rise to such pseudogenes. Indeed, a receptor protein that is non-functional to the expressing bacterium in principle could still be functional to an adsorbing phage, again selecting for receptor gene deletion over simply receptor gene mutation. The question I would have regarding this hypothesis, however, is one of just how important have deletion mutations been toward the evolution of phage resistance? Also, even if a deletion mutation is less costly, is there really also going to be, e.g., bacteria with point mutations to phage resistance that deletion mutants are going to be competing with, thereby demonstrating the selective superiority of those deletion mutations?

In addition, though deletions may be less costly on a per-gene basis, deletion mutations as noted also have a tendency to delete not just genes that are under selection to be removed, that is, deletion mutations often are not so 'clean' as to impact only a single open reading frame. For example, see Markwitz et al. (2021) for just such costs as associated with phage resistance achieved via deletion mutations in the range of 20,000 to 500,000 base pairs. Indeed, by impacting neighboring genes, deletion mutations themselves might result in the creation of still-expressed pseudogenes. With the Kuo and Ochman (2010) hypothesis, the need for these deletions would be less intense, essentially countering ongoing metabolic costs, than could be the need for bacterial evolution of phage resistance in the face of an abundance of lytically infecting phages. Thus, and overall, I have difficulty imagining that the existence of prophages in bacteria would select for increased rates of occurrence of deletion mutations and I have difficulty as well imagining that lytic phages would select for increased rates of occurrence of deletion mutations, but the idea that these greater than expected rates of occurrence of deletion mutations could exist in order to gradually remove more generally toxic bacterial proteins I find appealing.

Of additional relevance, while point mutations at least in principle are mutationally reversible, including potentially in the course of horizontal gene transfer, deletion mutations are much less easily reversed (reversion mutation). Thus, generating deletion mutations would seem to be a rather extreme means of achieving phage resistance, from a longer-term evolutionary perspective. Chromosomal rearrangements, by contrast, are in principle completely reversible, and as described by Scott et al. (2007), in fact also can give rise to resistance to phages. Indeed, reversion of these rearrangements, that is, recovery of the original phenotype upon further propagation and recombination, but in the absence of phage pressure, has been observed.

6.6 Bacterial Chromosomal Rearrangements and Prophages

In an interesting further twist, the genomic rearrangements observed by Scott et al. (2007) were themselves mediated by prophages, indeed in this case by Mu-like prophages. In fact, such intra-chromosomal recombination events that are mediated by similar sequences, as associated with more than one prophage being found within the same bacterial chromosome, are thought to be relatively common. Though involving homologous recombination (Brüssow et al. 2004, Menouni et al. 2015), they are not associated with sexual processes, so therefore can be described without qualification as phage-associated mutational processes. Just how commonly these rearrangements can lead to phage resistance, i.e., as observed by Scott et al. (2007), is however an open question.

References

Abedon ST (2008) Ecology of viruses infecting bacteria. In: Mahy BWJ, Van Regenmortel MHV (eds) Encyclopedia of Virology, 3rd edn. Elsevier, Oxford, pp 71–77

Abedon ST (2011) Size does matter – distinguishing bacteriophages by genome length (and 'breadth'). Microbiol Aust 32:95–96

Berenstein D (1986) Prophage induction by ultraviolet light in *Acinetobacter calcoaceticus*. J Gen Microbiol 132:2633–2636

Bobay LM, Touchon M, Rocha EP (2014) Pervasive domestication of defective prophages by bacteria. Proc Natl Acad Sci USA 111:12127–12,132

Brüssow H, Canchaya C, Hardt WD (2004) Phages and the evolution of bacterial pathogens: from genomic rearrangements to lysogenic conversion. Microbiol Mol Biol Rev 68:560–602

Hyman P, Abedon ST (2012) Smaller fleas: viruses of microorganisms. Scientifica 2012:734023

Iranzo J, Koonin EV (2018) How genetic parasites persist despite the purge of natural selection. EPL (Europhysics Letters) 122:58001

Kuo CH, Ochman H (2010) The extinction dynamics of bacterial pseudogenes. PLoS Genet 6: e1001050

Lawrence JG, Hendrickson H (2008) Genomes in motion: gene transfer as a catalyst for genome change. In: Schmidt H, Hensel M (eds) Horizontal Gene Transfer in the Evolution of Pathogenesis. Cambridge University Press, Cambridge, pp 3–22

Lawrence JG, Hendrix RW, Casjens S (2001) Where are the pseudogenes in bacterial genomes? Trends Microbiol 9:535–540

Markwitz P, Lood C, Olszak T, van Noort V, Lavigne R, Drulis-Kawa Z (2021) Genome-driven elucidation of phage-host interplay and impact of phage resistance evolution on bacterial fitness. ISME J 16:533–542

Menouni R, Hutinet G, Petit MA, Ansaldi M (2015) Bacterial genome remodeling through bacteriophage recombination. FEMS Microbiol Lett 362:1–10

Mira A, Ochman H, Moran NA (2001) Deletional bias and the evolution of bacterial genomes. Trends Genet 17:589–596

Pfeifer E, Hunnefeld M, Popa O, Frunzke J (2019) Impact of xenogeneic silencing on phage-host interactions. J Mol Biol 431:4670–4683

Scott AE, Timms AR, Connerton PL, Loc Carrillo C, Adzfa Radzum K, Connerton IF (2007) Genome dynamics of *Campylobacter jejuni* in response to bacteriophage predation. PLoS Path 3:e119

Shen M, Zhang H, Shen W, Zou Z, Lu S, Li G, He X, Agnello M, Shi W, Hu F, Le S (2018) *Pseudomonas aeruginosa* MutL promotes large chromosomal deletions through non-homologous end joining to prevent bacteriophage predation. Nucl Acids Res 46:4505–4514

Chapter 7
Genetic Drift and Phages

Contents

Abstract Genetic drift is a biological manifestation of the statistical concept of sampling error. That is, smaller-sized samples of populations display greater variance from expected values than larger-sized samples. Unlike natural selection, the resulting genetic drift, as is associated with small population sizes, is unbiased in terms of its impact on allele frequencies. Genetic drift thereby serves as a countering force to natural selection, with the action of natural selection weaker within smaller populations and the action of genetic drift weaker within larger populations. Phages can impose genetic drift on bacterial populations simply by reducing the sizes of bacterial populations, particularly as may be envisioned with the bacterial population being reduced in size in an unbiased manner, or at least with numbers of phage-sensitive bacteria not being reduced to zero. Discussed are general consequences of such reductions in bacterial numbers, i.e., as tending to result in the extinction of bacterial alleles, even potentially alleles that are selectively beneficial. Discussed as well is a special kind of genetic drift known as a founder effect, which potentially could be driven by the acquisition of prophages by bacteria.

> In other words, in small populations, the stochastic effects of random genetic drift overcome the effects of selection.—Laurent Duret (2008)

Mutations—that is, changes in the sequence of the genetic material making up an organism's genome—are the ultimate provider of the raw material upon which other evolutionary forces can act. Specifically, without mutation there can be no

S. T. Abedon, *Bacteriophages as Drivers of Evolution*,
https://doi.org/10.1007/978-3-030-94309-7_7

polymorphisms, i.e., allelic variation at individual loci, and thereby no genetic variation. Genetic migration and non-random mating can contribute to that genetic variation as well, with migration supplying mutations or combinations of alleles to a new population and non-random mating modifying the frequency of genotypes that are found within populations (these are discussed in Chaps. 9 and 10). Genetic drift and natural selection—contrasting mutation, migration, and non-random mating—instead are mechanisms of modification of existing allele or genotype frequencies, and in both cases this is accomplished in manners which explicitly have the effect of *reducing* the extent of genetic variation that is found within a population. Mutation along with migration thus add new variation to populations, non-random mating modifies to a degree how genetic variation is manifest going from individual to individual within a population, and then genetic drift and natural selection remove genetic variation from populations. This can almost be viewed as a cycle, at least if we ignore that all of the components can in fact be occurring simultaneously. That, is, the products of genetic drift and natural selection represent populations within which new mutations arise or are acquired via migration. These populations can then be subject to non-random mating, then to genetic drift and natural selection, and so on (Fig. 7.1). In this chapter we consider various aspects of the impact of phages on the occurrence of genetic drift in bacterial populations.

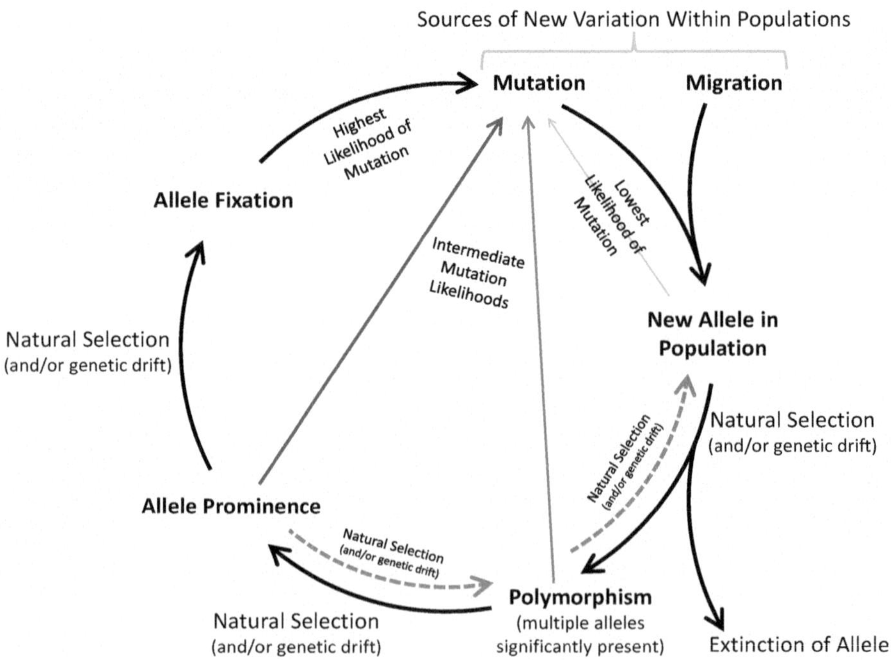

Fig. 7.1 Entrance and possible rise in frequency of new alleles within populations. Dashed lines represent reversals of the ascension of the frequency of a given allele. An assumption is made that the population under consideration otherwise remains relatively constant in size over time

7.1 Stochasticism Versus Determinism

What distinguishes genetic drift from natural selection is how their editing of genetic variation is manifest. With natural selection—as will be considered in substantially more detail starting especially with Chap. 23, but see also Chap. 3, Sects. 3.2.4 and 3.3 for an introduction—the forces involved can be described as deterministic. That is, knowing the specifics of how natural selection has acted in the past, we can predict with certainty what alleles or genotypes will increase in frequency in the future, otherwise holding environments constant and assuming an absence of genetic drift. This is the essence of the concept of Darwinian fitness as a predictive measure.

Thus, if we know that allele *A* has increased in frequency at the expense of allele *a* in the past, in environment X, then—assuming directional rather than stabilizing selection—we can describe carriers of allele *A* as possessing a higher fitness than carriers of allele *a*. That in turn means that we can predict that in the future allele *A* will increase in frequency at the expense of allele *a*, again holding the environment constant, since fitness measures can be environment specific. Indeed, these predictions typically are quantitative rather than simply qualitative. This is the essence of determinism: we can predict the specifics of future behavior based upon past behavior. Natural selection, as measured by its impact on the fitness of alleles, or on the fitness of whole genotypes, thus explicitly is a deterministic force.

Stochasticity by contrast is randomness. While we may be able to predict general behaviors of stochastic systems—e.g., allele *B* will go extinct over 100 generations with a probability of 0.25—in fact in a one locus, two-allele system operating solely as a function of stochasticity, so too allele *b* might go extinct after 100 generations with a probability of 0.25. Note that we are not predicting what the frequency of either allele may be after that 100 generations. Instead we are predicting only that one of the alleles, and we can't say with certainty which one, may no longer be present after that 100 generations. Such predictive power is equivalent to predicting that in tossing a coin three times in a row, heads will be thrown all three times. The odds of that happening are one-half raised to the third power, i.e., 0.125 or one eighth. We know therefore that throwing three heads might happen, and we can state what is our expectation that it will happen, but unlike with determinism, we cannot state with certainty the specifics of future outcomes. Genetic drift, contrasting the determinism of natural selection, thus is a stochastic process. We can calculate likelihoods, but unlike with natural selection, even given a perfect understanding of a system we cannot even in principle predict *specifically* what will occur in the future.

7.2 Sampling Error

Genetic drift is actually a specific example of a broader concept known as sampling error. Experimentally, often one samples a group rather than measuring the properties of all of the members of a group, e.g., determining the number of phages present within 100 μl rather than within the entire test tube. Since the entire group is not being measured, we cannot say with certainty that our sample is truly representative of the properties of all of the members of a group. That is, in not sampling the entire group, we are left with some degree of uncertainty. The greater the fraction of a group that we sample, however, then generally the lower that uncertainty will be, and particularly so to the degree that our sampling is not biased in some manner but instead is truly a random sampling of the larger group. The converse to this claim is that the less of the group that we sample, then the greater the uncertainty. Thus, if the group size is 100 and we measure all 100 individuals, then that obviously will provide less uncertainty than if we measure only 10. But if the group size were 1000 and we again measured 100 individuals, then that still would provide us with less uncertainty than if we only measured 10. This will not be as little uncertainty than had we sampled all 1000 members of the group, but still somewhat less uncertainty than if we measured only 10.

This is all somewhat equivalent to coin tossing as an example. If we throw three coins, then the odds of landing three heads is 0.125, but in fact 0.875 of the time we will not succeeding in throwing three heads. If you repeat these three coin tosses 10 times, however, you might throw three heads one-tenth (0.1) of the time. But if you repeat this 100 times, you will likely be closer to throwing three heads one-eighth of the time, and with 1000 throws closer still. This is all a manifestation of sampling error: You will be more likely to predict frequencies of occurrences the larger your sample size, but less likely the smaller your sample size. Even with deterministic processes, so long as some stochasticism also exists, then the odds of observing predicted outcomes will be greater the larger the sample size. Indeed, reduce the sample size far enough and stochastic processes can come to overwhelm deterministic processes (i.e., see the quotation at the start of this chapter). This is the essence of genetic drift: as we reduce a population in size, then fitness measures become less predictive. Note that this statement has nothing to do with the randomness of mutation, or of genetic migration. That is, it would hold even without those two forces operating. Thus, in small populations, or if a population is forced through a genetic bottleneck—which is a population size reduction—then allele frequencies will tend to change not just because one allele is fitter than another but instead, essentially, because one allele is luckier than another.

What has this to do with phages driving evolution? Clearly if bacterial population sizes are substantially reduced in an unbiased manner, e.g., to tens of individuals, then we can expect allele frequencies within that phage population, to the extent that genetic variation is present to begin with, to be modified mostly randomly rather than mostly deterministically (Blazanin and Turner 2021). Thus, we need to consider how phage actions might serve to reduce bacterial population sizes, thereby increasing the impact of sampling error. Four potential phenomena come to mind: (1) Most obvious is that lytically infecting phages can reduce bacterial population sizes in the course of phage replication and resulting phage population growth, thereby generating a genetic bottleneck. (2) Via the action of genetic hitchhiking, selection can randomly favor certain 'lucky' alleles that just happen to be found in association with phage-resistant bacterial mutants that have been selected due to the presence of lytically infecting phages. (3) There is a potential for phage-mediated genetic isolation of a bacterial subpopulation from its parental population, resulting in what is known as a founder effect. (4) Lastly is Muller's ratchet, which contrasting the first three listed phenomena, can be a product also of a *lack* of phage presence. Items (1) and (3) are addressed in this chapter, genetic bottlenecking and founder effects, whereas items (2) and (4), genetic hitchhiking and Muller's ratchet, are both considered instead in the following chapter (Chap. 8).

7.3 Reduction of Bacterial Population Size: Genetic Bottlenecking

The most obvious route toward phages impacting the size of bacterial populations, and therefore the likelihood of genetic drift, is as a consequence of phage-induced mortality of bacteria. This scenario we explore in this section, initially from a phage ecological perspective. That is, how might these reductions in bacterial population size occur. Some facile evolutionary consequences of such reductions in population size are then considered.

7.3.1 Ignoring (for Now) Natural Selection

To appreciate this scenario of phage-induced reductions in bacterial population size leading to genetic drift, it is best to first simplify it by removing the issue of natural selection. That is, some bacteria within a population will tend to possess genotypes that can serve to reduce their phage susceptibility (Chap. 18). Thus, exposure to phages will have the effect of selecting for those bacteria whose reproduction is least or, at least, less impacted by phage exposure. The result would be a biased propagation of those resistant bacteria versus ones that are more vulnerable to phages, i.e., positive selection for phage resistance. Of course, which bacterial genotypes will be less vulnerable will be dependent on what phage genotypes are present, a further complication, but as noted, at this point we are ignoring this issue of biases in what bacteria will be phage affected. Assume instead simply that all bacteria within a given population are equally genetically susceptible to the phage type in question, and further that these bacteria have no means of acquiring phage resistance and nor do they have any physiological differences that could impact their phage susceptibility.

With this simplification of ignoring natural selection, or biased bacterial reproduction generally, we then have the issue that a bacterial population that is exposed to a phage type to which it is vulnerable generally will be reduced in size, but to what degree? In particular, will that reduction in size be sufficient to result in sampling error becoming highly significant in bacterial evolution? The answer to that question from an ecological perspective has two components. First is the question of what fraction of the bacterial population is lost to phage exposure. Is this 1%, 10%, 90%, 99.9999%? Second is the question of the initial bacterial population size, which in turn is a function of a combination of bacterial density (of the affected bacterial population) along with the volume of the environment containing those bacteria. Thus, even substantial fractional reductions in bacterial densities will have less meaning, with regard to favoring genetic drift, if bacterial densities initially were quite high and volumes quite large, resulting in bacterial numbers even following phage exposure that remain relatively high. In addition is the question of just how few bacteria must remain for genetic drift in these populations to become especially important as a driver of bacterial evolution. First, though, more ecological considerations...

7.3.2 Bacterial Populations Can Display Numerical Vulnerabilities

Bacterial populations can display 'numerical vulnerabilities' with regard to the potential for phages to substantially reduce their population numbers. That phrase I present as a contrast to that of "Numerical refuge" (Chao et al. 1977) (Fig. 7.2). A numerical refuge is the density of a phage-susceptible bacterial population that is *insufficient* to support the growth of a phage population, especially to titers that can result in substantial reductions in the density of that bacterial population. Numerical vulnerability instead refers to a bacterial population possessing sufficient concentrations to support phage population growth to densities (titers) that in fact are able to substantially reduce the density of susceptible bacteria (see also Chap. 24, Sect. 24.6.1). These concepts, it should be noted, are simply considerations of standard Lotka-Volterra-type predator-prey type community dynamics, though for the sake of more realism in terms of system stability, an assumption of slower bacteria (as prey) population growth rates at higher bacterial densities often is made, e.g., (Wangersky 1978; Lenski 1988; Abedon 2009). With Lotka-Volterra-type community dynamics, prey populations (as 'victims') are expected to become more vulnerable to exploiter populations when prey population densities are higher, all else held constant. This is because victim population densities in these circumstances will tend to control what densities exploiter populations are able to reach. Thus, higher bacterial densities, so long as those bacteria (as prey/victims) are not entering into

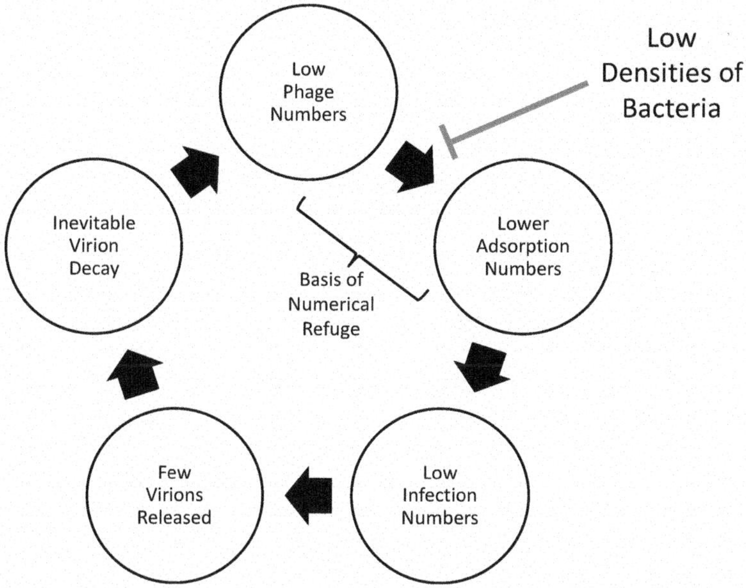

Fig. 7.2 Concept of numerical refuges as resulting from a damping effect of low bacterial densities on phage population densities (titers). Without sufficient numbers of bacteria around, then phages will not be able to increase in numbers sufficiently to result in substantial impacts on bacterial populations. Indeed, without sufficient numbers of bacteria around, then phage populations will tend to go extinct

stationary phase-like states at those higher densities, should as noted be able to support the growth of phage populations (as predators/exploiters) also to higher densities.

The degree to which a phage can reduce a bacterial population in number thus will in part be a function of the pre-phage bacterial population density, and this will be in combination with specifics of the phage type present along with environmental and phage-bacterial interaction details. These ideas can be summarized as: If you don't have enough bacteria around to support phage population growth to reasonably high titers, then the bacterial population will not be very vulnerable to decimation by the phage population (= numerical refuge). Alternatively, if the bacterial population is present at sufficiently high densities that the phage population is able to replicate to higher (inundative) densities, then to some degree the phage population will decimate the bacterial population (= numerical vulnerability). A variation on the latter, numerical vulnerability idea is that if bacterial populations are clumped, but still phage vulnerable, then even if bacterial densities on average across an environment are low (thereby seemingly resulting in a numerical refuge), they still can be high locally (thereby seemingly resulting in numerical vulnerability). This I have dubbed as a 'spatial vulnerability' (Abedon 2012), and see also Abedon (2017a). Note, though, that bacteria in many cases may also display reduced phage susceptibility in association with adaptations that they can display especially while in a clumped state, e.g., (Abedon 2016, 2017b; Ferriol-Gonzalez and Domingo-Calap 2020).

7.3.3 Consequences of Reducing Bacterial Population Sizes

Overall, from the above ecological considerations, we can see that even highly virulent phages may not automatically have a substantial impact on the size of bacterial populations. The question of whether phages will reduce bacterial populations sufficiently in size such that genetic drift becomes a substantial issue—whether across an environment or instead in terms of extremely localized bacterial sub-populations—therefore can be difficult to answer. Nevertheless, let's say that an otherwise isolated bacterial population has in fact been reduced substantially in number due to phage predation, such as to 100 bacteria in total. What will that mean? The first answer is that alleles which were rare in the pre-phage-exposure bacterial population, e.g., such as for example which had frequencies of less than 10^{-3}, will with high likelihood be lost from that population. Alternatively, randomly one or, less likely, a few of those previously rare alleles might instead end up making up a higher fraction of the resulting, bottlenecked population than they initially had, with the likelihood of that occurring a function of their initial allele frequency. For alleles that were more common in the population prior to the bottlenecking event, their frequency instead is likely to change in some random direction, assuming that more than one relatively common allele was originally present. Furthermore, the likelihood of extinction of these initially more common alleles will be dependent on just how small the resulting bacterial population has become, as well as whether more than one allele in fact was common prior to phage exposure.

Thus, if only a single allele were common at a given locus, then following phage exposure only that previously common allele may still be present. The population thereby can be said to have been purged of its genetic variation at that locus with the previously already common allele now fixed. Alternatively, by chance a previously rare allele may display a substantially greater frequency, though likely with purging of a majority or indeed all other rare alleles. If the phage population persists, or if the bacterial population retains a relatively small size such as due to competition with other not identically phage-vulnerable bacteria (Blazanin and Turner 2021), then the genetic bottlenecking may persist, resulting potentially in ongoing losses of alleles from the bacterial population as well as random changes in the frequencies of still common alleles. But, again, the magnitude of this impact will be dependent on just how small bacterial populations have been reduced to, and that in turn will be dependent on just how large the environment is that the bacterial population is present in along with the bacterial population's density and also what titers the attacking phages are able reach.

7.4 Potential Impact of Spatial Structure: Founder Effects

A specific scenario that can result in genetic bottlenecking, and therefore in genetic drift, are population founder events, resulting in founder effects, which basically is a form of short term genetic bottlenecking. Here new populations are established with relatively few individuals, thus representing a relatively small sample of the original population, that is, of the parental population. Due to sampling error, and thereby genetic drift, the new population can stochastically fail to resemble the parent population in terms of allele frequencies. How might phages drive founder events? Here we will consider prophage acquisition with a focus on lysogenic conversion (see also Chap. 15 for the latter). Specifically, lysogenic conversion can result in bacterial phenotypic changes that can promote invasion of new niches. For example, this can be conversion of a more commensal bacterium into a more tissue-invasive bacterium. As Brüssow et al. (2004) put it (p. 577): "its [lysogenic converting gene's] function allows the bacterium to conquer a new niche."

To the degree that such niche invasion events cut off a new population (one with the prophage effecting lysogenic conversion) from an old population (the pre-lysogenic conversion parental bacterial populations)—that is, given environments that are not well mixed but instead which contain what is known as spatial structure—then such niche invasions can give rise to founder events. The significance of such events as drivers of bacterial genetic drift, however, is perhaps questionable. This is due to a combination of the lack of obligate sexuality in bacteria and the potential, or lack of, for founder events to result in long-term population isolation. Thus, even without physical isolation, to a substantial degree individual bacteria are still somewhat genetically isolated from one another, and blocks on what gene exchange does occur may be ephemeral if physical isolation of a founder population is only short lived. In other words, it may be relatively rare that founder events have a lasting impact on bacteria.

Additional questions could include: How common is it in fact for phage genes to drive bacterial invasion of physically isolating niches? To what degree may this isolation scenario be legitimately distinguished from simply genetic hitchhiking (Chap. 8, Sect. 8.1)? That is, if niche invasion is selected, then this will indirectly select for all other genes linked with the converting gene. Thus, it seems at least possible that phages could drive bacterial lysogen founder events that result in the formation of genetically distinct subpopulations of bacteria, but how relevant this scenario might be should, in my opinion, remain an open question.

References

Abedon ST (2009) Deconstructing chemostats towards greater phage-modeling precision. In: Adams HT (ed) Contemporary trends in bacteriophage research. Nova Science Publishers, Hauppauge, pp 249–283

Abedon ST (2012) Spatial vulnerability: bacterial arrangements, microcolonies, and biofilms as responses to low rather than high phage densities. Viruses 4:663–687

Abedon ST (2016) Bacteriophage exploitation of bacterial biofilms: phage preference for less mature targets? FEMS Microbiol Lett 363:fnv246

Abedon ST (2017a) Active bacteriophage biocontrol and therapy on sub-millimeter scales towards removal of unwanted bacteria from foods and microbiomes. AIMS Microbiol 3:649–688

Abedon ST (2017b) Phage "delay" towards enhancing bacterial escape from biofilms: a more comprehensive way of viewing resistance to bacteriophages. AIMS Microbiol 3:186–226

Blazanin M, Turner PE (2021) Community context matters for bacteria-phage ecology and evolution. ISME J 15:3119–3128

Brüssow H, Canchaya C, Hardt WD (2004) Phages and the evolution of bacterial pathogens: from genomic rearrangements to lysogenic conversion. Microbiol Mol Biol Rev 68:560–602

Chao L, Levin BR, Stewart FM (1977) A complex community in a simple habitat: an experimental study with bacteria and phage. Ecology 58:369–378

Duret L (2008) Neutral theory: the null hypothesis of molecular evolution. Nat Educ 1:803–806

Ferriol-Gonzalez C, Domingo-Calap P (2020) Phages for biofilm removal. Antibiotics (Basel) 9

Lenski RE (1988) Dynamics of interactions between bacteria and virulent bacteriophage. Adv Microbial Ecol 10:1–44

Wangersky PJ (1978) Lotka-Volterra population models. Ann Rev. Ecol Syst 9:189–218

Chapter 8
Phages and Asexual Bacterial Populations

Contents

Abstract This chapter considers genetic hitchhiking and Muller's ratchet. Both are consequences of minimal gene exchange within populations, which in higher organisms would be described as a lack of mating but in bacterial populations is due to a lack of within-population horizontal gene transfer. Genetic hitchhiking is the elevation in frequency of an allele due to its being found in the same genome as a beneficial allele, essentially by luck, and thereby is genetically linked to that beneficial allele. The 'luck' is what ties this process to genetic drift, i.e., there is no bias in terms of what alleles are able to hitchhike other than the luck of being found in a certain, higher fitness genome. That beneficial allele, in terms of phages as drivers of evolution, could be one that bestows phage resistance. Muller's ratchet by contrast is truly a product of genetic drift, though of a special kind. Particularly, what is seen is an unbiased loss of whole genotypes including especially wild-type genotypes. This is rather than explicitly a loss of wild-type alleles from populations, however, as were gene exchange possible then the wild-type genotype with its wild-type alleles could be reestablished. Transducing phages, for example, could supply that within-population gene exchange.

> ...some bacilli show a greater aptitude than others to the acquisition of a resistance to the bacteriophage.—F. d'Herelle (1922, p. 76)

In the previous chapter (Chap. 7) we considered phenomena more classically attributed to genetic drift, i.e., as associated with genetic bottlenecking and founder events. In this chapter we consider two additional issues, as also mentioned in the previous chapter, that either provide genetic drift-like consequences or instead are a

S. T. Abedon, *Bacteriophages as Drivers of Evolution*,
https://doi.org/10.1007/978-3-030-94309-7_8

consequence of genetic drift. These are genetic hitchhiking and what is known as Muller's ratchet. Both are distinguishable from genetic bottlenecking and founder events in that one, genetic hitchhiking, actually is driven by natural selection, while the other, Muller's ratchet, though potentially caused by phages also instead can be prevented by phages.

8.1 Introducing Some Natural Selection: Genetic Hitchhiking

Genetic hitchhiking is a form of genetic drift-like stochastic evolution that nevertheless is driven by natural selection. It occurs when two genes are linked together, which essentially is the norm for all genes within a bacterial genome given clonal propagation. Thus, if an allele is present within a genome that bestows a selective advantage on its carrier, and that allele is not otherwise fixed within a population, then any increases in the prevalence of that selected allele will be associated with increases in the prevalence of any other not-yet-fixed allele that happens to be present within the same genome. This is equivalent to saying that this *other* allele is being indirectly selected by the direct selection acting on the first allele, i.e., the second allele is hitchhiking on the success of the first allele (Fig. 8.1).

8.1.1 Selection for Phage Resistance

There is more than one scenario by which phages could drive genetic hitchhiking within a bacterial population, scenarios that may be differentiated in terms of the nature of the directly selected bacterial allele. Thus, for example, there is where the selected allele confers phage resistance. In this case, any alleles that happen to be present as a component of phage-resistant genotype will tend to increase in frequency upon phage exposure (Fig. 8.2), e.g., (Harrison et al. 2015). Note again that such changes in allele frequency can occur only to the extent that an allele is not already fixed within the population. That is, there must be genetic variability upon which natural selection can act, and this is true even if natural selection acts indirectly on an allele's frequency rather than directly. In any case, in this example the result can be an increase in the prevalence of an allele, where that increase is a consequence of the phage-exposure of a bacterial population even if this other allele provides no selective advantage and otherwise has nothing to do with phage resistance. Indeed, this other allele might even bestow a selective disadvantage and still be subject to genetic hitchhiking, and this is so long as the selective advantage of phage resistance is and remains sufficiently high. What is not being referred to here, though, is an antagonistic pleiotropic effect (Chap. 22, Sect. 22.1), where one allele may provide both an advantage (e.g., phage resistance) and a

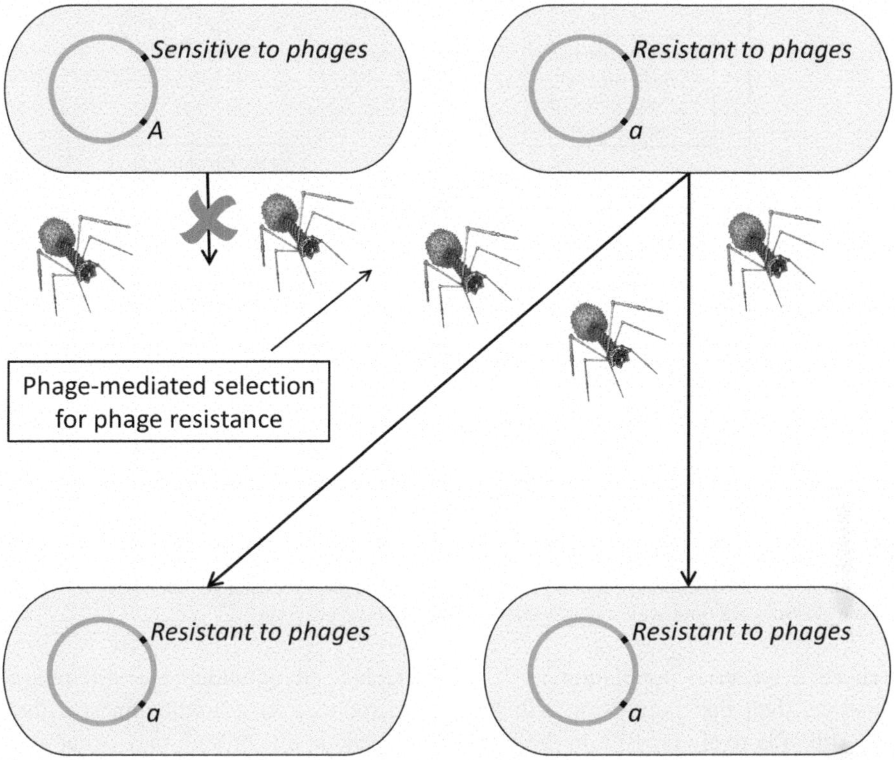

Fig. 8.1 Genetic hitchhiking, here as driven by phage exposure and resulting bacterial evolution of phage resistance. Allele *a* increases in frequency in this population only because it is genetically linked to the phage resistance-conferring bacterial allele

disadvantage (e.g., loss of a nutrient transport protein). Instead, this is a two-locus, two-mutation situation, where the advantage is found in one locus (that which is being directly selected for) while the possible disadvantage is found in the second locus, that which is being indirectly selected, i.e., which is hitchhiking on the success of the other allele.

Note how selection for phage resistance is different from phages simply reducing bacterial populations in size. This difference stems from the likelihood of bacterial mutation to phage resistance. Specifically, if the bacterial population is small enough and the rate of bacterial mutation to phage resistance low enough, then it is likely that only a single mutation event to successful phage resistance will have occurred in that population. The resulting bacterial mutant may then propagate, resulting in some number of clonally related mutants within that population even if the population has not yet been exposed to selecting phages (Luria and Delbrück 1943). Thus, if phages reduce a bacterial population to just that subpopulation of resistant mutants, even if the population consisted of 100 bacterial cells, the population in effect will have been bottlenecked through a single bacterial genotype. On the other hand, if phages

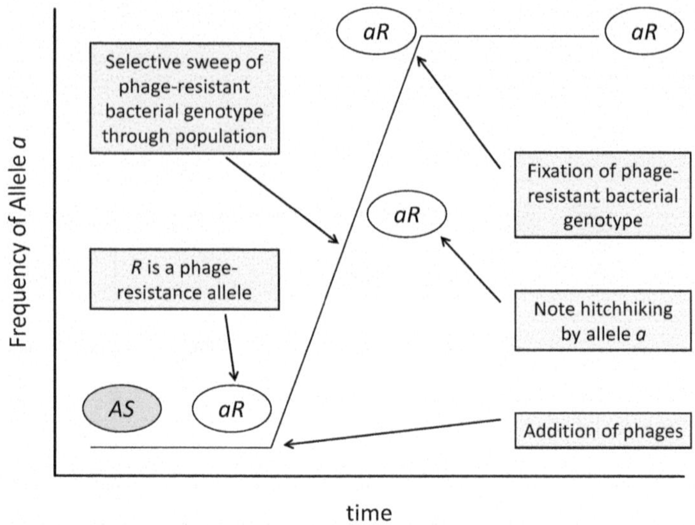

Fig. 8.2 Rise of bacterial allele *a* to fixation within a population while in the presence of phages due to linkage with a different bacterial allele (*R*) conferring phage resistance. Bacterial allele *A* by chance is associated only with a phage-sensitive (*S*) bacterial genotype

reduced a bacterial population to 100 individuals, all of which are still phage sensitive, then that population will have been reduced to a population size that may still effectively be 100. In the former case genetic hitchhiking may occur as a consequence of phage action, but in the latter case only sampling error will have been imposed, i.e., as due to genetic bottlenecking (Chap. 7, Sect. 7.3).

Interestingly, genetic hitchhiking will still have occurred in these scenarios even if more than one clonal population of phage-resistant genotypes were to exist, just not with as dramatic results as were only a single phage-resistant genotype to have existed prior to phage exposure. That is, the result may not be fixation but instead some form of polymorphism involving both selected alleles and any hitchhiking alleles associated with either. Furthermore, genetic drift would be imposed on these 'multi-clonal' phage-resistant populations only if further limitations on bacterial population sizes were to be imposed. That is, if the bacterial population were to simply recover in size, with no random losses of individuals, then we would have an expectation that natural selection rather than genetic drift would dominate their further evolution. Thus, explicitly, reductions in the size of bacterial populations which are not biased in terms of which individuals survive can have different stochastic consequences from reductions in bacterial population sizes which instead are biased in terms of which individuals survive.

8.1.2 *Selection Generally for Prophage Properties*

Another circumstance in which phages may drive genetic hitchhiking in a bacterial population is with the formation of a bacterial lysogen. In this case, to the extent that prophage presence provides a selective advantage (an issue that is addressed more fully in Chap. 15), then any alleles that are present in the host bacterium prior to prophage acquisition should be indirectly selected. This circumstance is more complicated than the one stemming from phage resistance, however, though the complication may be simplified to consideration only of the strength of the driving natural selection. That is, given sufficient densities of phages capable of exploiting a bacterial population within an environment, then selection for phage resistance may be rather robust, i.e., bestowing a large fitness advantage on resistant bacteria. By contrast, any growth or survival advantage bestowed by a newly acquired prophage may be relatively slight. Thus, an allele hitchhiking with the latter may not experience as substantial fitness gains as an allele hitchhiking with a resistance allele, but nevertheless will still be hitchhiking. You are still hitchhiking whether you catch a ride on a bicycle for a trip across a town, or on a motorcycle for a trip across a country.

8.1.3 *Selection Specifically for Superinfection Immunity*

There is a further complication to the bacterial lysogen-hitchhiking scenario, and that is that prophages themselves can supply phage-resistance alleles. These can be in the form simply of superinfection immunity, or in the form of carriage of phage-resistance systems as distinct from superinfection immunity, e.g., such as superinfection exclusion, restriction-modification systems, abortive infection systems, or CRISPR-Cas systems (Chap. 16). In the case of phage-resistance systems, note that a phage to which these systems bestow resistance must be present within the environment in order for the supplying prophage to, as a consequence, supply a selective advantage by encoding phage resistance. Here instead I will briefly focus on the most common of these circumstances, that of prophages supplying superinfection immunity.

Superinfection immunity is the expression of what is known as homoimmunity against a superinfecting phage (Chap. 4, Sect. 4.4 and Chap. 16, Sect. 16.1). Specifically, a prophage's repressor protein, which otherwise is responsible for retaining lysogenic cycles, interferes with both prophage induction and the display of successful infections by subsequently infecting phages, though only so long as those phages possess the same immunity type as the prophage. For a given phage population, e.g., phage λ, in fact all members should display the same immunity type. Thus, a lysogen that forms as a consequence of lysogenization by a member of a temperate phage population will be immune to all other members of that same phage population. If that phage population otherwise substantially reduces the size of the bacterial population, then surviving members of the bacterial population will be enriched for lysogens containing those prophages. The prophage, or at least its repressor gene, thus serves as selected allele, and therefore any other alleles carried by the newly formed lysogen will be able to hitchhike with that ‘phage-resistance’ allele. As with any phage-resistance motivated hitchhiking, that indirect selection will occur only so far as selecting phages are sufficiently prevalent, though super-infection immunity is special in this regard since the same phage type is providing both selection for phage resistance and the selected phage resistance.

8.2 Muller’s Ratchet

> ...gene loss may be detrimental, but insufficiently so to be prevented...—Jeffrey G. Lawrence and Heather Hendrickson (2008, p. 11)

Genetic drift represents random changes in the frequencies of alleles or genotypes, as can result in extinction of specific alleles or genotypes. This extinction can occur in some cases particularly given sufficiently small population sizes, and under these circumstances this is even if what is lost is fitter than what remains. Such losses, that is, can occur simply due to sampling error. In Chap. 7 our interest was primarily in the fate of individual alleles, with the fate of whole genotypes in the face of genetic drift not quite the emphasis. In this section, by contrast, it is the fate of whole genotypes that is the primary interest. A special example of especially genotype loss rather than allele loss due to genetic drift is known as Muller’s ratchet. In Muller’s Ratchet what is lost is the fittest genotype within a population, again as due to the randomness of sampling error, but this loss is special in the sense that though the fittest genotype is lost from a population, the alleles making up that genotype are not also lost from the population.

A Muller’s ratchet scenario goes something like this, upper case letters indicating wild-type alleles and lower case letters indicating mutated and thereby potentially detrimental alleles: We can start with a population that consists of a fixed genotype, *AB*. Over time mutations occur, resulting in the population consisting instead of a combination of *AB*, *Ab*, and *aB*. Due to genetic drift, e.g., the *Ab* genotype comes to dominate, ultimately resulting in the random loss of the *AB* genotype, i.e., wild type. The population still possesses both allele *A* and allele *B*. Therefore the *AB* genotype could in principle be reconstituted through sexual processes. Alternatively, the wild alleles could be reestablished via reversion mutations, but an assumption usually is made that reversion mutations are sufficiently rare that they do not occur, and indeed would be even rarer in the smaller populations within which genetic drift robustly operates. Thus, particularly unless there is sex, then the wild-type and presumably most-fit genotype is lost from the population and this is solely due to the sampling error seen with ongoing genetic bottlenecking. The population, that is, has been ratcheted down one click toward lower fitness. These ideas, only involving three rather than only two alleles, are illustrated in Fig. 8.3.

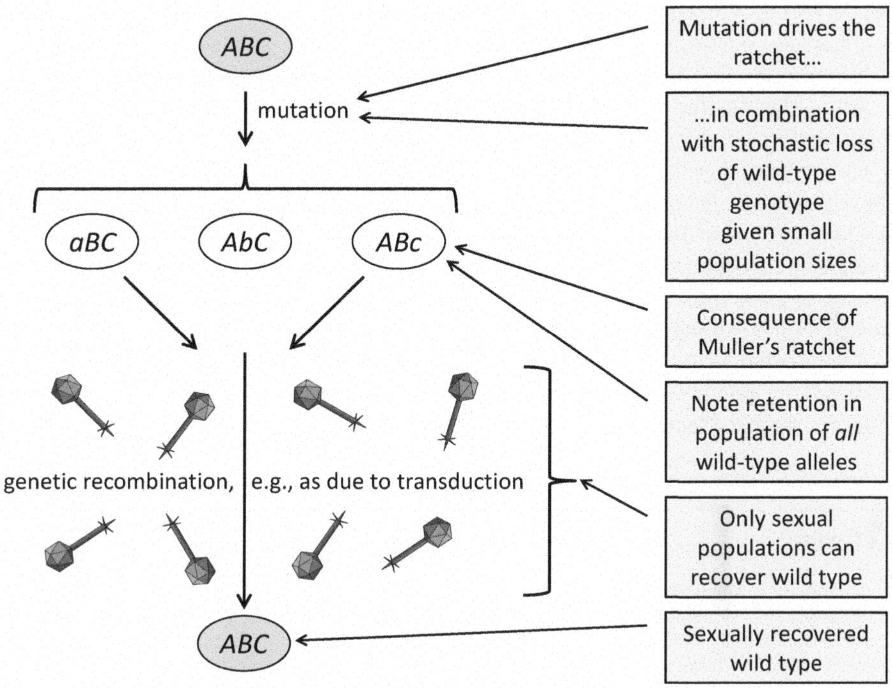

Fig. 8.3 Schematic representation of Muller's ratchet. Note the loss of the wild-type genotype from the population due to genetic drift but the failure to lose, at least proximately, any wild-type alleles from the population (here *A*, *B*, and *C*). Recovery of the wild-type genotype is possible only via reversion mutation (low likelihood) or genetic recombination, the latter here suggested as following transduction. If there is no recombination between genotypes, however, then there is little potential for recovering the wild-type genotype and thus the population ratchets toward an increasingly reduced fitness

We can extend the above scenario to include additional loci, still with each click of the ratchet resulting in loss of that genotype carrying the most wild-type alleles. Again, the population as a whole can still contain all of the wild-type alleles, but no one individual genome contains all of them. And over time, the number of wild-type alleles carried by any one individual declines. The result thus is an ongoing 'ratcheting' down of the collective fitness of a population as the remaining most-fit genotypes are sequentially lost.

As indicated, an important assumption in a Muller's ratchet scenario is that gene exchange between individual members of a population is rare. That is, it is not specific alleles that are being lost from populations in Muller's ratchet but instead specific genotypes, but with gene exchange fittest genotypes in principle can be reconstituted. For example, gene exchange between an *Ab* individual and an *aB* individual can reconstitute an *AB* individual. As one means of gene exchange among bacteria is via transduction, i.e., as mediated by phages, Muller's ratchet may be countered by transducing phages, particularly generalized transducing phages, as well as by other mechanisms of horizontal gene transfer such as transformation, e.g., as potentially occurring within an "Arena" of an otherwise isolated, small population of bacteria (Bordenstein and Wernegreen 2004; Bordenstein and Reznikoff 2005).

Altogether, to the extent that phage actions can result in bacterial populations experiencing genetic drift, then a result of that action could be Muller's ratchet, hence inclusion here of discussion of this phenomenon. Alternatively, it seems more likely that those bacterial populations that would tend to prone to Muller's ratchet would not be those that are more exposed to phages but instead those less exposed to phages, i.e., as due to the noted role that particularly transducing phages can play in horizontal gene transfer, as discussed in Chaps. 10 and 13 but also Chap. 12. First, though, in the following chapter (Chap. 9), we consider the potential for randomness in gene exchange as it can occur among especially homospecific bacteria and as mediated by phages. Also at the end of that chapter we return to the subject of Muller's ratchet (Sect. 9.4).

References

Bordenstein SR, Reznikoff WS (2005) Mobile DNA in obligate intracellular bacteria. Nat Rev Microbiol 3:688–699

Bordenstein SR, Wernegreen JJ (2004) Bacteriophage flux in endosymbionts (*Wolbachia*): infection frequency, lateral transfer, and recombination rates. Mol Biol Evol 21:1981–1991

d'Herelle F (1922) The Bacteriophage: Its Role in Immunity. Williams and Wilkins Co./Waverly Press, Baltimore

Harrison E, Wood AJ, Dytham C, Pitchford JW, Truman J, Spiers A, Paterson S, Brockhurst MA (2015) Bacteriophages limit the existence conditions for conjugative plasmids. MBio 6:e00586–e00515

Lawrence JG, Hendrickson H (2008) Genomes in motion: gene transfer as a catalyst for genome change. In: Schmidt H, Hensel M (eds) Horizontal Gene Transfer in the Evolution of Pathogenesis. Cambridge University Press, Cambridge, pp 3–22

Luria SE, Delbrück M (1943) Mutations of bacteria from virus sensitivity to virus resistance. Genetics 28:491–511

Chapter 9
Phage Impact on Non-random Mating Among Bacteria

Contents

Abstract The genetic structure of populations can be modified by mutation, genetic drift, genetic migration, and natural selection, but also by violations of panmixis, that is, as due to blocks on random mating. With bacteria, that 'mating' occurs via processes of horizontal gene transfer, which in terms of phages as drivers of evolution predominantly involves the processes of transduction. Non-random mating among bacteria can occur due to a variety of factors therefore involving phages. This can be seen as a consequence of transduction events being rare, which naturally will result in some pairings of bacterial genotypes occurring but not others. Another route to non-random mating involving phages are limitations on phage movement, which will tend to bias matings to between closer rather than more distantly located bacterial pairs. In addition, of course, are biases in transduction occurrence between bacteria that are a function of phage host ranges. Discussed as well is the importance of DNA retention within bacteria, once that DNA has been delivered into their cytoplasms, as well as consequences associated with a complete absence of mating. I suggest that phages can counter the latter even within well-isolated bacterial populations should those bacteria carry transducing phages as prophages.

> Adaptive transfer of genes is limited to those that can be transferred as a functional unit, provide a niche-transcending adaptation, and are compatible with the architecture and physiology of other organisms. Horizontally transferred adaptations may bring about fitness costs, and natural selection may ameliorate these costs."—Jane Wiedenbeck and Frederick M. Cohan (2011) (p. 957)

S. T. Abedon, *Bacteriophages as Drivers of Evolution*,
https://doi.org/10.1007/978-3-030-94309-7_9

Of the five assumptions necessary to attain Hardy-Weinberg equilibrium—these being (1) no mutation, (2) large population sizes (i.e., no genetic drift), (3) no migration, (4) no selection, and (5) random mating—random mating can be viewed also as the odd-concept out. The reason for this latter statement is that the randomness of mating does not directly impact allele frequencies. Mutation, genetic drift, genetic migration, and natural selection by contrast all are evolutionary because they explicitly impact allele frequencies. Non-randomness of mating instead, traditionally, i.e., as according to violations of Hardy-Weinberg assumptions, impacts the genetic structure of populations by affecting the frequency of *genotypes*, that is, collections of alleles making up individual organisms, rather than the frequency of individual alleles themselves. Thus, in addition to the frequency of alleles helping to define a population's genetic structure, so too does the frequency with which alleles are found together in individuals help to define a population's genetic structure.

9.1 Transduction and Panmixis

By 'random mating', what is meant, statistically, is that any one individual has the same likelihood of mating with any other one individual within a population—yes, taking into account that matings, given the existence of genders, still are not completely random since males will be limited to mating with females, and vice versa. Spatially or biogeographically, we would say, at a minimum, that the population therefore must be well mixed and thus that the likelihoods of any one individual physically encountering any other one individual are equivalent. Of course, no population can be perfectly well mixed, but given small populations within environments that contain little internal spatial structure, then such mixing and thereby a potential for random mating at least can be approximated.

This randomness of mating traditionally is described as panmixia or panmixis, meaning 'all' 'mingling'. Non-random mating that is due, for example, to limitations on organism movement, would therefore be an example of an absence of panmixia. The question posed by this chapter is how phages can contribute to an absence of panmixia, i.e., particularly in terms of limiting the randomness of 'matings' within a bacterial population. That is, what is the impact of phage-mediated transduction on the non-randomness rather than the randomness of gene exchange among bacteria? We especially will consider a variety of mechanisms that could contribute to biases in what bacteria are impacted by transduction.

I note at the outset that the idea of mating in a Hardy-Weinberg sense certainly is not equivalent to 'mating' as mediated via transduction, since not only do complete genomes not intermingle given transduction but so too the intermingling of genomes given transduction is not reciprocal. That is, unlike with meiosis-associated sex, and owing to both of those qualities—incompleteness and lack of reciprocation—with transduction there is a both a donor and a recipient of genetic material. In addition, with transduction donors die and recipients are relatively rare. Thus, the question being asked still is one of 'who' might be mating with 'whom', and why, but the actual matings differ somewhat from the matings that we are used to observing among obligately sexual creatures. With regard to the impact of transduction on the randomness of mating, therefore the key question is who is *not* mating with whom, qualitatively speaking, and why not?

9.2 Various Limitations on Random Mating

Why might matings *not* occur randomly between individuals, that is, instead of occurring non-randomly? Three possible, non-mutually exclusive answers are discussed in this section. These are sampling error, environmental spatial structure, and assortative mating, all of which can be relevant also to phage-mediated transduction.

9.2.1 Sampling Error and Non-random Mating

Sampling error can result in specific alleles being more likely to be found associated together within the genomes of individuals than would be predicted based solely on the frequency of those alleles within populations. In this case, the 'sampling' is the mating itself, e.g., as undertaken by two individuals chosen randomly from a population. Specifically, when populations are small, then the pool of mates is small as well. Consider at an extreme a population consisting of four individuals each with a different genotype, *W*, *X*, *Y*, and *Z*. If matings occurred once and involved pairings, then the combinations would either be *WX* and *YZ*, *WY* and *XZ*, or *WZ* and *XY*. At the other extreme, were the population infinite and consisting of equal numbers of each genotype, then all combinations would be represented, each making up one-sixth of matings (obviously, in this example requirements for gender compatibility are being ignored). At the extreme of the first example, only two combinations of pairings are even possible, versus six as population size moves toward the other extreme. With small sample sizes, genotype frequencies thus can fluctuate away from what would be predicted statistically.

By analogy, the rarer that transduction events are, then the less likely that they will collectively give rise to panmixis. Instead, certain allele combinations will tend to be randomly created and thereby become overrepresented in a population, at least among possible recombinant products. Importantly, however, the rarer that transduction events occur and therefore the greater the random divergence of a population from panmixis, so too the less that transduction will contribute to the genetic structure of a population. Thus, at the same time that panmixis may not be occurring because of sampling error, more importantly panmixis will not be occurring because for the most part mating is not occurring much at all, at least via transduction. In addition, note that due to the nature of transduction, it still should not in and of itself modify allele frequencies even if rare, and this is because for every gain of an allele by a recipient, that allele will have also been lost by a donor, i.e., as due to the latter's death.

9.2.2 Spatial Structure and Non-random Mating

The presence of some allele combinations at higher frequencies in specific geographic regions can be maintained—ignoring both natural selection and genetic drift—as a consequence of incomplete mixing of populations. This is literally an absence of 'all mingling', with random mating potentially still possible locally while still not occurring across a population as a whole. What genotypes will be present explicitly would therefore reflect local allele frequencies rather than global allele frequencies.

Phages, like bacteria, tend to be limited in their geographical ranges, with populations of specific phage types not evenly distributed, that is, not necessarily equivalently prevalent (same titers) across all portions of all environments, particularly spatially structured environments such as soils and sediments or as found in association with metazoans. Phage movement within environments also often is not necessarily unimpeded (e.g., poor diffusion through soils), and certainly phages are limited in terms of what ecosystems they tend to thrive in. Thus, the ability of phages to transduce specific bacterial alleles to other specific bacteria found in the same bacterial population will tend to be limited by a combination of uneven spatial distributions of bacterial alleles and an inability of transducing phages to readily span distances between these alleles. I know these statements may strike some as flying in the face of, "Everything is everywhere, but, the environment selects"; however, nowhere in that statement is a claim of spatial homogeneity within environments (de Wit and Bouvier 2006).

9.2.3 Assortment and Non-random Mating

Another contribution to non-random mating is described as assortative, which is where more similar individuals tend to mate with each other rather than with less similar individuals. Assortative mating especially results in greater-than-expected frequencies of specific combinations of alleles, i.e., especially whatever it is that makes two individuals more similar. Non-random mating also can be a consequence of matings that instead are disassortative, i.e., with like individuals tending to avoid each other—in less technical terms: disassortative mating would be 'opposites attract'. This also has the effect of increasing frequencies of specific combinations of alleles above what would be expected based solely on random mating, particularly increasing combinations of those alleles that make individuals different.

The trouble with considerations of assortative mating with regard to transduction is the question of just what is a bacterial population and how exactly do phage host ranges coincide with bacterial populations. For instance, are all members of a bacterial population, save for a few random mutants, equivalently susceptible to the same set of phages? If so, then it would be difficult to see how disassortative transduction could actually occur. Alternatively, one can readily envisage different bacterial strains, as making up different bacterial populations, possessing differences in phage susceptibility. Issues of assortative or disassortative mating I therefore consider further in the following chapter (Chap. 10, Sect. 10.2), which focusses on gene exchange between bacterial populations rather than within bacterial populations, i.e., with the latter instead the emphasis of the current chapter.

9.3 Bacterial Retention of Transduced DNA

Limitations on phage host ranges as well as limitations on the geographical ranges of bacteria mean that a transducing particle produced by any one bacterium will have a limited number of bacteria to which virion-carried DNA might be transduced. Yet other factors are relevant to alleles becoming associated within the same genome, however, particularly as resulting in that DNA being abortively transduced. Particularly are limitations on the biological processes that result in an allele becoming incorporated into a genome. Generally, however, the more similar a transduced allele is to an already existing allele, thereby allowing for homologous recombination into the recipient organism's genome, then the greater the likelihood that transduction will not be aborted. That similarity would be the expectation given gene exchange occurring *within* populations. Alternatively, an allele might possess a means of assuring its segregation without homologously recombining into bacterial DNA. That, again within bacterial populations, generally will require that more genetic material be found in association with the transfered allele, such as its being found in association with a plasmid, genomic island, or prophage.

A second additional factor is natural selection. That is, will a new allele combination once formed tend to persist even given effective segregation into daughter cells? This then has two components, i.e., is the newly acquired allele beneficial in its own right, on the one hand (that is providing some sort of beneficial function) (Lawrence 1999), and is the newly acquired allele not excessively antagonistic to its new genetic/physiological background, on the other hand (being difficult to retain due, e.g., to toxicities to its new organism). Popa and Dagan (2011) describe these as "Functional barriers" to the acquisition of new DNA and discuss a number of categories of such barriers. Newly acquired alleles that provide synergistic benefits in combination with alleles already present within a genome, while at the same time integrating well into the physiology of their new organism, could provide the greatest selective benefits, but such alleles probably are only relatively rarely acquired transductively. Alternatively, within a bacterial population, how many especially homologous alleles that are subject to transduction are likely to be toxic to their recipients?

Thus, what will interfere with the potential of all possible genotypes being generated in bacteria by transductive processes? Factors will include the rarity of transduction itself, limitations on both phage movement and the mixing together of different bacterial strains across environments, limitations on the potential for bacteria to segregate new DNA into daughter cells, and the potential for natural selection to not favor probably a majority of especially somewhat divergent new genotypes should they transductively arise. In addition are limitations on phage host range that tend to give rise to more assortative matings, though as noted this likely applies more to between-population transduction than within-population transduction, unless phage resistance has arisen to a substantial extent within a bacterial population. Non-random mating between bacteria as mediated by phages in any case is likely to be violated in the real world.

9.4 What Happens If There Is No Horizontal Gene Transfer Within Populations?

> If the population has recombination, the fixation of mutants at different loci will be more or less independent. Favorable mutants which arise in different individuals can ultimately be combined into the same genome by recombination. But if there is no recombination in the population, two mutants can both succeed in fixing only if the second occurs in one of the offspring of the first. Otherwise their offspring can at best compete with one another, and only one of the two mutants can ultimately succeed in fixing. Thus, many of the newly-occurring mutants must be lost, more than would be the case in the presence of recombination. A population with recombination can therefore evolve faster than one without it.—Joseph Felsenstein (1974) (p. 738)

What happens if a bacterial population has no phages and otherwise no horizontal gene transfer? This might be viewed as a special case of non-random mating, indeed amixis. We began to explore that question in Chap. 8 (Sect. 8.2) when considering Muller's ratchet and in this section we continue that discussion. Specifically, here we are referring to an absence of gene exchange particularly from more closely related bacteria. Basically, without horizontal gene transfer not only do bacteria have difficulty moving forward evolutionary (i.e., see the quotation, above), but they can have trouble even staying in the same place.

In the absence horizontal gene transfer, bacterial populations become fully clonal (Hauck and Maiden 2018). An absence of phages and transduction is a necessary though not sufficient criterion for the maintenance of such pure clonality, however, as other routes of horizontal gene transfer exist as well, i.e., transformation and conjugation. With full clonality, all adaptations that an organism possesses must either have already been present in the lineage at the point of establishment of full clonality or instead must be a consequence of genetic changes that occurred only within each individual clonal lineage, with any mutational changes that occur passed down solely from parent to offspring. Furthermore, should a lineage accumulate pseudogenes through especially genetic drift, including genetic hitchhiking, then there will be no means of recovering wild-type versions of those genes except via reversion mutation, or pseudoreversion, which as noted (Chap. 8) tend to be inefficient processes. Thus, pathways to new evolutionary innovation are extremely curtailed in these organisms, as too is the potential to retain current adaptations, particularly retention of genes that are not otherwise essential, or nearly so, to organism functioning.

A perhaps interesting twist on these ideas could be were a bacterial lineage to carry its own gene exchange-mediating adaptations, which in terms of phages would be one or more prophages that generate transducing particles. Prophages certainly are known to produce transducing particles, e.g., (Schicklmaier et al. 1998; Stanczak-Mrozek et al. 2017; Schneider 2021), and indeed one could easily argue that this could be an advantage of encoding such prophages for bacteria possessing lifestyles that cause them to otherwise not be exposed to horizontal gene transfer mechanisms, including to exogenously sourced phages, e.g., such as highly invasive obligate pathogens. This, I believe, could be an interesting scenario to explore from theoretical, experimental, and observational perspectives. That is, to determine if there is some potential for such phages to counter the occurrence of Muller's ratchet especially during long-lived, otherwise clonal bacterial infections. I imagine that *Mycobacterium tuberculosis* could represent just such a pathogen (Dos Vultos et al. 2008) for which this mechanism would be useful, or perhaps "strictly maternally transmitted endosymbionts" (Bobay and Ochman 2017, p. 499), but do forgive me if I have overlooked studies that have already looked at this possibility. Note also that the potential for prophages to generate potentially transformable eDNA is mentioned in Chap. 12 (Sect. 12.4).

References

Bobay LM, Ochman H (2017) Biological species are universal across life's domains. Genome Biol Evol 9:491–501

de Wit R, Bouvier T (2006) 'Everything is everywhere, but, the environment selects'; what did Baas Becking and Beijerinck really say? Environ Microbiol 8:755–758

Dos Vultos T, Mestre O, Rauzier J, Golec M, Rastogi N, Rasolofo V, Tonjum T, Sola C, Matic I, Gicquel B (2008) Evolution and diversity of clonal bacteria: the paradigm of *Mycobacterium tuberculosis*. PLoS One 3:e1538

Felsenstein J (1974) The evolutionary advantage of recombination. Genetics 78:737–756

Hauck S, Maiden MCJ (2018) Clonally evolving pathogenic bacteria. In: Molecular mechanisms of microbial evolution. Springer, Cham, pp 308–325

Lawrence JG (1999) Gene transfer, speciation, and the evolution of bacterial genomes. Curr Opin Mirobiol 2:519–523

Popa O, Dagan T (2011) Trends and barriers to lateral gene transfer in prokaryotes. Curr Opin Microbiol 14:615–623

Schicklmaier P, Moser E, Wieland T, Rabsch W, Schmieger H (1998) A comparative study on the frequency of prophages among natural isolates of *Salmonella* and *Escherichia coli* with emphasis on generalized transducers. Antonie van Leeuwenhoek J Microbiol 73:49–54

Schneider CL (2021) Bacteriophage-mediated horizontal gene transfer: transduction. In: Harper DR, Abedon ST, Burrowes B, McConville M (eds) Bacteriophages: biology, technology, therapy. Springer, New York City, pp 151–192

Stanczak-Mrozek KI, Laing KG, Lindsay JA (2017) Resistance gene transfer: induction of transducing phage by sub-inhibitory concentrations of antimicrobials is not correlated to induction of lytic phage. J Antimicrob Chemother 72:1624–1631

Wiedenbeck J, Cohan FM (2011) Origins of bacterial diversity through horizontal genetic transfer and adaptation to new ecological niches. FEMS Microbiol Rev. 35:957–976

Part III
Genetic Migration

Chapter 10
Genetic Migration and Phages

Contents

Abstract Contrasting random mating, which is a within-population phenomenon, genetic migration by definition occurs between populations. For bacteria, this 'mating' in both cases occurs via horizontal gene transfer, and phages can mediate this horizontal gene transfer via the various forms of transduction. As too with random mating, there exist biases in terms of what bacteria can serve as transductive donors and recipients, as can be dependent on bacterial as well as transducing phage properties. Discussed in particular are how these biases can result in the potential for bacteria to reciprocate transduction events, that is, the potential for bacterium type A to serve as a donor to bacterium type B and then B back to A. Alternatively, A and B types may not be capable of displaying reciprocal transduction, but still could be connected by transduction via intermediary bacterial types, thereby potentially resulting in *indirect* reciprocal transduction. I discuss as well how transduction can serve to either increase the similarity of bacteria toward their better approximating a single population or instead can result in the divergence of bacteria from their parent population.

> …lineages are no longer restricted to exploring logically accessible niches; rather, they could acquire any set of genes—from any other organism, living at any place or time—to thrust them into completely novel, previously unavailable ecological contexts.—Jeffrey G. Lawrence and Heather Hendrickson (2008, p. 4)

The concept of migration can be confusing because biologically the word has multiple meanings. One of those meanings is in terms of the movement of an entity from one place to another, such as birds flying south for the winter. Proteins migrate

S. T. Abedon, *Bacteriophages as Drivers of Evolution*,
https://doi.org/10.1007/978-3-030-94309-7_10

as well, such as within gels during electrophoresis, etc. The concept of migration being used here, by contrast, is genetical. This too involves movement, but in this case it is movement of DNA from one population to another.

Genetic migration can occur either vertically or horizontally. Vertical movement is from parent to offspring, which in the case of genetic migration involves matings between would-be parents that are members of different populations. Horizontal movement instead occurs between individuals in a manner that does not directly result in reproduction and for bacteria movement of genetic material between individuals occurs exclusively by this horizontal route. Chap. 9 emphasized phage-mediated movement of genetic material as that can occur *within* bacterial populations, which strictly speaking is not genetic migration. For movement of DNA to also result in genetic migration, then there has to instead be some potential for transfer of new alleles from one population into another, which as mediated by phages is the emphasis of this chapter.

In terms of phages as drivers of genetic migration, the acquisition of new prophages by bacteria within a population represents an obvious example of how this can occur (Bartoli et al. 2016). Prophage acquisition also is an important contributor to bacterial genome size expansion as that also occurs via genetic migration (Gao et al. 2019), though bacterial genome expansion is not an emphasis of this chapter, nor really anywhere in this book. Furthermore, I will not be cataloging specific functions that may be transferred between bacteria; for some examples of that, especially as delivered within prophages, see instead references provided in Chap. 15 on lysogenic conversion. Rather, in this chapter we consider more generically the movement of DNA from one distinct bacterial lineage into a second distinct bacterial lineage, as can occur both within and between bacterial species, focusing on some basic principles of that movement.

10.1 Phage-Mediated Introduction of New Alleles into Bacterial Populations

Classically, DNA transfer to bacteria is considered to occur via transformation, conjugation, or transduction. The latter, of course, is phage mediated, but phage-mediated transfer of DNA between bacteria can occur via a number of processes, not all of which are necessarily described strictly as transduction. Of those that are so described, there is specialized versus generalized transduction. Those concepts were introduced in Chap. 1, Sect. 1.1. Movement of DNA by phages in association with temperate phage genomes can also occur as products of illegitimate recombination (Chap. 13, Sect. 13.2). Such genes have come to be described as 'morons' (Chap. 14), many of which are also lysogenic converting genes (Chaps. 15 and 16). These and other genes carried by prophages ultimately may become domesticated as bacterial genes rather than remaining encoded within phage genomes (Chap. 17). The different pathways toward phage-mediated DNA transfer

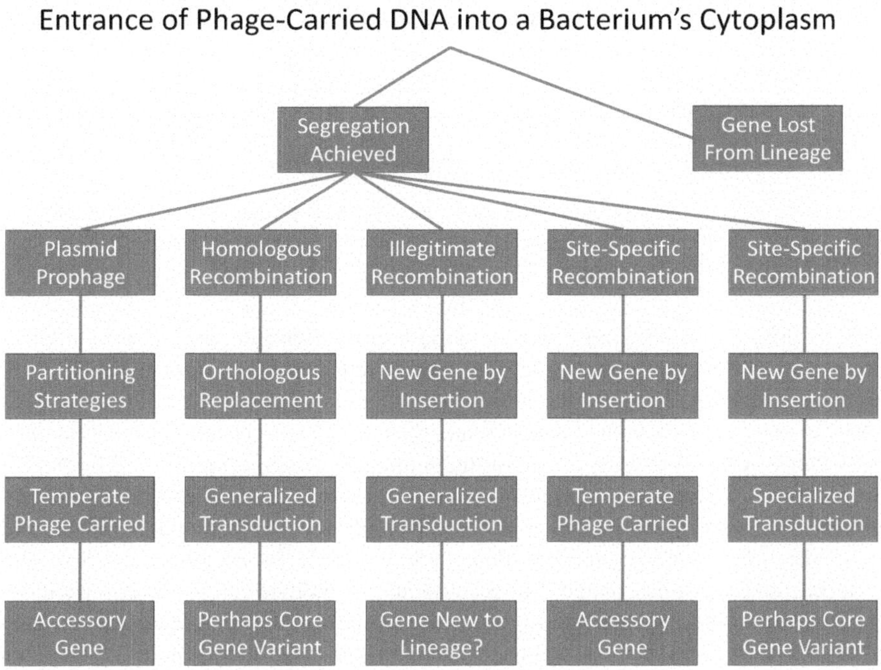

Fig. 10.1 Phage-mediated horizontal gene transfer of bacterial genes. Not all possible combinations are shown. For example, homologous recombination could occur following any of the indicated mechanisms that result in gene segregation, i.e., everything except "Gene lost from lineage", though traditionally in phage biology homologous recombination of DNA into the recipient bacterium following phage-mediated transfer is most closely associated with generalized transduction. Though not explicitly indicated, "Site-specific recombination" refers to the acquisition by a bacterium of an integrated prophage. "Perhaps core gene variant" is meant to imply that the new gene either is or could be related to an allele already found in the recipient bacterium. A "Zone of paralogy" (not explicitly indicated) would be associated especially with illegitimate recombination (Chap. 11, Sect. 11.2). All acquired genes, even if segregation is achieved, will be subject to natural selection along with genetic drift as well as mutation and potentially also further genetic migration

to bacteria—plasmid prophage-mediated transfer, generalized transduction (two entries), carriage of morons or lysogenic conversion genes, and specialized transduction—are summarized in Fig. 10.1.

For all of these mechanisms, it is relevant to distinguish phage-mediated horizontal gene transfer from phage-mediated genetic migration, and thus distinguish between whether or not genes have been moved between members of the same bacterial population. Here, though, we will not attempt to grapple with the question of just what constitutes a single bacterial population. Instead, we will differentiate between phage-mediated transfer of genetic material to bacteria that either does or does not represent new alleles, with this chapter addressing particularly phage-mediated transfer of new alleles into a given bacterial lineage, i.e., as could have come from a different bacterial population. To reduce clutter, I will use the term 'transduction' and its variants generically to describe phage-mediated movement of bacterial DNA between bacteria regardless of the mechanism.

10.2 Assortative Mating Introduced by Phages

> Recombination that depends on vectors, such as bacteriophage-mediated transduction or plasmid-mediated conjugation, is limited by the host ranges of the respective vectors.—Frederick M. Cohan (2001, p. 514)

Though not always the case, phages nevertheless often are somewhat limited in terms of what bacterial strains they are able to infect (Hyman and Abedon 2010, Koskella and Meaden 2013, de Jonge et al. 2019, Ross et al. 2016, Hyman 2019). This can serve to bias what transductive 'matings' can occur among bacteria. The result can be viewed as a form of assortative mating (Chap. 9, Sect. 9.2.3) because it is typically similarity among bacteria that will tend to determine whether those bacteria are found within a phage's host range. Usually, for example, this would mean that specific phage receptor molecules (Chap. 1, Sect. 1.1) are displayed by bacteria found within a phage's host range along with fewer *dissimilar* bacteria-encoded phage-resistance mechanisms, or at least a lack of such mechanisms in both bacteria which are able to efficiently inactivate the transducing phage. Thus, the more similar two bacteria, then the more likely that a phage capable of transducing bacterial genetic material will be able to infect both.

If indeed that latter statement is true, then at a minimum transduction should be biased away from substantially modifying bacteria in terms of those genes which underlie factors that phages must interact with, or instead must avoid, in order to successfully infect. That is, bacterial genes encoding factors that phages need to infect would tend to be shared among bacteria making up a phage's host range and thereby may be less altered in their distribution via transduction than genes encoding factors that phages don't need in order to successfully infect. Exceptional, however, would be bacterial factors that individual phage types can successfully interact with despite a diversity in their forms. And, as noted, phage-resistance genes that substantially impact a given phage would not tend to be shared among bacteria found within that phage's host range. There likely is a way around these constraints on what can be transduced and that involves something that can be described as transduction that is unable to be directly reciprocated.

10.3 Reciprocal Versus Non-reciprocal Transduction

This scenario of bacteria assortatively having their genes transferred back and forth by phages, albeit at much lower rates than is seen when gene exchange is mediated instead by meiotic sexual processes, comes with a small caveat. Specifically, there is no reason to imagine that all possible transduction processes between two bacterial

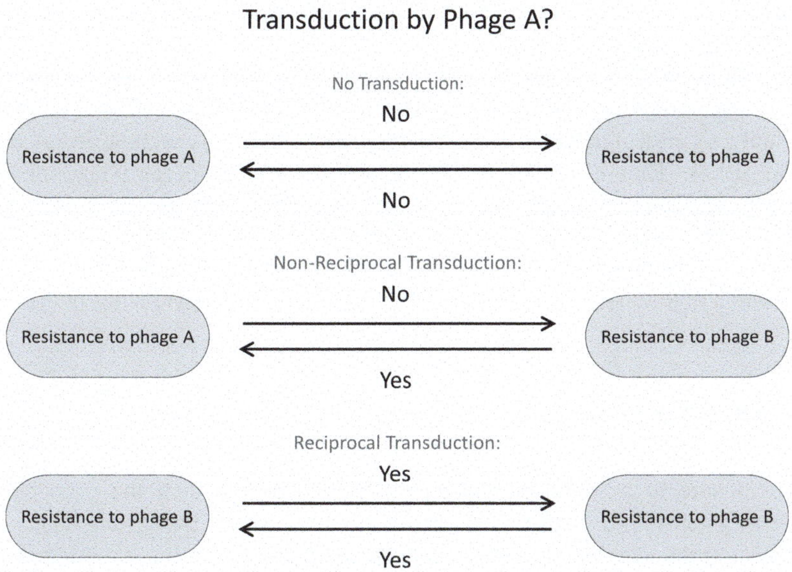

Fig. 10.2 The potential for transduction of mechanisms of phage-infection resistance, such as restriction-modifications systems, requires a lack of that phage resistance while infecting the donor bacterium but not necessarily also a complete lack of phage resistance diplayed by the recipient bacterium (middle example)

strains are necessarily reciprocal (Ozeki 1959). It is possible, in other words, for transduction to occur between two bacterial strains, only one of which is able to support a productive infection by the phage in question. Such transduction relationships could be viewed as non-reciprocal, with one strain within a given transductive pairing serving solely as the donor and the other bacterial strain serving solely as the transduction recipient (Fig. 10.2). Phage production of virions, on the one hand, is a minimum requirement for a bacterium to serve as a transductive donor. On the other hand, phage delivery of DNA to a bacterium's cytoplasm is not dependent on subsequent production of virions in that cytoplasm. Successful reception of phage-mediated horizontal transfer of DNA instead minimally only requires some degree of avoidance of abortive transduction once that DNA has been received.

Thus, a bacterium that can support a productive phage infection by a transducing phage could serve as a DNA donor to a bacterium that cannot support a productive phage infection by the same transducing phage, but not vice versa. In addition, generally the productive host ranges of phages likely are narrower than the penetrative host ranges of phages. That is, there likely tend to be fewer bacteria which a given phage is able to productively infect versus able only to successfully deliver DNA to (Hyman and Abedon 2010, Kenzaka et al. 2010, Penades et al. 2015), or at least kill as is the case for the transducing phage P1, which is able to kill *P. aeruginosa* strains despite that bacterial species not, as tested, serving as a productive host of this phage (Amati 1962). With non-reciprocal transduction, bacterial matings therefore could be less assortative than would be case were only reciprocal transductions considered.

Note that in the following section (Sect. 10.4) I will describe reciprocal transduction instead as *direct* reciprocal transduction, with the latter designation equivalent to how simply reciprocal transduction is being considered in this section (lower example, Fig. 10.2). That is, direct reciprocal transduction involves phage-mediated movement of DNA between two bacteria, both of which are able to support productive infections by the transducing phage. Non-reciprocal transduction, by contrast, involves phage-mediated movement of DNA between two bacteria, only one of which able to support a productive infection by the transducing phage.

10.4 Direct and Indirect Reciprocal Transduction

Given exposure to a sufficient diversity of phage types, then the properties of individual phages may matter less in determining which bacterium mates with which bacterium, and this particularly may be the case to the extent that non-reciprocal transduction in one direction by one phage type may to a degree be balanced by non-reciprocal transduction in the other direction by another phage type. This scenario I'm describing thus would consist of a form of *indirect* reciprocal transduction (Fig. 10.3). This suggests a possibility of networks of reciprocal and non-reciprocal transductive gene exchange among a diversity of bacterial strains making a single species. Given enough time and enough phages, then various forms of overlapping host ranges should result in transductive gene exchange throughout a species. Nonetheless, though panmixis (Chap. 9, Sect. 9.1) might be better approximated, still biases should exist such that some bacterial strains can more readily exchange genes with certain other bacterial strains, as dependent on a combination of (1) bacterial properties, (2) the extent to which their geographic ranges overlap both locally and globally, and (3) also the properties of what phages are present.

Notwithstanding these various considerations of inter-strain gene exchange, the most likely transductive gene exchange between bacteria still would be intra-strain or nearly so, i.e., with "much of the nearly homologous recombination taking place between more closely related strains", Hendrickson (2012), p. 70, and see also Popa and Dagan (2011). That is, no two bacteria should be more similar than those that make up the same bacterial strain, including especially those individual bacteria that are clonally related. So too the same strains of bacteria, and particularly clonally related individuals, should be more likely to occupy the same geographical ranges. Despite this, however, within-strain transductive exchanges would be expected to be the least meaningful evolutionarily since the power of gene exchange as an evolutionary force is a function of the extent to which new genetic variation may be introduced into individuals. The more closely related two individuals, however, then the less genetic variation that is present between them and therefore the less likely that we could equate that gene exchange with genetic migration, though this would still constitute horizontal gene transfer (Chap. 9).

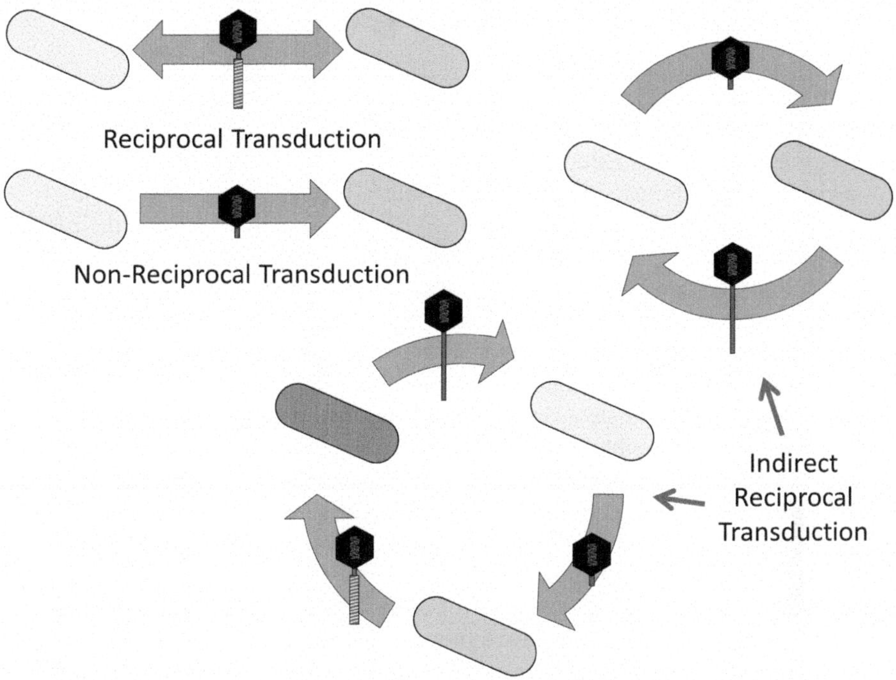

Fig. 10.3 Reciprocal transduction, non-reciprocal transduction, and indirect reciprocal transduction. With reciprocal transduction the same phage type can transduce DNA back-and-forth between two different bacterial strains. With non-reciprocal transduction, movement is possible from one strain to another, but not back again via the same phage type. Reciprocality, however, can be achieved even as based on non-reciprocal transduction if more than one phage is involved, here called indirect reciprocal transduction

Gene exchange between more closely related bacteria still should modify bacterial population structure, to the extent that more genotypically similar bacteria will tend to remain more genotypically similar as a consequence of assortative transduction. Nonetheless, the more similar two bacteria, then the more likely that they will be able to participate in *direct* reciprocal transduction. Thus, we can predict that as genetic migration becomes more relevant as a potential force of evolutionary change, then indirect reciprocal transduction as well as non-reciprocal transduction, rather than direct reciprocal transduction, may be especially relevant toward effecting that change.

10.5 Stabilizing Versus Disruptive Transduction

To the extent that transduction can be reciprocal, whether direct or indirect, then it can serve as a mutually stabilizing force, that is, of the structure of a bacterial population. For bacteria, this naturally leads to the question of just what is and what is not a population? The easy answer can be that a bacterial population is simply a group of more or less clonally related individuals, i.e., for which most of their DNA can be traced back to a single ancestor cell. Alternatively, to the extent that gene exchange is strong enough between two otherwise distinct lineages of bacteria, then some greater degree of genetic cohesiveness may be achieved, resulting perhaps in multiple lineages coming to represent a single population. Whether or not that scenario is even possible, we can predict that multiple bacterial lineages can come to at least better approximate a single population the more that reciprocal transduction among them is possible, again, whether achieved by direct or instead indirect means. Whether that results in transduction between the different lineages ever becoming *not* genetic migration, that is, becoming other than transduction occurring between otherwise distinct bacterial populations, is an open question. Perhaps gene exchange becoming *not* genetic migration is more likely to be achieved, however, given the occurrence of direct rather than just indirect reciprocal transduction.

Alternatively, with purely non-reciprocal transduction, transduction between different lineages should always consist of genetic migration. Furthermore, it should always result in disruption or at least modification of a population's genetic structure, particularly that of the recipient population. This disruption will not necessarily be substantial, i.e., if it consists of the reception of only a gene or two. Alternatively is the potential for phage-mediated movement of whole genomic islands, as considered in Chap. 13, Sect. 13.3. Furthermore, once acquired by a recipient population, with low likelihood of transduction back to the original donor population, then there could be greater potential for those newly acquired genes or genomic islands to evolutionarily diverge. In addition, those new populations might themselves be able to transduce DNA to yet additional bacterial populations, resulting in phage-mediated transmission of DNA from bacterium to bacterium that greatly extends the transductive reach of individual phages, i.e., as resulting in a transductive equivalent to "All the world's a phage" (Chap. 11, Sect. 11.3).

References

Amati P (1962) Abortive infection of *Pseudomonas aeruginosa* and *Serratia marcescens* with coliphage "Plant". J Bacteriol 83:433–435

Bartoli C, Roux F, Lamichhane JR (2016) Molecular mechanisms underlying the emergence of bacterial pathogens: an ecological perspective. Mol Plant Pathol 17:303–310

Cohan FM (2001) Bacterial species and speciation. Syst Biol 50:513–524

de Jonge PA, Nobrega FL, Brouns SJJ, Dutilh BE (2019) Molecular and evolutionary determinants of bacteriophage host range. Trends Microbiol 27:51–63

Gao NL, Chen J, Wang T, Lercher MJ, Chen WH (2019) Prokaryotic genome expansion is facilitated by phages and plasmids but impaired by CRISPR. Front Microbiol 10:2254

Hendrickson H (2012) The lion and the mouse: how bacteriophages create, liberate, and decimate bacterial pathogens. In: Hyman P, Abedon ST (eds) Bacteriophages in health and disease. CABI Press, Wallingford, pp 61–75

Hyman P (2019) Phages for phage therapy: isolation, characterization, and host range breadth. Pharmaceuticals (Basel) 12:35

Hyman P, Abedon ST (2010) Bacteriophage host range and bacterial resistance. Adv Appl Microbiol 70:217–248

Kenzaka T, Tani K, Nasu M (2010) High-frequency phage-mediated gene transfer in freshwater environments determined at single-cell level. ISME J 4:648–659

Koskella B, Meaden S (2013) Understanding bacteriophage specificity in natural microbial communities. Viruses 5:806–823

Lawrence JG, Hendrickson H (2008) Genomes in motion: gene transfer as a catalyst for genome change. In: Schmidt H, Hensel M (eds) Horizontal gene transfer in the evolution of pathogenesis. Cambridge University Press, Cambridge, pp 3–22

Ozeki H (1959) Chromosome fragments participating in transduction in Salmonella typhimurium. Genetics 44:457–470

Penades JR, Chen J, Quiles-Puchalt N, Carpena N, Novick RP (2015) Bacteriophage-mediated spread of bacterial virulence genes. Curr Opin Microbiol 23:171–178

Popa O, Dagan T (2011) Trends and barriers to lateral gene transfer in prokaryotes. Curr Opin Microbiol 14:615–623

Ross A, Ward S, Hyman P (2016) More is better: selecting for broad host range bacteriophages. Front Microbiol 7:1352

Chapter 11
Bacterial Reproductive Isolation and its Violation by Phages

Contents

Abstract This chapter considers aspects of the extent to which DNA can be broadly moved by phages between bacteria. One issue is the potential impact on bacterial evolution to limitations on this movement of genetic material particularly between different bacterial strains. A second consideration is how it is, especially in to terms of bacteria biology rather than phage biology, that these limitations can occur. The chapter then transitions to considering the extent to which movement of DNA can be more extensive than traditionally thought. Particularly, this is with regard to the existence of networks of movement of DNA among phages infecting different bacterial taxa, though also relevant to phages as drivers of bacterial evolution is the ability of DNA to move from bacteria into phages. In both of these cases, to the extent that especially temperate phage genes can ultimately become bacterial genes, then there is a potential for genes first to move into phages from bacteria, then to move broadly between phages, and then from these phages back to being bacterial rather than phage genes. This chain of horizontal gene transfer opens the door to a possibility of vast phage-mediated movement of bacterial genes between disparate bacteria taxa, particularly by routes that are unlike traditional considerations of phage-mediated DNA transduction.

All the world's a phage—Roger W. Hendrix et al (1999, p. 2192)

In this chapter we consider limitations on phage-mediated movement of DNA between especially different bacterial strains and what happens if those limitations are relaxed especially over evolutionary time scales. Related to the latter idea are important concepts associated with the late Roger Hendrix and his collaborators, i.e.,

S. T. Abedon, *Bacteriophages as Drivers of Evolution*,
https://doi.org/10.1007/978-3-030-94309-7_11

the just mentioned "All the world's a phage" along with what they dubbed as "Moron accretion". Both stem from work on the comparative genomics of temperate phages such as and especially of lambdoid phages, the latter as reviewed by Hendrix and Casjens (2006). Both of these concepts, however, are more relevant to phage evolution than to the impact of phages on bacterial evolution. Nevertheless, to the extent that phages can acquire new genes and then pass those new genes on to other phages, which in principle might then pass those genes on to new bacterial hosts, then these ideas are not exactly irrelevant to bacterial evolution. Indeed, as above (Chap. 10), they are simply phage-mediated horizontal gene transfer except over more complex paths than normally we consider. Thus, in this chapter we consider both limitations on phage-mediated horizontal gene transfer and violation of those limitations, starting with limitations.

11.1 Barriers to Transduction as Defining Bacterial Species?

The ideas discussed in Chap. 9 on limitations to non-random mating may be framed instead in terms of what are known as reproductive isolation mechanisms, i.e., blocks on gene exchange between different species or at least between would-be different species. Roles of blocks on transduction are discussed by Shapiro (2018) in terms of their potential to contribute to bacterial speciation, but who also notes that (p. 32), "rates of gene flow are higher within than between species, but cross-species gene transfer can still occur". In this section I expand on these ideas. Particularly, I am interested in how blocks on transduction may be considered in terms of reproductive barriers, which in the literature concerning obligately sexual organisms is how blocks on gene exchange often are described, and then where that sort of thinking might lead us, especially in terms of how reproductive barriers might contribute to the existence of distinct bacterial species.

Traditionally, reproductive barriers have been differentiated into those that occur prior to DNA entering the same cytoplasm versus those which act following that event. Barriers that interfere with the entrance of DNA from different parents into the same cytoplasm are described as prezygotic. Postzygotic barriers, by contrast, act after such cytoplasmic mixing of DNA. Blocks on virion attachment or on phage DNA translocation following attachment (Chaps. 18 and 19) I suggest may be viewed—in terms of phage-mediated transduction of genetic material—as prezygotic barriers. This is because they occur prior to phage-mediated DNA delivery into a bacterium's cytoplasm, thereby directly interfering with cytoplasmic mixing of DNA. The existence of these barriers result in a lower potential for species barriers to be broken down by transduction, i.e., somewhat dissimilar bacteria will tend to resist becoming more similar because of blocks on phage delivery of DNA into cytoplasms. This would conform with one aspect of the biological species concept, where members of a species are defined in terms of the likelihood of successful matings occurring between them, i.e., with prezygotic barriers serving to reduce that likelihood.

One might thereby consider a concept of 'transductive species' within bacterial communities as analogous to biological species. These would be populations that are defined by their potential to share transducing phages and therefore to share DNA via transduction, i.e., with ability to share transducing phages depending on both phage and bacterial properties. That is, a transductive species would consist, perhaps *maximally*, of a collection of bacteria for which substantial prezygotic barriers to transduction by at least one phage type do not exist. This would be different from all of the bacteria that may be transduced from one bacterial lineage necessarily also coming to make up the same *population*, i.e., as considered at the end of Chap. 10 (Sect. 10.5), as that instead would require some form of reciprocal transduction. Thus, in a sense a maximal transductive species might be viewed as more encompassing than a biological species, where the latter generally is defined in terms of reciprocal gene exchange, i.e., as the biological species concept comes from the world of obligate sexuality and obligately sexual reproduction is inherently a reciprocal form of gene exchange, whereas a transductive species could be maximally defined by the occurrence of both reciprocal transduction and non-reciprocal transduction. That is, all of those bacteria that may be transduced by at least a single phage type.

A concept of transductive species would not be without ambiguity, however. For example, and as noted, we might define such a 'species' based upon only a single type of transducing phage. In that case, we might have a *core* of bacteria that this phage is able to effectively transduce *from*, and a *periphery* of bacteria that this phage is only able to transduce *to*. Furthermore, we have the potential for there to exist multiple transducing phages with only partially overlapping transductive cores as well as only partially overlapping transductive peripheries. Perhaps, then, it would be preferable to replace 'species' with a concept that is more amorphous, such as 'group'. Thus, included in a core group could be those bacteria for which reciprocal transduction can occur with reasonably high probability, whether directly or indirectly (Chap. 10, Sect. 10.4), and a peripheral group for which only non-reciprocal transduction, to those bacteria, can occur. A further difficulty we would run into, however, is that there is more to transduction than just avoiding prezygotic barriers as postzygotic reproductive barriers can exist as well.

First among these postzygotic barriers are lower potentials for retention of DNA once it has entered into a bacterium's cytoplasm, particularly such as due to reduced likelihoods of homologous recombination the greater the phylogenetic distance between bacteria. That is, especially peripheral transductive groups could be defined not just by the potential for phages to deliver DNA into cytoplasms but also in terms of the potential for those would-be transduction events to not be abortive, that is, resulting in retention (not abortive transduction) rather than loss (abortive transduction) of the transferred DNA. In addition, and also acting as a postzygotic barrier to transduction, is the impact of natural selection on bacterial transductants, i.e., on recombinants or hybrids should they form. Arguably, at a minimum, for the transductive species to be analogous to a biological species, then there might be an expectation that at least homologous recombination should be able to occur once DNA has been transductively received, and then also for that DNA to have at least a reasonable likelihood of having a neutral impact on the fitness of the recipient bacterium, i.e., just as is the case for most matings that occur within an obligately sexual species. This then might eliminate most members of peripheral transductive groups from being full members of transductive species, leaving the core group as perhaps somewhat biological species-like, at least in terms of there existing some degree of stabilizing transduction occurring (Chap. 10, Sect. 10.4).

Natural selection as a postzygotic reproductive barrier also should be compatible with bacterial species being defined in terms of the ecological species concept, as also highlighted by Shapiro (2018), i.e., the idea of incompatibilities between genomes that are consequences of differences in the ecological niches that different species occupy and thereby the different adaptations they need to possess. In metazoans, we can expect such incompatibilities to select for the evolution of more robust prezygotic reproductive barriers, which for bacteria would be anti-phage defenses, particularly what I describe as phage 'avoidance' mechanisms (Chaps. 18 and 19). My guess, though, is that transduction is both sufficiently rare and sufficiently not costly when it does occur that as a selective force for prezygotic barriers it pales in comparison to costs associated with simply being killed by adsorbing phages; this is particularly since barriers preventing the initiation of phage infections, if sufficiently broadly acting as to be useful, likely would be effective against non-transducing phages as well. Thus, it is unlikely that bacterial species exist because of the evolution of specifically transduction-limiting mechanisms. Consequently, there may be little selection against being part of a peripheral transduction group, even if homologous recombination given transduced-DNA delivery is relatively unlikely. Successful phage-mediated gene exchange may still occur without homologous recombination, however, or without even prophage acquisition, as the following section considers (Sect. 11.2).

11.2 Zones of Paralogy

Lawrence and Hendrickson (2003) suggest a means by which postzygotic barriers to successful transduction might be avoided, what they describe as zones of paralogy. They posit that greater novelty in association with non-homologous, i.e., illegitimate recombination—basically DNA insertion events that assure subsequent DNA segregation to daughter cells (Sect. 13.2)—could result in the acquisition of novel functions by more distantly related bacteria. These literally new genetic loci might then be retained by natural selection due to that novelty. That is, to become a new genetic locus via horizontal gene transfer, generally that gene must be physically acquired (e.g., enter a bacterium's cytoplasm such as via transduction), must achieve segregation due to some means other than homologous recombination with existing bacterial DNA (not orthologous replacement), must supply a new function to the recipient bacterium (or at least be linked to a gene supplying a new function, which at a minimum means that a new gene must be expressed), and that new function must then be useful in some manner to the recipient organism.

As an aside, note that "Paralog" is being defined by Lawrence and Hendrickson as a reuniting of "previous orthologues in the same genome, where they would appear as paralogues" (p. 746), whereas the original definition of paralogs (Sonnhammer and Koonin 2002) is, p. 619, "genes that derive from a single gene that was duplicated within a genome." That is, the "Paralogy" of "Zone of paralogy" represents a reuniting of two genes that have diverged from each other while occupying different genomes, though more broadly this "Zone" might be viewed simply as the acquisition of a gene by illegitimate rather than homologous recombination and which then is retained by natural selection. It is an open question how often the process described by Lawrence and Hendrickson (2003), as based on such reuniting, would occur as a consequence of transduction. As analogous to the concept of biological species, however, if postzygotic barriers to transduction between bacterial strains can *only* be breached via illegitimate recombination, rather than in the course of homologous recombination, then a transductive donor and recipient should be viewed as definitively *not* falling within the same at least transductive species.

There is a third way that genes can be moved by phages and then established in new bacterial species given an absence of absolute prezygotic barriers. That is, besides as involving either illegitimate or homologous recombination. Rather, it involves the establishment of lysogenic cycles. Lysogenic cycle-establishing temperate phages, however, do not inherently consist of bacterial genes and therefore are not necessarily transductive. Nevertheless, toward initiating my support of the importance of this idea of temperate phages being important drivers of bacterial evolution in terms bacterial acquisition of new genes, in the next two sections (Sects. 11.3 and 11.4) I briefly return first to the issue that phages perhaps really can impact a periphery of their host-range breadth (i.e., all the world's a phage) and then to the idea that these phages also inherently may be carriers of acquired genes, including of bacterial genes.

11.3 All the World's a Phage

"All the world's a phage" comes from the Hendrix et al. (1999) publication titled, "Evolutionary relationships among diverse bacteriophages and prophages: all the world's a phage". These relationships are ones that are driven by recombination between diverse phage types, resulting in what is known as mosaic evolution. The idea that phages display mosaic evolution (Chaps. 4 and 17) predates their publication, with the contribution of the article instead the idea of networks of relationships between phages infecting disparate bacterial taxa (Fig. 11.1). Thus, phage genes, perhaps particularly as associated with temperate phages, have a potential to move far beyond even 'transductive species', and thereby may be able to drive genetic migration to more bacterial types than one might imagine based solely on typical knowledge of the host ranges of individual transducing phage types.

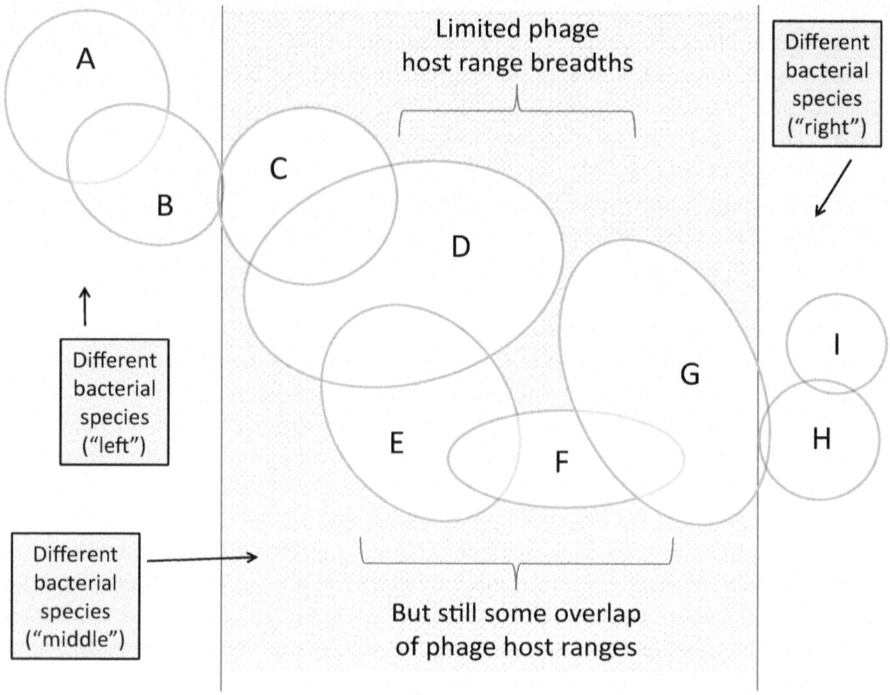

Fig. 11.1 All the world's a phage! Illustration of the idea of overlapping phage host-range breadths. The area of the figure represents genotype space for three different bacterial species (left, right, and middle) as reduced to two-dimensional planes. Capital letters are individual phage types, with host ranges (ovals) encompassing some number of distinct bacterial strains within each species. That is, each 'pixel' represents essentially a different bacterial genotype with more closely related genotypes represented as more closely associated pixels. Thus, the various ovals, representing each phage type's host range, encompass a collection of bacterial genotypes that together make up a portion of at least one bacterial species but in some cases also strains from relatively closely related bacterial species (the latter as crossing the vertical dashed lines). These host ranges can be broader for phages particularly if only DNA transfer to individual hosts is required for a bacterium to be found in its host range (penetrative host range) and not also an ability for a phage to successfully infect (e.g., productive host range)

All the world's a phage thus would seem to be driven by a combination of recombination occurring between phages that happen to be coinfecting the same bacterium, the host ranges of different phages not completely overlapping and thereby allowing genes that have recombined into a phage to move beyond the hosts available to the previous phage carrier of those genes, and at least some of those phages being able to successfully infect, even if only occasionally, more than one bacterial species. For this to result in the movement especially of bacterial genes between bacteria, rather than just phage genes, then there has to exist a means by which phages can acquire these bacterial genes. That concept we consider further especially in Chap. 14. Fundamentally, however, it is an issue of genetic migration along with violations of reproductive isolation, though involving phages as the evolving organisms rather than phages only as drivers of that evolution. Specifically, this has been described as the noted, "Moron accretion".

11.4 Moron Accretion

. . .a common view of the origin of phages is that they arose by accretion and adaptation of functional modules originally of host origin (Campbell and Botstein 1983; Casjens et al. 1992).—Robert A. Weisberg (1996, p. 2446)

. . .in the host might replace, by recombination, a virus copy or vice versa.—Allan Campbell and David Botstein (1983, p. 367)

Presumably during the evolutionary construction of viruses, genes have been appropriated by viruses from their hosts' genomes and evolutionarily modified to suit their own purposes.—Sherwood Casjens et al. (1992, p. 393)

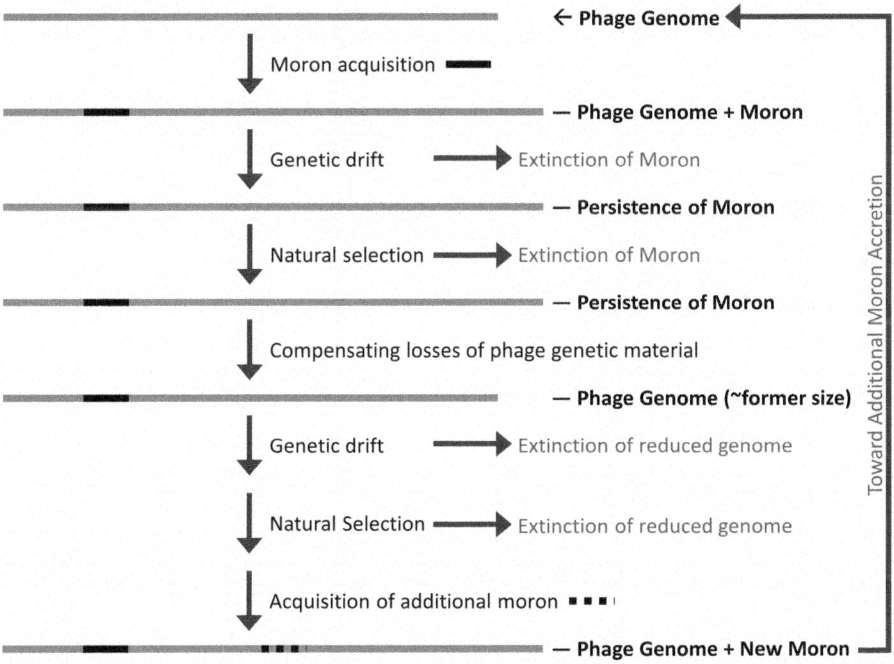

Fig. 11.2 Moron accretion. Morons are acquired via especially illegitimate recombination. They are then subject to genetic drift and natural selection, both of which can result in extinction of the moron-containing phage genome. Losses of genetic material also can occur, potentially allowing phage genomes to not grow excessively in size despite continued moron acquisition. These losses too are subject to genetic drift and natural selection. If the moron-containing genome survives, then additional morons may be acquired, and so on

The idea behind what can be described as a "Moron-accretion hypothesis" (Hendrix et al. 2000) is that phages may be assembled one genetic component at a time via a combination of horizontal gene transfer and illegitimate recombination (Fig. 11.2). Thus, phages are exposed to the DNA of other phages as well as to bacterial DNA. New DNA can insert into nascent phage genomes from random places, into random locations found within these phage genomes, forming new phage genetic loci, with 'new' to phages generally meaning that the DNA comes from non-phage sources such as bacteria. Occasionally a useful gene or two inserts into a promising location, and this new gene may then be retained over evolutionary time (Campbell and Botstein 1983; Hendrix 2005, 2008). Expanding this idea to beyond phages, as phages incorporate new genes, it is possible for some of those new genes to eventually transition into becoming bacterial genes, e.g., Chap. 17. To the extent that "All the world's a phage", then bacterial genes acquired as phage morons could be transferred well beyond specific bacterial species.

Going one step further, phage genes may be evolutionarily refined within the phage genome before transitioning to (or back to) becoming bacterial genes. As Comeau and Krisch (2005) put it (p. 492),

> We suggest that the war between phage and bacteria serves as a 'proving ground' for the biosphere, where evolutionary innovation is created, evaluated and validated. Only the most successful of these innovations are retained within the phage genome pool and those that are more generally useful are eventually co-opted by their bacterial hosts. The use of phage as an evolutionary proving ground has crucial advantages: high turn-over rates, high mutation rates, and high throughput rates. As a result, diversity can be rapidly created on a massive scale at a relatively low biological cost and can then be distributed by horizontal transfer to places in the biosphere where it can be used.

See Wahl and Pattenden (2017) for similar thoughts but from the phage rather than bacterial perspective and also Ren et al. (2017) for an example of a eukaryote gene that appears to have descended from a phage gene.

It is with auxiliary metabolic genes that the suggestion of Comeau and Krisch (2005) may be most applicable. This is because these genes by definition are ones that produce products that are able to interact directly with host metabolic processes, either replacing host gene products or directly augmenting the functioning of host factors. The modern concept of these auxiliary metabolic genes dates from the early 2000s (Mann et al. 2003; Lindell et al. 2004; Hurwitz and U'Ren 2016; Warwick-Dugdale et al. 2019). As reviewed by Comeau and Krisch (2005), however, the idea that phage genes might functionally replace bacterial genes at least slightly predates the photosynthetic-gene origins of that concept. Going even further back, however, note that Campbell (1988) suggested that (p. 6), "Relatedness between host and phage genes might mean that the host genes are descended from phage genes rather than the reverse."

References

Campbell A (1988) Phage evolution and speciation. In: Calendar R (ed) The bacteriophages, vol 1. Plenum Press, New York, pp 1–14

Campbell A, Botstein D (1983) Evolution of the lambdoid phages. In: Hendrix RW, Roberts JW, Stahl FW, Weisberg RA (eds) Lambda II. Cold Spring Harbor Laboratory, Cold Spring Harbor, pp 365–380

Casjens S, Hatfull GF, Hendrix RW (1992) Evolution of dsDNA tailed-bacteriophage genomes. Semin Virol 3:383–397

Comeau AM, Krisch HM (2005) War is peace—dispatches from the bacterial and phage killing fields. Curr Opin Mirobiol 8:488–494

Hendrix RW (2005) Bacteriophage evolution and the role of phages in host evolution. In: Waldor MK, Friedman DI, Adhya SL (eds) Phages: their role in bacterial pathogenesis and biotechnology. ASM Press, Washington DC, pp 55–65

Hendrix RW (2008) Phage evolution. In: Abedon ST (ed) Bacteriophage ecology. Cambridge University Press, Cambridge, pp 177–194

Hendrix RW, Casjens S (2006) Bacteriophage λ and its genetic neighborhood. In: Calendar R, Abedon ST (eds) The bacteriophages. Oxford University Press, Oxford, pp 409–447

Hendrix RW, Lawrence JG, Hatfull GF, Casjens S (2000) The origins and ongoing evolution of viruses. Trends Microbiol 8:504–508

Hendrix RW, Smith MCM, Burns RN, Ford ME, Hatfull GF (1999) Evolutionary relationships among diverse bacteriophages and prophages: all the world's a phage. Proc Natl Acad Sci U S A 96:2192–2197

Hurwitz BL, U'Ren JM (2016) Viral metabolic reprogramming in marine ecosystems. Curr Opin Microbiol 31:161–168

Lawrence JG, Hendrickson H (2003) Lateral gene transfer: when will adolescence end? Mol Microbiol 50:739–749

Lindell D, Sullivan MB, Johnson ZI, Tolonen AC, Rohwer F, Chisholm SW (2004) Transfer of photosynthesis genes to and from *Prochlorococcus* viruses. Proc Natl Acad Sci USA 101: 11013–11018

Mann NH, Cook A, Millard A, Bailey S, Clokie M (2003) Marine ecosystems: bacterial photosynthesis genes in a virus. Nature (London) 424:741

Ren Q, Wang C, Jin M, Lan J, Ye T, Hui K, Tan J, Wang Z, Wyckoff GJ, Wang W, Han GZ (2017) Co-option of bacteriophage lysozyme genes by bivalve genomes. Open Biol 7:160285

Shapiro BJ (2018) What microbial population genomics has taught us about speciation. In: Population genomics: microorganisms. Springer, Cham, pp 31–47

Sonnhammer EL, Koonin EV (2002) Orthology, paralogy and proposed classification for paralog subtypes. Trends Genet 18:619–620

Wahl LM, Pattenden T (2017) Prophage provide a safe haven for adaptive exploration in temperate viruses. Genetics 206:407–416

Warwick-Dugdale J, Buchholz HH, Allen MJ, Temperton B (2019) Host-hijacking and planktonic piracy: how phages command the microbial high seas. Virol J 16:15

Weisberg RA (1996) Specialized transduction. In: Neidhardt FC, Curtiss R III, Ingraham JL, Lin ECC, Low KB, Magasanik B, Reznikoff WS, Riley M, Schaechter M, Umbarger HE (eds) *Escherichia coli* and *Salmonella* cellular and molecular biology. ASM Press, Washington, D.C., pp 2442–2448

Chapter 12
Phage-Provided Environmental DNA and Superspreading

Contents

Abstract Carriage of encapsidated DNA between bacteria, as ultimately may add to the genetic repertoire of recipient bacteria, is only one means by which phages can contribute to horizontal gene transfer. Though transduction along with prophage encoding of bacteria-useful genes is probably the most important means by which phages can contribute to this process, it is also possible for phages to contribute to bacterial transformation, particularly by converting donor bacterial DNA into environmental DNA, or eDNA for short. Some phages, however, are more efficient eDNA generators than others. This consideration has led to an idea that these more efficient eDNA generating phages can serve as eDNA 'superspreaders'. I do not question that possibility, but do address whether 'superspreader' is an appropriate descriptor of the role of phages in this process, though alternatively highly effective transducing phage-bacterium combinations certainly could serve as *transduction* superspreaders. Independent of their transduction superspreader status, phages capable of generalized transduction are likely also efficient generators of highly intact eDNA. Particularly in the absence of generalized transduction, temperate phages capable of generating highly intact eDNA could serve as a means fostering intra-bacterial strain recombination toward fending off Muller's ratchet

> Lysis of dead bacterial cells, e.g., following phage infection, and the spontaneous release of both chromosomal and plasmid DNA during different phases of bacterial growth have been

S. T. Abedon, *Bacteriophages as Drivers of Evolution*,
https://doi.org/10.1007/978-3-030-94309-7_12

suggested to be among the sources of such environmental DNA.—C. Vettori et al (1996, p. 251)

An important reservoir of genetic material within environments is free DNA. This DNA, known as environmental or eDNA, forms largely as a consequence of the breakdown of formerly living organisms, or formerly living cells shed by still-living multicellular organisms. The prevalence of eDNA is a function in part of the rapidity of separation of that DNA from source organisms versus rates of environmental decay of the now mostly naked DNA. One route to eDNA dispersal is in the course of lysis of cells, such as phage-induced lysis of bacteria. Also involving phages is the breakdown of virion particles, in which case viral DNA becomes eDNA, as well as the release of unencapsidated virus DNA following phage-induced cell lysis. Evolutionarily, eDNA is a source DNA for transformation. This chapter considers especially the role of phages in generating eDNA, with some emphasis placed on what have been described as bacteriophage 'superspreaders'.

12.1 Uptake of Naked DNA

There is a hypothesis that is known as "You are what you eat" (Doolittle 1998). This in turn has been expanded (Keeling and Palmer 2008) to "You are what you eat, what you live on, what lives on you, and what lives in you." The general idea is that exposure to DNA—in whatever form so long as ultimately it exists as somewhat intact free DNA—will result in some likelihood of that DNA becoming incorporated into the so-exposed organism. This seems to be especially the case with eukaryotic organisms that habitually bring DNA into their cells, whether in the course of phagocytosis or instead because of breakdown of entities already existing in their cytoplasms (the latter, i.e., endosymbionts or instead abortively invading pathogens). Alternatively, especially in the absence of cell walls, DNA as eDNA may simply slip into cells and thereby inadvertently find its way into the nucleus and thus potentially become incorporated into the nuclear genome. Virus-induced lysis of cells in turn can make DNA, as eDNA, more available for acquisition by other cells.

For organisms possessing cell walls, such as bacteria, or which do not possess endosymbionts or intra-cytoplasmic cellular parasites, routes to DNA acquisition generally are less promiscuous than can be the case for eukaryotes, especially cell wall-less, phagotrophic eukaryotes. As has been noted (Chap. 10), virus-mediated movement of genetic material via transduction is one way around the cell wall barrier, as too is conjugation. These mechanisms, however, do not represent routes toward acquisition of eDNA. Instead, eDNA is taken up explicitly by a horizontal gene transfer process known transformation (Domingues et al. 2012; Overballe-Petersen et al. 2013). An ability to take up eDNA varies among bacterial taxa and some organisms clearly are more promiscuous than others in terms of what eDNA they are likely to take up (Johnsborg et al. 2007; Mell and Redfield 2014). But those lineages that have been found to be able to take up eDNA into their cytoplasms are

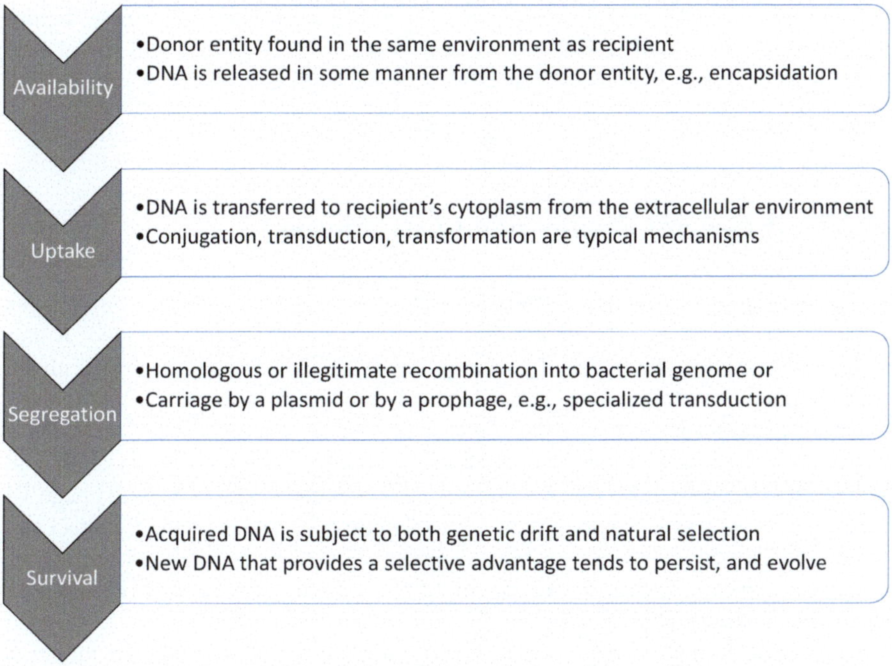

Fig. 12.1 What does it take for a bacterium to acquire and retain new DNA via horizontal gene transfer? The DNA must be available in a recipient's environment, must be taken up into a recipient's cytoplasm, must segregate to daughter cells upon binary fission, and otherwise the recombinant bacterium must persist, including despite often starting out in very low numbers

said to be naturally competent. Furthermore, as with any form of horizontal gene transfer, the reception of genetic material must be followed by its retention both over the course of subsequent cell division and with regard to natural selection (Fig. 12.1; also Chaps. 9 and 11).

12.2 Phage-Generation of eDNA

For phages to drive bacterial evolution that stems from transformation, then phage actions must give rise to the generation of eDNA. In the course of infecting a bacterium, phage's both take over the gene expression program of the host bacterium and redirect resources to phage gene expression and virion production. One such resource is nucleotides, and various phages are able to degrade their host's genome, such as to provide additional nucleotides for phage gene expression and genome replication. Different phages, however, break down host DNA to varying degrees. The less such breakdown, then the more such DNA that can then be released extracellularly to contribute to the eDNA pool. Furthermore, differences in environments will determine how far that DNA, once released, must travel before encountering a potentially transformable cell, how readily that DNA may travel over these distances, and how stable that DNA will be such as in terms of exposure to nucleases.

12.2.1 A General Feature of all Lytic Phages?

Even with phage T4, the paradigm of host chromosome destruction among phages (Chap. 2, Sect. 2.1), host DNA upon infection is not completely degraded. Kutter et al. (2018) provides estimates of resulting host DNA sizes in megadaltons, with the average molecular weight of a DNA nucleoside monophosphate being 327 daltons. During a prolonged phage T4 infection of *E. coli* B at 37 °C, the size of these fragments is roughly 1 megadalton, or about 3 kilobases. This does not mean that all of the host DNA is found at this size. It does, however, mean that even with phage T4 infections, some host DNA should be released upon lysis that is relatively intact and thereby presumably available for transformation as eDNA. By comparison, eDNA over time typically comes to be no more than about 100 bp (0.1 kb) in size (Overballe-Petersen et al. 2013).

Keen et al. (2017) found less plasmid degradation with phage SUSP2 infection (also, like phage T4, a myovirus) than with phage T4 infection. In terms of transformation potential, most of the other phages they tested (20 in total) were more like phage T4 in this regard then SUSP2, and all of the phages tested were presumptively non-temperate. Phage SUSP1, which also is a myovirus, by contrast was more like SUSP1 in preserving transformation ability. Phage T7, a podovirus, instead was one of the 20 that included phage T4. Thus, from these examples, 20 presumably strictly lytic phages appear to be somewhat destructive of host DNA whereas 3, including the *B. subtilis* myovirus, phage SPO1 as mentioned also in Chap. 2, Sect. 2.1 (Yehle and Ganesan 1972), are not. Plasmids aside, based on the T4 model it seems likely that most or all lytic phages still may readily supply transformable bacterial eDNA to environments. In addition, and with reasonably high certainty, as yet unencapsidated phage genomic nucleic acid should also be released upon phage-induced lysis from within.

12.2.2 A Role for Biofilms?

Whether phage-generated eDNA will actually result in transformation in nature, even of naturally competent bacteria, is an open question. We can imagine, for example, that DNA that is released within a biofilm can rapidly encounter other cells, whereas DNA released into well-mixed planktonic environments may rapidly disperse. Another issue, however, is the degree of genetic relatedness of eDNA donors and eDNA recipients. To the extent that immediately neighboring cells within biofilms are clonally related to a lysed, eDNA-releasing cell, i.e., as making up bacterial microcolonies (Kreft 2005; Abedon 2015b), then even if transformation is readily achieved, the potential for changes to the genetic makeup of eDNA recipients may be slight. On the other hand, bacterial biofilms often consist of closely associated multiple species (Kvich et al. 2020), so lack of diversity between donor and recipient organisms within biofilms may be not always be an issue (Molin and Tolker-Nielsen 2003).

That is, evolutionarily a key aspect of transformation is the ability of a cell to be 'transformed', meaning changed by incorporating new genetic material, and reception by a bacterium of DNA that is identical to existing DNA does not leave much room for evolutionary change (recalling that key to evolutionary change is genetic variability, and here the argument is that genetic variability may be lacking). On the other hand, within mixed-species biofilms the potential for reception of eDNA by somewhat genetically divergent organisms could be profound. Thus, it certainly seems reasonable that any mixed-species biofilm that is being successfully attacked lytically by phages may have an elevated potential for supporting the genetic migration/horizontal gene transfer of eDNA between bacteria. In addition, it is possible that with prophage induction, as leading to lysis of the host bacterium, eDNA may be released as well, potentially allowing for further opportunities for recombination between even well-isolated clonal bacteria (Chap. 9, Sect. 9.4). This latter possibility is discussed further below, in the last section of this chapter (Sect. 12.4).

12.3 Phages as Mediators of 'Superspreading'

Epidemiologically, a superspreader is an individual or type of individual that contributes especially to the dissemination of an infectious disease. Thus, the spread of pathogens between individuals instead of being 'smooth' or linear can be much more episodic, with infections being transmitted at relatively low rates until it is a superspreader that becomes infected. At that point, the rate of spread is greatly accelerated, particularly because the superspreader infects so many more individuals than the typical member of an affected population. Topically, as I write, we are still in the midst of the COVID-19 pandemic (2020–2021) and we have all heard stories of certain individuals who are responsible for transmitting the virus to far more others than the typical person infects. Those individuals are superspreaders, as too of course was 'Typhoid Mary'.

12.3.1 Horizontal Gene Transfer Superspreaders?

In this section I consider the concept of superspreader as applied to horizontal gene transfer, particularly as mediated by phages. It is important to recognize, however, that horizontal gene transfer ‘superspreader’ events are not necessarily described using that term. Dionisio et al. (2002) specifically described what could be dubbed as conjugation-mediated superspreading, where certain hosts to these plasmids happen to be more adept at mediating conjugation events with other bacteria. Thus, spread of the plasmid occurs at some relatively slow ongoing rate until the plasmid reaches these ‘superspreader’ bacteria, at which point the rate of spread of the plasmid increases dramatically. Dionisio, however, described this as an “Amplification effect” rather than as a superspreader effect, though clearly were we to apply ‘superspreader’ to horizontal gene transfer then that term could have been used by Dionisio et al. as well. Indeed, Fellous et al. (2012) make that very point. Certainly as well, Dionisio et al. used the word “Spread” in their article, and indeed that word is found in both their title and their running title.

12.3.2 Phages as Transformation Superspreaders?

The above all serves as an introduction to an article by Keen et al. (2017) on the potential for phages to at least mediate ‘superspreader’ events. In this case, rather than the perhaps anticipated transduction (more on that in a bit), it instead is transformation that is the mechanism of horizontal gene transfer. As also was the case with Dionisio et al. (2002), it is specifically the transfer of plasmids that Keen et al. considered.

The specific premise of the Keen et al. (2017) study is the above-noted idea that phage-mediated release of potentially transformable bacterial DNA should be enhanced to the extent that a given lytic phage infection does not severely degrade bacterial DNA. Explicitly, if a bacterium is infected with a phage that does not degrade bacterial plasmids, or indeed does not degrade any other bacterial DNA, then following phage-induced lysis that DNA should be available for transformation of naturally competent bacteria found in the same environment. Thus, by this logic, phages that do not degrade bacterial DNA during infections might be viewed as ‘superspreaders’, though really they are more ‘superspreading-mediating’ factors rather than superspreaders themselves, i.e., as considered from an epidemiological perspective. By contrast, in the Dionisio et al. (2002) study the permissively conjugating bacteria really are superspreaders, or at least they are to the extent that we are willing to apply that term to horizontal gene transfer events.

Of interest, this idea of phages mediating transformation was also suggested by Udden et al. (2008), p. 11952: “The transformation frequency of various recipient strains in the presence of the phage was $>$100-fold higher than that in the absence of the phage. Adding DNase to the culture reduced the transformation efficiency by $\approx$50-fold, suggesting that phage-mediated release of cellular DNA from the donor cells was an important step in the transformation.” These experiments were done using *V. cholerae* strains versus the *E. coli* strains employed by Keen et al. (2017). In addition, the eDNA in the Udden et al. (2008) experiments were provided following prophage induction events rather than upon the addition of virions to bacteria as was the case in the Keen et al. (2017) experiments. We will shortly return to possible roles of prophages in phage-mediated eDNA generation.

12.3.3 *Phages as Transduction Superspreaders?*

The various ideas considered above should lead to a question of whether phages themselves could serve as horizontal gene transfer superspreaders. In this case, it would be superspreading via transduction rather than phages indirectly contributing to horizontal gene transfer by providing eDNA for transformation events. Presumably, then, phages which are more adept at transducing bacterial DNA—that produce more transducing particles per infection, and perhaps that also can subsequently transduce a wider diversity of bacteria (wide genome 'penetrative' host range; Fig. 12.2)—could be viewed as superspreaders relative to less adept phages. Indeed, phages that are highly effective at transducing bacterial DNA, DNA which otherwise must not be excessively degraded for especially generalized

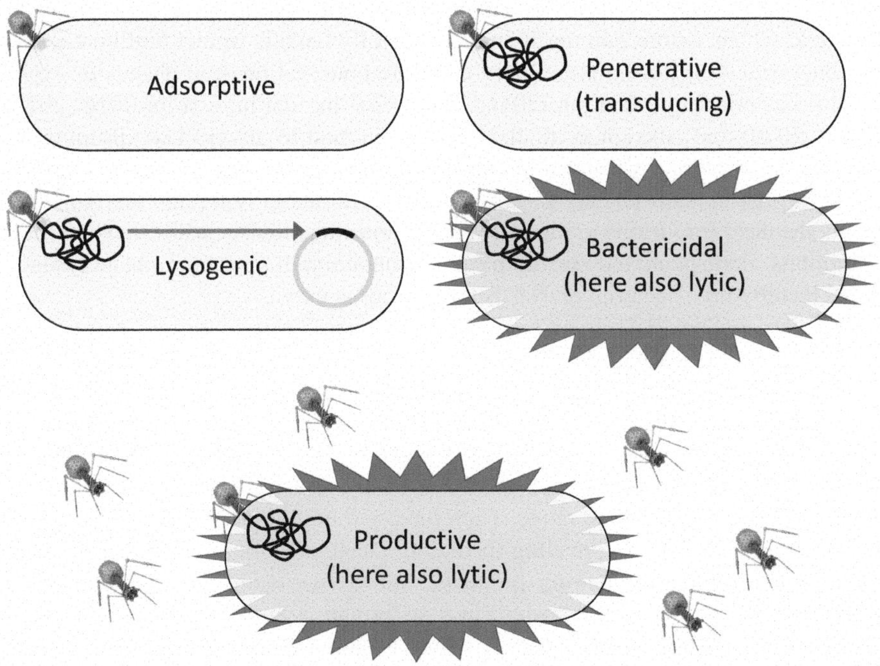

Fig. 12.2 Various host-range categories. Virions are able to attach (adsorptive host range; upper-left). Virions are able to deliver their genomes into the adsorbed bacterium (penetrative host range and also potentially transductive host range; upper-right). Phages are able to display lysogenic cycles (lysogenic host range; right-middle). Phages are able to kill the bacteria they have infected (bactericidal host range; left-middle). Phages are able to produce and release new virions (productive host range; bottom)

transduction to take place, ought also to be highly effective at generating transformable DNA, i.e., as Keen et al. (2017) considered.

Notwithstanding both of those possibilities, facilitating transduction as well as transformation, I am not too comfortable with making a claim that transducing phages can serve as superspreaders, and this is specifically because it should be bacteria that are described as the superspreaders in this context, i.e., 'superspreaders of infectious DNA', rather than describing as superspreaders the means by which the superspreading is mediated. Thus, perhaps bacteria which are more susceptible to being infected by transducing phages and/or at having their DNA packaged into transducing particles could be viewed as possible transductive superspreaders, rather than transducing phages themselves, or the high transformation-mediating phages described by Keen et al. (2017).

12.4 Temperate Phages as Generators of eDNA

Though mixed-species biofilms could serve as a means by which donor and recipient bacteria can be in close proximity, it continues to be an interesting consideration, the degree to which virions can reach and successfully lytically infect biofilm-bacteria. Leaving aside that I am a great believer in the potential for free phages to exploit biofilm bacteria, e.g., Abedon (2011, 2012a, 2012b, 2015a, 2015b, 2016, 2017a, 2017b, 2020) and Abedon et al. 2021, it is important to always keep in mind that lysogenized bacteria by definition are already phage infected. Therefore, biofilms can form that contain already-phage-infected bacteria, as lysogens, thereby eliminating the need for virions to reach those bacteria exogenously while they are found in biofilms. Perhaps mixed-species biofilms containing transducing-phage lysogens could thereby host bacterial eDNA 'superspreading events'.

In Chap. 9 (Sect. 9.4), I suggested that prophages encoding transducing phages could help to maintain gene exchange within well-isolated bacterial populations. Indeed, that presumably could be the case within biofilms as well. Whether or not those phages might be able to transduce DNA to other bacterial species, however, would at best be an open question and at worst probably is unlikely. Transformation, however, is different in that uptake of eDNA by transformation recipients is not dependent on a phage's host range. Therefore, within biofilms, perhaps lysis caused by the above-noted superspreading-mediating phages—but in this case as mediated by temperate phages following prophage induction—could provide a means of amplifying horizontal gene transfer via transformation among diverse bacteria. So too, such phages could supply eDNA within bacterial populations toward reducing clonality, e.g., as suggested by Naito and Pawlowska (2016), though those authors do not explicitly include a phage-induced lysis component in the scenario they present (p. 6): "...perhaps through the uptake of DNA from dead [mycoplasma-related endobacterial] cells present in the same host cytoplasmic niche." As Naito and Pawlowska note, however, it is vitally important that the bacteria involved retain rather than evolutionarily lose their recombination systems for natural competence to result in retention of taken-up DNA. They note as well that the bacteria they studied both possess phages and explicitly do succeed in staving off Muller's ratchet despite their confinement within the cytoplasms of their fungal hosts.

A variation on this idea is that it is not just phages lysing their parental lysogens that could provide this eDNA, but also released temperate phages infecting and lysing neighboring, presumably more closely related bacteria that also could be releasing eDNA. This possibility has been suggested by Molina-Quiroz et al. (2020) as based on experiments also employing *V. cholerae*. I should point out, however, that this idea of endogenously produced superspreading-mediating phages infecting and lysing neighboring, related bacteria may be no more plausible ecologically (but certainly not impossible!) than exogenously supplied superspreading-mediating phages also reaching and lysing biofilm bacteria. That is, a key requirement for endogenously supplied superspreading-mediating phages doing this is that closely related but not clonally related bacteria—bacteria found within the released phage's host range but which are not identical to the originally phage-producing lysogen—in fact are found nearby to the originally lysing lysogens, just as a key requirement for exogenously supplied superspreading-mediating phages is that they are able to also reach bacteria that they are able to infect. By contrast, with simply lysis of bacterial lysogens as initiated by superspreading-mediating prophages, no constraining steps to eDNA horizontal gene transfer potential should be equivalently present.

Perhaps consistent with the above observation of Molina-Quiroz et al. (2020), Haaber et al. (2016) show that certain prophages, in this case as shown in a *S. aureus* system, can produce virions that are released from lysogens that then serve as transducing phages after having infected neighboring bacteria. This would suggest that in addition to occasionally producing transducing particles, these lytic phages while infecting neighboring bacteria also probably are causing the release of relatively intact eDNA from those bacteria. In any case, there remains in this scenario a requirement for close association of two different bacterial strains that are both found within the same phage's host range, the original lysogen and the phage-amplifying neighboring bacteria.

Molina-Quiroz et al. (2020) do not actually mention biofilms in their study but rather "Chitin microcosms" were used. These, though, involved inclusion of shrimp shell (1% of the media) upon which biofilm perhaps grew. Also, their emphasis was on the potential for phage-generated eDNA to transform clone mates of the original, lysing lysogen. Contrasting within biofilms, phages finding new, planktonic bacteria to infect within liquid media and thereby generating eDNA may be less effective in leading up to transformation events. This is because eDNA once generated from planktonic bacteria likely would not be as concentrated within liquid media in the immediate vicinity of potentially transformable cells as should be eDNA that instead is generated within a biofilm. Alternatively, within liquid media whatever constraints biofilms may impose on the ability of phage virions to reach bacteria of course would not be present either. Note that we will return to the idea of lysogens releasing phages that can lyse neighboring, related, but not clonally related bacteria when we consider the concept of 'lysogen allelopathy' in Chap. 24, Sect. 24.4.

References

Abedon ST (2011) Bacteriophages and biofilms: ecology, phage therapy, plaques. Nova Science Publishers, Hauppauge

Abedon ST (2012a) Spatial vulnerability: bacterial arrangements, microcolonies, and biofilms as responses to low rather than high phage densities. Viruses 4:663–687

Abedon ST (2012b) Thinking about microcolonies as phage targets. Bacteriophage 2:200–204

Abedon ST (2015a) Ecology of anti-biofilm agents I: antibiotics versus bacteriophages. Pharmaceuticals (Basel) 8:525–558

Abedon ST (2015b) Ecology of anti-biofilm agents II: bacteriophage exploitation and biocontrol of biofilm bacteria. Pharmaceuticals (Basel) 8:559–589

Abedon ST (2016) Bacteriophage exploitation of bacterial biofilms: phage preference for less mature targets? FEMS Microbiol Lett 363:fnv246

Abedon ST (2017a) Active bacteriophage biocontrol and therapy on sub-millimeter scales towards removal of unwanted bacteria from foods and microbiomes. AIMS Microbiol 3:649–688

Abedon ST (2017b) Phage "delay" towards enhancing bacterial escape from biofilms: a more comprehensive way of viewing resistance to bacteriophages. AIMS Microbiol 3:186–226

Abedon ST (2020) Bacteriophage-mediated biocontrol of wound infections, and ecological exploitation of biofilms by phages. In: Shiffman M, Low M (eds) Biofilm, pilonidal cysts and sinuses. recent clinical techniques, results, and research in wounds, vol 1. Springer Nature, Cham, pp 121–158

Abedon ST, Danis-Wlodarczyk KM, Wozniak DJ, Sullivan MB (2021) Improving phage-biofilm in vitro experimentation. Viruses 13:1175

Dionisio F, Matic I, Radman M, Rodrigues OR, Taddei F (2002) Plasmids spread very fast in heterogeneous bacterial communities. Genetics 162:1525–1532

Domingues S, Harms K, Fricke WF, Johnsen PJ, da Silva GJ, Nielsen KM (2012) Natural transformation facilitates transfer of transposons, integrons and gene cassettes between bacterial species. PLoS Pathog 8:e1002837

Doolittle WF (1998) You are what you eat: a gene transfer ratchet could account for bacterial genes in eukaryotic nuclear genomes. Trends Genet 14:307–311

Fellous S, Duncan AB, Quillery E, Vale PF, Kaltz O (2012) Genetic influence on disease spread following arrival of infected carriers. Ecol Lett 15:186–192

Haaber J, Leisner JJ, Cohn MT, Catalan-Moreno A, Nielsen JB, Westh H, Penades JR, Ingmer H (2016) Bacterial viruses enable their host to acquire antibiotic resistance genes from neighbouring cells. Nat Commun 7:13333

Johnsborg O, Eldholm V, Havarstein LS (2007) Natural genetic transformation: prevalence, mechanisms and function. Res Microbiol 158:767–778

Keeling PJ, Palmer JD (2008) Horizontal gene transfer in eukaryotic evolution. Nat Rev. Genet 9: 605–618

Keen EC, Bliskovsky VV, Malagon F, Baker JD, Prince JS, Klaus JS, Adhya SL (2017) Novel "superspreader" bacteriophages promote horizontal gene transfer by transformation. MBio 8: e02115–e02116

Kreft J-U (2005) Conflicts of interest in biofilms. Biofilms 1:265–276

Kutter E, Bryan D, Ray G, Brewster E, Blasdel B, Guttman B (2018) From host to phage metabolism: hot tales of phage T4's takeover of *E. coli*. Viruses 10:387

Kvich L, Burmolle M, Bjarnsholt T, Lichtenberg M (2020) Do mixed-species biofilms dominate in chronic infections?—Need for *in situ* visualization of bacterial organization. Front Cell Infect Microbiol 10:396

Mell JC, Redfield RJ (2014) Natural competence and the evolution of DNA uptake specificity. J Bacteriol 196:1471–1483

Molin S, Tolker-Nielsen T (2003) Gene transfer occurs with enhanced efficiency in biofilms and induces enhanced stabilisation of the biofilm structure. Curr Opin Biotechnol 14:255–261

Molina-Quiroz RC, Dalia TN, Camilli A, Dalia AB, Silva-Valenzuela CA (2020) Prophage-dependent neighbor predation fosters horizontal gene transfer by natural transformation. mSphere 5:e00975-20

Naito M, Pawlowska TE (2016) Defying Muller's ratchet: ancient heritable endobacteria escape extinction through retention of recombination and genome plasticity. MBio 7:e02057-15

Overballe-Petersen S, Harms K, Orlando LA, Mayar JV, Rasmussen S, Dahl TW, Rosing MT, Poole AM, Sicheritz-Ponten T, Brunak S, Inselmann S, de Vries J, Wackernagel W, Pybus OG, Nielsen R, Johnsen PJ, Nielsen KM, Willerslev E (2013) Bacterial natural transformation by highly fragmented and damaged DNA. Proc Natl Acad Sci USA 110:19860–19865

Udden SM, Zahid MS, Biswas K, Ahmad QS, Cravioto A, Nair GB, Mekalanos JJ, Faruque SM (2008) Acquisition of classical CTX prophage from *Vibrio cholerae* O141 by El Tor strains aided by lytic phages and chitin-induced competence. Proc Natl Acad Sci USA 105:11951–11,956

Vettori C, Paffetti D, Pietramellara G, Stotzky G, Gallori E (1996) Amplification of bacterial DNA bound on clay minerals by the random amplified polymorphic DNA (RAPD) technique. FEMS Microbiol Ecol 20:251–260

Yehle CO, Ganesan AT (1972) Deoxyribonucleic acid synthesis in bacteriophage SPO1-infected *Bacillus subtilis*. I. Bacteriophage deoxyribonucleic acid synthesis and fate of host deoxyribonucleic acid in normal and polymerase-deficient strains. J Virol 9:263–272

Chapter 13
Transduction of Large Amounts of DNA

Contents

Abstract Generalized transduction allows phages to move phage genome-sized segments of DNA between bacteria. Not all phages are able to so-transduce bacterial DNA and variations exist also in terms of both what DNA is transferred and how likely it is that bacterial DNA will be encapsidated at all. Encapsidation and then movement of bacterial DNA to a new bacterium, as mediated by phage transducing particles, is also only the first step of the process of generalized transduction. Important as well is what happens to that DNA once it arrives in the cytoplasm of the recipient bacterium, and there are three possible fates of this DNA. These are lack of subsequent integration into the bacterial genome (known as abortive transduction), homologous recombination resulting especially in the orthologous replacement of segments of similar DNA (as generalized transduction traditionally was detected experimentally), and illegitimate recombination, which can provide new genes to transduction recipients. In addition, it is possible for certain forms of bacterial DNA to parasitize phages, increasing their likelihood of encapsidation but also possessing means of better assuring their insertion into recipient bacterial genomes than illegitimate recombination can provide. Considered as well is the potential for the generalized transduction of plasmids between bacteria.

> Transduction can occur by several different mechanisms... the simplest and likely most common mechanism is simply DNA packaging error in which a phage genome sized fragment of the chromosome of the infected cells is packaged instead of phage DNA. ... [Considering] the fraction of tailed phages that can perform transduction... 10% is a

S. T. Abedon, *Bacteriophages as Drivers of Evolution*,
https://doi.org/10.1007/978-3-030-94309-7_13

reasonable and conservative estimate…—Roger W. Hendrix and Sherwood R. Casjens (2008, p. 89)

The primary evolutionary consequence of horizontal gene transfer is that adaptations need not develop solely linearly within lineages, that is, where each adaptation must be built up more or less one mutation at a time, serially, on top of the suite of existing adaptations already present in an organism (Fig. 13.1). At its most powerful, however, horizontal gene transfer can result in the movement of multiple genes, such as whole operons underlying adaptations, or even genomic islands underlying whole new life styles. That in fact is the power of generalized transduction (Casas and Maloy 2018), which allows the horizontal gene transfer of relatively vast pieces of DNA, consisting of tens of thousands of base pairs. Not my emphasis here, though definitely of interest to many, phages also can move antibiotic-resistance genes via generalized transduction (Casas and Maloy 2018, Gilcrease and Casjens 2018, Villa et al. 2019, Balcázar 2020). Whether phages also often encode antibiotic resistance genes, though, remains an open question (Enault et al. 2016). In this chapter, we consider generalized transduction, along with the horizontal gene transfer of large segments of DNA between bacteria more generally.

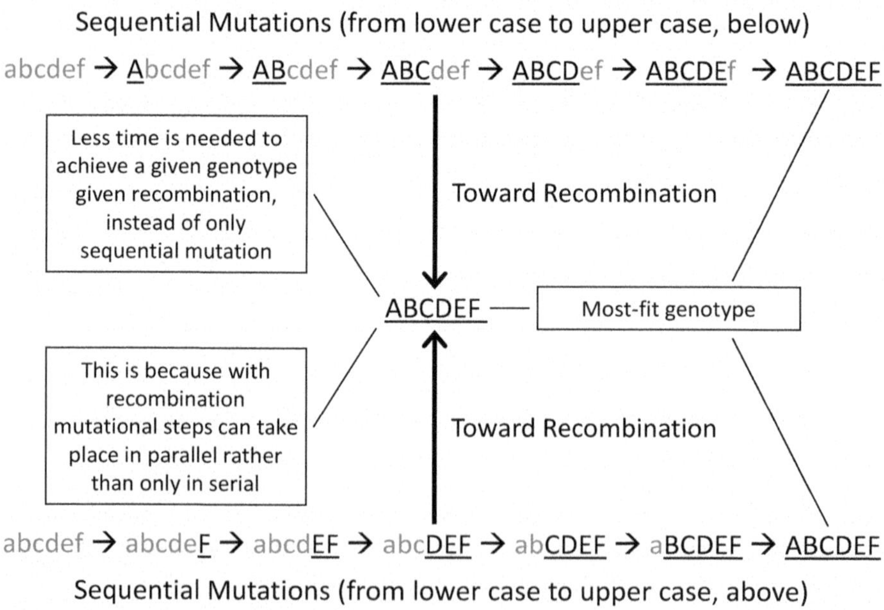

Fig. 13.1 Generation of most-fit genotype via sequential mutation-only versus sequential mutation in association with recombination

13.1 Generalized Transduction

Generalized transduction is the encapsidation of DNA that is not of viral origin, or at least not of the virus doing the encapsidating (Harrison and Brockhurst 2017, Chiang et al. 2019, Schneider 2021). The encapsidation of non-phage DNA by phages instead constitutes an error on the part of the virus (Fig. 13.2), since with generalized transduction capsids are wasted with regard to virion production. Also, as pointed out in Chap. 10, Sect. 10.3, the breadth of bacteria that a given phage may be able to transduce can be broader than the breadth of bacteria they are able to productively infect. Generalized transduction thus represents something of a DNA-delivery promiscuity on the part of the virus, but is not observed with all phages. Specifically, there are two criteria necessary for phages to accidentally encapsidate bacterial DNA with relatively high frequencies. These and other issues associated with generalized transduction are as discussed as follows.

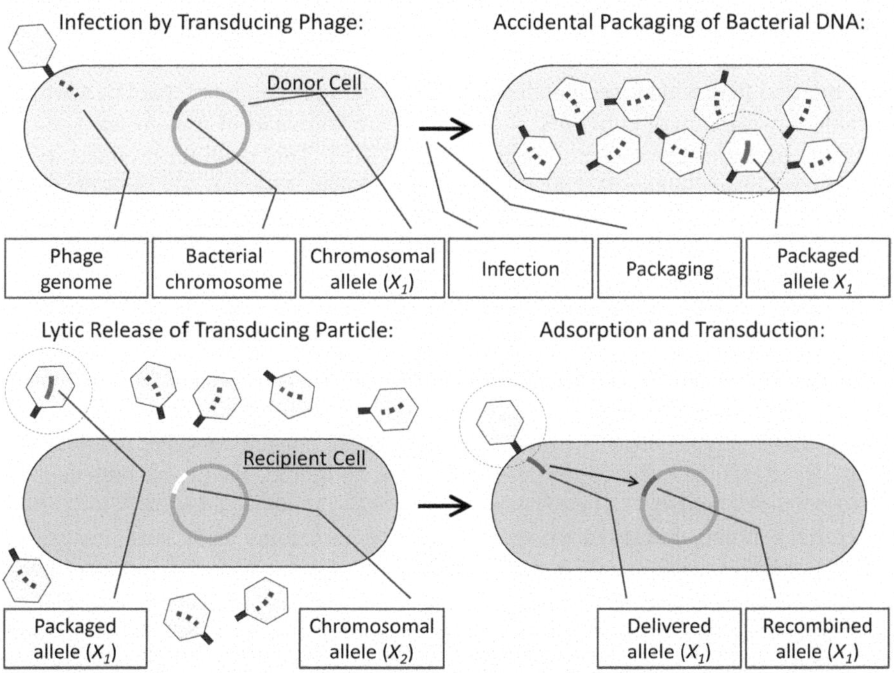

Fig. 13.2 Generalized transduction. Note that the transducing phage first does not yet carry bacterial DNA (upper-left), second has accidentally packaged bacterial DNA (upper-right), third has lytically released a transducing particle carrying that bacterial DNA (lower-left), and fourth, that transducing particle has then delivered that bacterial DNA to a new bacterium (lower-right). That new bacterium does not necessarily contain DNA that differs in sequence from the transduced DNA, but here this new DNA is being shown as allelically distinct from the otherwise homologous chromosomal recipient DNA

13.1.1 Criteria for Generalized Transduction

The first criterion for generalized transduction is that the means by which phages package DNA into virions must display sufficient promiscuity to result in the packaging of bacterial DNA rather than solely packaging phage DNA. There are three mechanisms which can result in this packaging of bacterial DNA. One mechanism might be described as having high promiscuity, where no packaging signal is required for successful bacterial DNA packaging. This results in there being little bias in terms of what DNA is packaged. High-promiscuity DNA packaging is exemplified by coliphage P1 as a mostly though not entirely unbiased transducing phage, though this is despite phage P1 otherwise use of a *pac* site for packaging (Hanks et al. 1988, Lehnherr 2006, Dutra et al. 2007). Indeed, phage P1 is apparently able to non-specifically package more than one piece of DNA at the same time (Coren et al. 1995). The other mechanism is the recognition of a single sequence as found on a phage genome but that also can be found in bacterial genomes with varying degrees of homology. This we might describe as resulting in a generalized transduction involving a 'medium' level of promiscuity since there also can be diversity in terms of what bacterial DNA may be packaged. Still, there is substantial bias, more so than what is seen with phage P1, toward packaging of that DNA which is found downstream of either *pac* sites or what are known as of *pseudo-pac* sites as found on bacterial chromosomes (Chen et al. 2018). This medium promiscuity is exemplified by coliphage P22 as a transducing phage. Alternatively, and the third mechanism, specialized transduction may be viewed as a form of particularly low-promiscuity transduction, to a point where it is not considered to be a form of generalized transduction.

Ignoring for now specialized transduction, note that it is those phages that require recognition sequences for DNA packaging, but don't employ *pac* sites, that tend to *not* be generalized transducers. With *pac* sites, that is, a single recognition site is utilized, with phage DNA otherwise packaged by a headful mechanism (the latter as discussed in the next subsection). Alternatively, when phage packaging involves the use of *cos* sites, then two recognition sequences are required, as found at both ends of the to-be-packaged DNA. The existence of two recognizable packaging sites within a bacterial genome, that also are not too far apart, is something that is sufficiently rare that these phages, such as coliphage λ, generally are not considered to be generalized transducing phages. Bacterial chromosomal DNA still can evolve for the sake of promoting transfer by these *cos*-containing phages (Penades et al. 2015), though again that would not be an example of generalized transduction given the substantially lower diversity of bacterial DNA that may be so transduced. Coliphage λ nonetheless is an effective specialized transducer of bacterial DNA, though this is possible because its *cos* sites are not found at the ends of the integrated prophage but instead in the middle (Fig. 13.3).

The second criterion for generalized transduction is that bacterial DNA must not be substantially degraded in the course of phage infection (Chapters 2 and 11). This is because encapsidation requires sufficiently intact DNA. Thus, those phages that

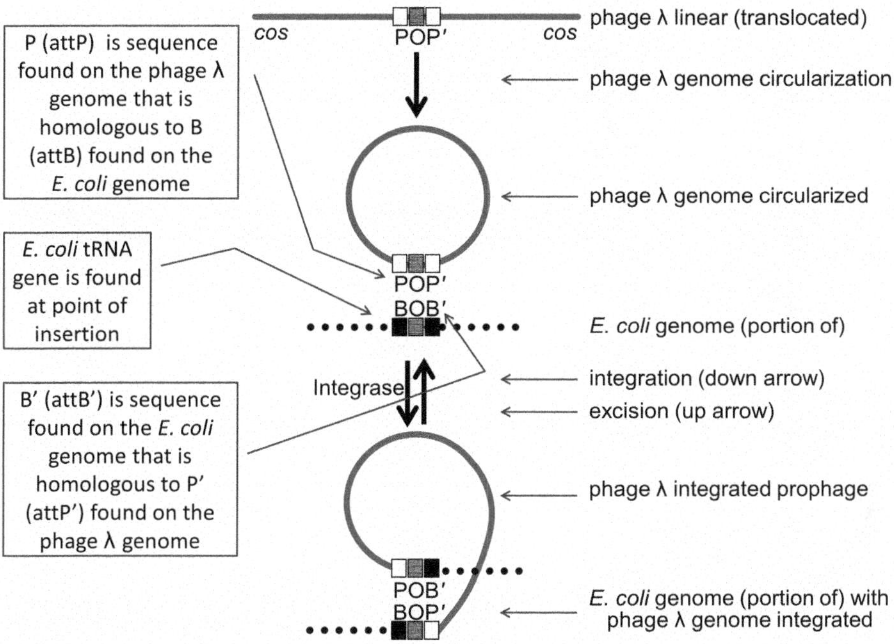

Fig. 13.3 Temperate phage integration, with phage λ (solid gray lines) and *E. coli* (dotted black lines) as models. This figure is derived from that of Brown and Cox (2021) which in turn was derived from Weisberg and Landy (1983). The various genome features are not drawn to scale

don't use recognition sites for packaging and also don't degrade their host's DNA should be exceptional transducers of bacterial DNA, a 'prediction' that I will return to shortly. How common wild-type phages might be which possess both of these qualities, however, is an open question.

13.1.2 Headful Packaging

DNA uptake into virions without dual signaling—with or without use of *pac* sites but not with use of *cos* sites—occurs in combination with headful packaging. Here, phage genome replication results in the generation of concatemers of DNA that are then packaged in terms of total lengths of DNA. To make sure that all of the phage genome is packaged, the amount of DNA that is packaged is greater than the actual genome length of the phage. With phage T4, for example, headful packaging is approximately 102% of the genome (Rao and Black 2005) or 110% for phage P1 (Lehnherr 2006). This results in what is known as genome circular permutation (Fig. 13.4), which is to say that, for example, the phage T4 genome as packaged can be represented as a circle, with neither a fixed beginning nor a fixed end. The actual genome, both as packaged and as functional during infections, is however linear. As

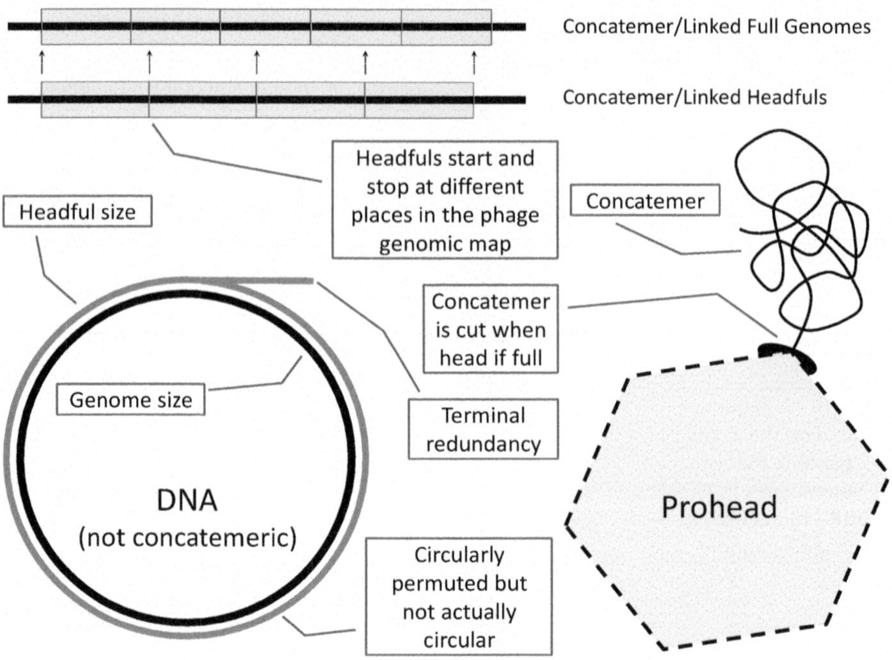

Fig. 13.4 Circular permutation and headful packaging. Concatemers are long pieces of covalently linked genomes. With headful packaging, more than 100% of a phage genome is packaged as cut from these concatemers. As a result, packaged genomes repeat themselves at their ends, which is described as terminal redundancy. Because of terminal redundancy, headful packaging results in genomes with neither consistently placed beginnings nor ends of their linear chromosomes. Nonetheless, any one genome can be arranged metaphorically into a circle, with terminal redundancies overlapping. As the points of overlap will vary from genome to genome, this is known as circular permutation, with the circularity of these genomes therefore a concept rather than a physical actuality

Wilson et al. (1979) summarize (p. 81), "Possession of a circularly permuted phage chromosome and the ability to carry out generalized transduction often go hand in hand."

Phage T4, despite using headful packaging, is a terrible transducing phage. There, however, is an article, published in *Nature* in 1979 (Wilson et al. 1979), that has as its title, "High-frequency generalized transduction by bacteriophage T4". I mention this to prove the point that it is signal-less packaging that is crucial for generalized transduction. Specifically, in the Wilson et al. article they employed a T4 strain that possesses mutations in multiple genes that ultimately prevent the phage from degrading host DNA. Indeed, they reported transduction of frequencies that (p. 80) "are higher than those observed for other transducing phages." Thus, as with 'superspreaders' discussed previously (Chap. 12), for a phage to contribute substantially to horizontal gene transfer, a key criterion is to not substantially degrade bacterial DNA, while to contribute transductively there also needs to be a potential to promiscuously package bacterial DNA should it remain intact. Another thing to keep in mind, and as also considered in Chap. 12 (Sect. 12.4), is that temperate phages, depending on the phage, may transduce DNA, some of it via generalized transduction (Schneider 2021), following prophage induction (Colomer-Lluch et al. 2014).

13.1.3 Selection Against Rather Than for Generalized Transduction

One thing to always keep in mind when considering especially generalized transduction—as a mechanism of DNA transfer that does not involve also the transfer of phage DNA—is that it is unlikely that the transducing phage is under selection to maintain this mechanism. Alternatively, one could posit a group-selection mechanism for the existence of transduction, e.g., from Davidson (2018, p. 153), in discussing the results of Chen et al. (2018), "High-frequency transduction may actually be an evolutionary adaptation for phages to increase the general fitness of their host."

Such a scenario would require that the same phages carrying alleles that are promoting higher frequencies of transduction are also (1) infecting bacteria that are recipients of those transduction events, (2) are infecting bacteria at higher rates than phages that do not carry these phage alleles, and/or (3) that the transduced bacteria have not become fitter in ways that improve the fitness of phages other than the transducing phages. While the first of these criteria seems to be reasonably plausible, the second is less so unless the phages carrying the transduction-promoting allele are isolated during selection for that allele away from competing phages, e.g., as a form of kin selection. The third criterion is perhaps even less reasonable given time lags between transduction and resulting phenotypic modification of the transduced bacterium. That is, if a phage needs a specific gene function upon infection, then why not just encode that function itself? This is all to say that, as a general rule, be careful to not fall into group-selectionist thinking when pondering the evolution of traits. More narrowly, in the case of generalized transducing phages, it seems somewhat more likely that an ability to transduce is a product of insufficient selection *against* that trait rather than necessarily selection directly for it. With specialized transduction and/or the acquisition of phage morons, on the other hand, it is quite reasonable to posit that *still-viable* phages carrying new genes could be benefiting from carrying those genes (see Chaps. 11, 14, and 15).

13.2 Homologous Recombination Versus Illegitimate Recombination

Homologous recombination is that which we are taught about in introductory biology class. It is the crossing-over process observed during meiosis, though it occurs in bacteria as well. The process involves pairing up of similar DNA sequences, with then one sequence replacing the other, or indeed for meiotic crossing over, with the generation of now two recombinant strands, one for each chromosome homolog, rather than just one. In either case, key is that it is DNA that is highly similar in terms of DNA sequence that is being either replaced or swapped. This allows for a replacement of alleles following horizontal gene transfer, or allele swapping, which is also known as orthologous replacement. Would-be phage-mediated transfer of bacterial DNA to a new bacterium that does not result in homologous recombination can be described as an aborted transduction.

Illegitimate recombination, also known as non-homologous recombination, is much less well understood mechanistically than homologous recombination. It either involves recombination based on little sequence homology between the recombining strands, or no sequence homology at all. Furthermore, rather than replacing alleles, illegitimate recombination results, at least so far as we are concerned, in insertion of new DNA (though not to be confused with the insertions associated with prophage acquisition; Chap. 5). Lastly, the rate at which illegitimate recombination occurs will tend to be many orders of magnitude lower than rates of homologous recombination. Illegitimate recombination, that is, is both very rare and results, again so far as we are concerned, in gains in the amount of DNA carried by an organism. That new DNA can be explicitly of phage origin or it can be as carried by transducing particles. For this new DNA to be functional, generally it must consist of complete genes, or even complete pathways, e.g., whole operons.

It is primarily via generalized transduction that many bacterial genes or whole operons may be carried by transducing particles. Hence, generalized transduction, illegitimate recombination, and gains by bacteria of large amounts of potentially functional DNA all should be viewed as somewhat co-occurring phenomena (though note that plasmids, too, can insert into chromosomes). Another phenomenon that may be associated with the encapsidation of large amounts of non-phage DNA is acquisition by bacteria of so-called genomic islands (Sect. 13.3).

13.3 Genomic Islands

Genomic islands, often which serve as pathogenicity islands, are clearly products of horizontal gene transfer (Lawrence and Hendrickson 2008, Juhas et al. 2009, Busby et al. 2013, Carraro et al. 2017, Maguire et al. 2020). Indeed, they are 'islands' of relatively newly acquired DNA sequence which lie within seas of more stably present genetic material. Given their sizes, ranging from 10 to 200 kb, genomic

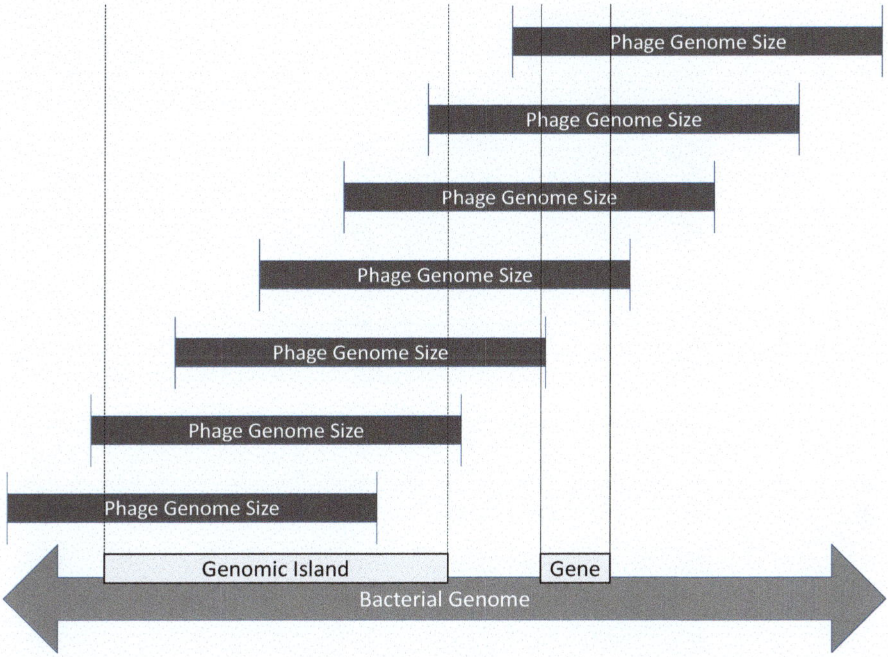

Fig. 13.5 Randomly packaging regions of DNA becomes less likely the larger the region of DNA to be packaged and the smaller the piece of DNA this is being packaged. In this case, only the second-from-the-left randomly packaged region of DNA by chance encompasses the whole genomic island. The individual gene, as shown further to the right near the bottom of the figure, would by contrast be packaged here as part of four different randomly chosen regions of bacterial genomic DNA (as indicated in the figure by the top four "Phage genome size" fragments)

island that are not capable of being transferred via conjugation must be carried from bacterium to bacterium by transducing phage particles. Note, though, that for a genomic island to be transferred as a complete entity with high likelihood, then it must have evolved a means of assuring that complete packaging, thereby with genomic islands serving as parasites of specific phage types in terms of their horizontal transfer (see also Sect. 4.5 of Chap. 4 as well as below). Alternatively, a genomic island would need to be somewhat smaller than the total piece of DNA being transferred so that ends of the DNA that is being transduced are not located with too high probability in the middle of a genomic island (Lawrence and Hendrickson 2008) (Fig. 13.5).

Possessing packaging signals or not, as a general rule the larger the genomic island, then the fewer transducing phages will possess genome sizes that allow them to transfer that island in full, with stringencies only increased if that transfer is to be accomplished with reasonably high likelihood via promiscuous generalized transduction alone (versus the genomic island encoding *pac* sites or even *cos* sites). The likelihood of transfer thus must decline with increasing genomic island size if only because fewer phages would exist which are able to package the DNA in full, but

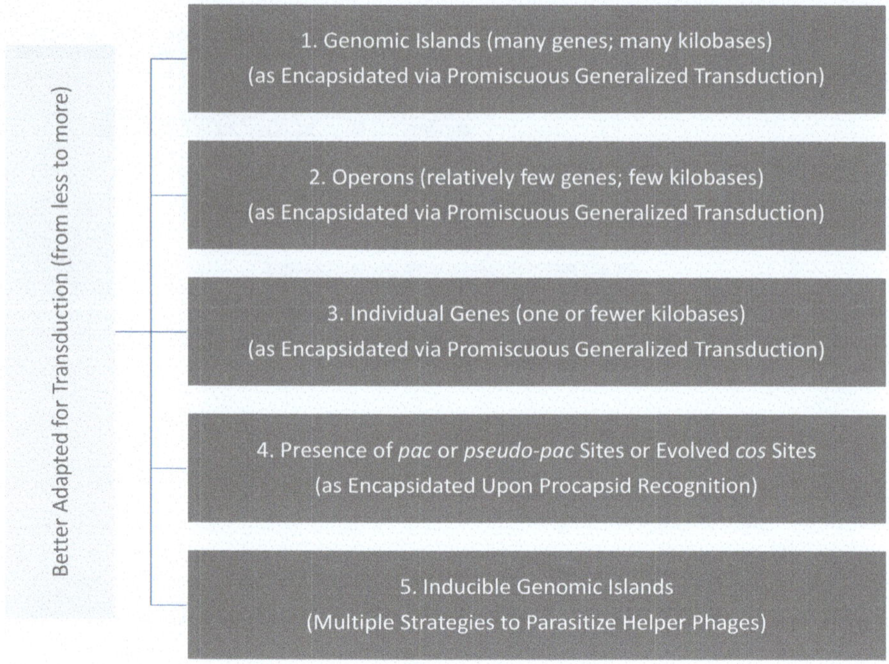

Fig. 13.6 Hierarchy of transducibility. Generalized transduction without packaging signals (1. through 3.), i.e., as involving more 'promiscuous' transduction, is least likely, with potential for encapsidation as a whole decreasing as the size of the DNA segment (here number of kilobases) increases (3. > 2. > 1.). Packaging signals can increase the likelihood of transduction (4. > 3.) but are still relatively passive means of assuring transduction. Phage-inducible genomic islands (5.) employ additional, active mechanisms to increase the amount of encapsidation achieved per transducing- or helper-phage-infected bacterium

even more so to the extent that packaging signals are lacking. Alternative, it is possible for some genomic islands to be transduced in parts rather than in full (Moon et al. 2015, 2016).

Among genomic islands that are packaged in full using specific mechanisms of phage packaging are what are known as phage-inducible chromosomal islands (as introduced in Chap. 4, Sect. 4.5) (Feaga and Dworkin 2019). These are not phages but instead are parasites of phages (Penadés and Christie 2015). Specifically, they are mobile genetic elements found in Gram-positive as well as Gram-negative bacteria (Fillol-Salom et al. 2019, Barth et al. 2020) that parasitize helper phages to effect their transmission to new bacteria. They are distinct from genomic islands that simply possess *pac* sites or *cos* sites in that they replicate to increase their rates of packaging by helper phages (Fig. 13.6), often to the great detriment of the helper phage infection (Chap. 20, Sect. 20.2.3).

As Penadés and Christie (2015) point out (p. 194):

> There is a growing body of evidence that [*S. aureus* phage-inducible pathogenicity islands] can undergo horizontal gene transfer at high frequency not only between different

> staphylococcal species but also across long phylogenetic distances. . . . Such transfer does not require that the helper phage be able to propagate in the recipient strain. As long as the appropriate receptor is present for transducing particle adsorption and subsequent DNA ejection. . .

The same of course can be said for the generalized transduction of any DNA segment (Chap. 10), though with the caveat that the further the phylogenic distance with generalized transduction of random bacterial chromosomal DNA, then the less likely that homologous recombination upon acquisition will occur and therefore the less likely that segregation of the acquired DNA will be achieved (Chap. 11, Sect. 11.2). With phage-inducible genomic islands, the limitation instead, once they have entered a bacterium's cytoplasm, is an ability to successfully integrate their genomes into their host bacterium's chromosome, i.e., as equivalent to the requirement for integration by many temperate phages to display lysogenic cycles. Subsequently, of course, as in all of these cases, it is also helpful for natural selection to favor their presence within the genomes of recipient bacteria.

These inducible genomic islands can also be described as satellite viruses (or satellite phages). As parasites of other, helper phages, there also can be arms races whereby helper phages evolve resistance to satellite viruses while phage-inducible chromosomal islands evolve enhancements in their abilities to parasitize helper phages (Frigols et al. 2015), i.e., as representing a form of antagonistic coevolution (Chap. 25). Remarkably, phage-inducible genomic islands also can effect transduction processes of bacterial DNA in a manner that their helper phages are unable, and this occurs through a process that requires *pac*-site equivalents (*pseudo-pac* sites; see above) that are found on bacterial chromosomes.

13.4 Generalized Transduction of Plasmids

Generalized transducing phages not only can encapsidate fragments of bacterial chromosomal DNA, they can transduce plasmids as well. As a result, plasmids may be transferred between bacteria not just via conjugation, or transformation (Chap. 12, Sect. 12.3.2), but also via transduction, e.g., (Raya et al. 1989, Jiang and Paul 1998, Ammann et al. 2008).

Saye et al. (1987) provided evidence that the efficiency of this packaging can be greater to the extent that the size of a plasmid's DNA is similar to the size of the transducing phage's genome rather than much smaller. Specifically, phage F116, with a genome size of 65 kb (Byrne and Kropinski 2005) was able to transduce a plasmid of size 67.5 kb with 4 times the efficiency of a plasmid of size 55 kb and was not able to transduce at detectible levels plasmids of size 7 kb. It is notable, though, that variables besides size might have resulted in this correlation instead. One means by which plasmids may get around size inefficiencies with regard to generalized transduction, however, can instead to be encapsidated as concatenated DNA (Hahn et al. 1991). Plasmids present at higher copy numbers also may be subject to higher rates of transduction. So far as I can tell, though, this has been demonstrated particularly for non-standard means of transduction rather than because phages are simply promiscuously transducing higher copy-number pieces of DNA (Valero-Rello et al. 2017, Rodriguez-Rubio et al. 2020). Still, it seems logical that higher copy numbers should correlate, all else held equal, with higher rates of accidental packaging.

To gain some perspective on how many phages might be capable of generalized transduction, Shousha et al. (2015) tested 243 phages of *E. coli* that were isolated from chicken meat and nearly 25% of those phages were able to transduce antibiotic resistance markers. These abilities varied between phages in terms of what antibiotic resistance marker was being tested, and I speculate that this is because the resistance markers were associated with plasmids (from p. 4605: "without the need for homologous recombination") and the phages tested varied in terms of their abilities to transfer different plasmids. Note that we can speculate further that the same phages that are effective at transducing these antibiotic markers may also be able to serve as transformation 'superspreaders' (Chap. 11). Still, 25% is an impressively high number such as when compared to the estimate of 10% suggested by Hendrix and Casjens (2008).

Thus, overall, it is not unusual to find that transducing phages can also transduce plasmids, though more evidence of such transfer has been obtained with more tractable antibiotic-marked plasmids than with natural systems. An expectation therefore should exist that not just conjugation but so too phages can be responsible for moving plasmids between bacteria. Furthermore, we may actually expect that transduction of plasmid DNA could be more achievable between otherwise somewhat dissimilar bacteria than transduction of non-plasmid DNA, that is, in terms of likelihood of retention of the transferred DNA, because plasmids possess autonomous mechanisms for DNA retention. Specifically, plasmids need not undergo recombination with recipient-host DNA to achieve segregation. Overall, phage-mediated plasmid movement between bacteria likely plays an important and indeed complementary role to conjugation in their horizontal gene transfer, perhaps particularly in terms of transferring plasmids between bacteria for which conjugation is not achievable.

References

Ammann A, Neve H, Geis A, Heller KJ (2008) J Bacteriol 190:3083–3087

Balcázar JL (2020) Implications of bacteriophages on the acquisition and spread of antibiotic resistance in the environment. Int Microbiol 23:475–479

Barth ZK, Silvas TV, Angermeyer A, Seed KD (2020) Genome replication dynamics of a bacteriophage and its satellite reveal strategies for parasitism and viral restriction. Nucl Acids Res 48:249–263

Brown N, Cox C (2021) Bacteriophage use in molecular biology and biotechnology. In: Harper DR, Abedon ST, Burrowes B, McConville M (eds) Bacteriophages: biology, technology, therapy. Springer, New York, pp 465–506

Busby B, Kristensen DM, Koonin EV (2013) Contribution of phage-derived genomic islands to the virulence of facultative bacterial pathogens. Environ Microbiol 15:307–312

Byrne M, Kropinski AM (2005) The genome of the *Pseudomonas aeruginosa* generalized transducing bacteriophage F116. Gene 346:187–194

Carraro N, Rivard N, Burrus V, Ceccarelli D (2017) Mobilizable genomic islands, different strategies for the dissemination of multidrug resistance and other adaptive traits. Mob Genet Elements 7:1–6

Casas V, Maloy S (2018) The role of phage in the adaptation of bacteria to new environmental niches. In: Rampelotto PH (ed) Molecular mechanisms of microbial evolution. Springer, Cham, pp 267–306

Chen J, Quiles-Puchalt N, Chiang YN, Bacigalupe R, Fillol-Salom A, Chee MSJ, Fitzgerald JR, Penades JR (2018) Genome hypermobility by lateral transduction. Science (New York, N Y) 362:207–212

Chiang YN, Penades JR, Chen J (2019) Genetic transduction by phages and chromosomal islands: the new and noncanonical. PLoS Pathog 15:e1007878

Colomer-Lluch M, Jofre J, Muniesa M (2014) Quinolone resistance genes (*qnrA* and *qnrS*) in bacteriophage particles from wastewater samples and the effect of inducing agents on packaged antibiotic resistance genes. J Antimicrob Chemother 69:1265–1274

Coren JS, Pierce JC, Sternberg N (1995) Headful packaging revisited: the packaging of more than one DNA molecule into a bacteriophage P1 head. J Mol Biol 249:176–184

Davidson AR (2018) A common trick for transferring bacterial DNA. Science (New York, N Y) 362:152–153

Dutra BE, Sutera VA Jr, Lovett ST (2007) RecA-independent recombination is efficient but limited by exonucleases. Proc Natl Acad Sci U S A 104:216–221

Enault F, Briet A, Bouteille L, Roux S, Sullivan MB, Petit MA (2016) Phages rarely encode antibiotic resistance genes: a cautionary tale for virome analyses. ISME J 11:237–247

Feaga HA, Dworkin J (2019) A wolf in sheep's clothing: chromosomal pathogenicity islands co-opt phage capsids to facilitate horizontal spread. Mol Cell 75:889–890

Fillol-Salom A, Bacarizo J, Alqasmi M, Ciges-Tomas JR, Martinez-Rubio R, Roszak AW, Cogdell RJ, Chen J, Marina A, Penades JR (2019) Hijacking the hijackers: *Escherichia coli* pathogenicity islands redirect helper phage packaging for their own benefit. Mol Cell 75:1020–1030

Frigols B, Quiles-Puchalt N, Mir-Sanchis I, Donderis J, Elena SF, Buckling A, Novick RP, Marina A, Penades JR (2015) Virus satellites drive viral evolution and ecology. PLoS Genet 11:e1005609

Gilcrease EB, Casjens SR (2018) The genome sequence of *Escherichia coli* tailed phage D6 and the diversity of *Enterobacteriales* circular plasmid prophages. Virology 515:203–214

Hahn DR, McHenney MA, Baltz RH (1991) Properties of the streptomycete temperate bacteriophage FP43. J Bacteriol 173:3770–3775

Hanks MC, Newman B, Oliver IR, Masters M (1988) Packaging of transducing DNA by bacteriophage P1. Mol Gen Genet 214:523–532

Harrison E, Brockhurst MA (2017) Ecological and evolutionary benefits of temperate phage: what does or doesn't kill you makes you stronger. BioEssays 39:1700112

Hendrix RW, Casjens SR (2008) The role of bacteriophages in the generation and spread of bacterial pathogens. In: Hensel M, Schmidt H (eds) Horizontal gene transfer in the evolution of pathogenesis. Cambridge University Press, Cambridge, pp 79–112

Jiang SC, Paul JH (1998) Gene transfer by transduction in the marine environment. Appl Environ Microbiol 64:2780–2787

Juhas M, van der Meer JR, Gaillard M, Harding RM, Hood DW, Crook DW (2009) Genomic islands: tools of bacterial horizontal gene transfer and evolution. FEMS Microbiol Rev 33:376–393

Lawrence JG, Hendrickson H (2008) Genomes in motion: gene transfer as a catalyst for genome change. In: Schmidt H, Hensel M (eds) Horizontal gene transfer in the evolution of pathogenesis. Cambridge University Press, Cambridge, pp 3–22

Lehnherr H (2006) Bacteriophage P1. In: Calendar R, Abedon ST (eds) The bacteriophages. Oxford University Press, Oxford, pp 350–364

Maguire F, Jia B, Gray KL, Lau WYV, Beiko RG, Brinkman FSL (2020) Metagenome-assembled genome binning methods with short reads disproportionately fail for plasmids and genomic islands. Microb Genom 6:mgen000436

Moon BY, Park JY, Hwang SY, Robinson DA, Thomas JC, Fitzgerald JR, Park YH, Seo KS (2015) Phage-mediated horizontal transfer of a *Staphylococcus aureus* virulence-associated genomic island. Sci Rep 5:9784

Moon BY, Park JY, Robinson DA, Thomas JC, Park YH, Thornton JA, Seo KS (2016) Mobilization of genomic islands of *Staphylococcus aureus* by temperate bacteriophage. PLoS One 11: e0151409

Penades JR, Chen J, Quiles-Puchalt N, Carpena N, Novick RP (2015) Bacteriophage-mediated spread of bacterial virulence genes. Curr Opin Microbiol 23:171–178

Penadés JR, Christie GE (2015) The phage-inducible chromosomal islands: a family of highly evolved molecular parasites. Annu Rev Virol 2:181–201

Rao VB, Black LW (2005) DNA packaging in bacteriophage T4. In: Viral genome packaging machines: genetics, structure, and mechanism. Springer, Cham, pp 40–58

Raya RR, Kleeman EG, Luchansky JB, Klaenhammer TR (1989) Characterization of the temperate bacteriophage ϕadh and plasmid transduction in *Lactobacillus acidophilus* ADH. Appl Environ Microbiol 55:2206–2213

Rodriguez-Rubio L, Serna C, Ares-Arroyo M, Matamoros BR, Delgado-Blas JF, Montero N, Bernabe-Balas C, Wedel EF, Mendez IS, Muniesa M, Gonzalez-Zorn B (2020) Extensive antimicrobial resistance mobilization via multicopy plasmid encapsidation mediated by temperate phages. J Antimicrob Chemother 75:3173–3180

Saye DJ, Ogunseitan O, Sayler GS, Miller RV (1987) Potential for transduction of plasmids in a natural freshwater environment: effect of plasmid donor concentration and a natural microbial community on transduction in *Pseudomonas aeruginosa*. Appl Environ Microbiol 53:987–995

Schneider CL (2021) Bacteriophage-mediated horizontal gene transfer: transduction. In: Harper DR, Abedon ST, Burrowes B, McConville M (eds) Bacteriophages: biology, technology, therapy. Springer, New York, pp 151–192

Shousha A, Awaiwanont N, Sofka D, Smulders FJ, Paulsen P, Szostak MP, Humphrey T, Hilbert F (2015) Bacteriophages isolated from chicken meat and the horizontal transfer of antimicrobial resistance genes. Appl Environ Microbiol 81:4600–4606

Valero-Rello A, Lopez-Sanz M, Quevedo-Olmos A, Sorokin A, Ayora S (2017) Molecular mechanisms that contribute to horizontal transfer of plasmids by the bacteriophage SPP1. Front Microbiol 8:1816

Villa TG, Feijoo-Siota L, Rama JR, Sánchez-Pérez A, Viñas M (2019) Horizontal gene transfer between bacteriophages and bacteria: antibiotic resistances and toxin production. In: Horizontal gene transfer. Springer, Cham, pp 97–142

Weisberg RA, Landy A (1983) Site-specific recombination in phage lambda. In: Hendrix RW, Roberts JW, Stahl FW, Weisberg RA (eds) Lambda II. Cold Spring Harbor Laboratory, Cold Spring Harbor, NY, pp 211–250

Wilson GG, Young KKY, Edlin GJ, Konigsberg W (1979) High-frequency generalized transduction by bacteriophage T4. Nature (London) 280:80–82

Part IV
Prophage-Encoding of Bacterium-Expressed Genes

Chapter 14
Phage Morons

Contents

Abstract An important means by which phages can drive the evolution of bacteria is by providing new genes that are encoded within the genomes of especially non-defective phages. If those phages are strictly lytic, then association of those genes with a viable bacterium may occur only under circumstances where the phage infection is not successful. Alternatively, if the phage is temperate, then those genes can be associated with both successful phage infections and still-viable bacteria, i.e., in the course of lysogenic cycles. Particularly, genes that bioinformatically are suspected to be newly associated with a given phage have been dubbed "Morons" and this is due their representing 'more' phage DNA. Considered in this chapter are various evolutionary ecological issues associated with phage morons including what they are, how they likely come into existence, and problems for phages that can be associated with their acquisition. The concept of morons as lysogenic converting genes is then briefly considered, though lysogenic conversion—also from an evolutionary ecological perspective—is considered more fully in the succeeding chapter.

S. T. Abedon, *Bacteriophages as Drivers of Evolution*,
https://doi.org/10.1007/978-3-030-94309-7_14

> Whereas there is as yet no case in which a phage becomes a (non-defective) converting phage by recombination with the bacterial chromosome, there seems to be no reason to think that this cannot happen.—Allan Campbell (1961, p. 157)

> . . .we propose to give them the name, moron, to indicate the fact that when one is present in the genome there is more DNA than when it is not present.—Robert J. Juhala et al. (2000, pp. 35–36)

A more detailed description of "Moron" is provided by Hendrix et al. (2000, p. 505), as published in the same year are Juhala et al. (2000) (the bracketed text is an insertion of theirs):

> Consider the late operons of lambdoid phages, whose virion morphogenetic and lysis genes are very well studied. In pairwise comparisons of the head and tail gene regions of these genomes, there is occasionally an 'extra' gene in one of the phages, inserted between two genes that are adjacent in related phages; this gene appears to have nothing to do with head or tail assembly or function. These extra genes are typically preceded by a transcription promoter sequence and followed by a factor-independent (stem-loop) transcription terminator, an arrangement that allows the genes to be transcribed autonomously, even from a repressed prophage. These genetic elements [the coding region(s) plus transcription control sequences] have been termed 'morons' in accordance with the fact that their addition to the genome means that there is 'more DNA' than there is without the element. The nucleotide and dinucleotide composition of moron genes is typically substantially different from that of adjacent genes, arguing that the moron has recently entered the genome from an outside source.

The term 'moron' thus was playfully created to refer to phage DNA sequences that represent more DNA than is required for basic phage functioning, with this extra DNA either identified or speculated as having been acquired horizontally, particularly originally from non-phages sources. Though the sources of individual morons is not necessarily known, moron genes are found between flanking sequences that are homologous to sequences (i.e., genes) which are found in other phages (Hendrix 2005). Thus, a moron is extra DNA because it is not commonly found in phage genomes, it is extra DNA because it is not necessarily of phage origin, and it is extra DNA because it is not found even in relatively closely related phages.

Already considered, in Chap. 11, is the moron accretion hypothesis, though there the concept of moron is more broadly defined than will be the case in this chapter. Instead, in this chapter, 'more' refers not just to extra DNA but generally also to 'more' phage phenotype than is required to sustain or directly optimize phage lytic or lysogenic cycles. Optimization of lysogenic cycles via the expression of additional phage genes is a topic we will consider especially in Chap. 15, however. Here we focus explicitly on considering the concept of phage moron in a narrower sense than that described under the heading of moron accretion. Narrow-sense morons are relevant to phages driving evolution in part because morons can supply additional phenotypes to bacteria (lysogenic conversion; Chaps. 15 and 16) and in part because morons represent another means by which strictly bacterial genes can enter into new bacterial lineages (Chap. 17).

14.1 What Is a Moron? (Narrow Sense)

> We speculate that many moron genes benefit the phage indirectly, by providing a direct benefit to the host cell from the prophage and therefore a disincentive for the cell to delete the prophage.—Roger W. Hendrix and Sherwood R. Casjens (2008, p. 82)

> In principle morons can also provide a selective benefit to the phage by enhancing its lytic growth. . .—Roger W. Hendrix and Sherwood R. Casjens (2008, p. 83)

My interpretation of the Hendrix et al. (2000) study is that there are two ways of viewing morons, one broader and the other narrower. The broader definition is, as noted, associated with the moron accretion hypothesis (Chap. 11): more DNA in a quantitative sense, but not more DNA in a more specific, qualitative sense. In a narrower sense, a moron instead can represent not just 'extra' phage DNA but extra DNA that potentially can supply a phage with a new function. Perhaps not all new phage functions may be consistently ascribable to morons, however. Specifically, a key issue is the question of just what might be meant by 'new' phage function? From that perspective, this section attempts to answer the question of just what isn't a moron in a narrower sense, taking as a given that morons are horizontally acquired segments of DNA that typically will have sequence signatures that are consistent with their being relatively new components of phage genomes.

14.1.1 Not Useful Especially During Productive Cycles

One possible answer to the question of what specifically is a moron, in a narrow sense, is that it would be something that is *not* directly involved in a phage productive cycle, that is, which is only useful and perhaps even only expressed during lysogenic cycles. Consistently, mutational knock out of a moron (and moron alone) should result in no harm to the various key phage growth parameters: adsorption characteristics, latent period, burst size, and virion durability following release.

Ideally, this consideration would extend to beyond just the laboratory environment. That is, no negative impact on phage productive infection growth parameters in the phage's natural environment. Thus, though not all phage genes are essential to phage virion production, many nevertheless may be helpful such as in terms of increasing burst sizes, extending host range, or reducing phage latent periods (the latter particularly without otherwise negatively impacting burst sizes; Abedon 2022). Such genes may not be essential, but certainly could be described as being useful, including while infecting different bacterial hosts, and yet useful in this case especially in terms productive rather than necessarily also lysogenic cycles.

My sense is that any gene that contributes to the effectiveness of a phage productive cycle, or which even is expressed solely during a productive cycle, most authors would not consider to be morons in this narrower sense. That is, a moron should be more than just more phage DNA, even if a gene does indeed appear to have been newly acquired.

14.1.2 Newly Acquired Lysogenic Converting Gene

On the other hand, and why I suggest that there should be both a narrower and broader definition of 'moron', Hendrix et al. (2000) explicitly note that (p. 505), "Morons might also increase phage fitness by assisting directly in lytic cycle functions of temperate or virulent phages." That quotation, however, immediately follows the words, "Moron accretion", which as I have noted, appears to be associated with a broader rather than narrower concept of moron. By contrast, the discovery of morons involved especially the study of temperate phages (Hendrix et al. 2000). Furthermore, in their initial description of just what are morons, Hendrix et al. (2000) explicitly suggest that these genes are associated in terms of their function especially with lysogenic cycles (p. 505):

> It seems likely that in some or all of these cases the function provided by the moron (from its prophage location) gives the host cell increased fitness, and that this in turn provides a selective disincentive for the cell to lose the prophage carrying the moron. Thus, the presence of the moron increases phage fitness by increasing prophage fitness.

Explicitly in its narrower sense and as considered particularly in this chapter, a moron therefore is a lysogenic converting gene, or at least potentially one, that has the hallmarks of being a newly acquired phage gene rather than one that is well established within phage genomes.

That definition should lead to a question of just when or how often are lysogenic converting genes not also morons, but that is not a question that I will be attempting to address here. Instead, I consider in this chapter various aspects of morons as extra phage genes that (1) are not *directly* involved in productive infections, (2) are not directly *required* by lysogenic cycles, but which (3) are more likely studied as components of temperate phages rather being components of phages that are unrelated to temperate phages. At the end of this chapter as well as in the following chapter (Chap. 15), however, I look explicitly at lysogenic conversion.

14.2 What Isn't a Moron? (Narrow Sense)

If we can accept that a moron in a narrow sense is a gene associated with a temperate phage that (1) is not directly useful to phage productive cycles, (2) is not directly required for phage lysogenic cycles, and (3) appears to be relatively newly acquired by phages, that still leaves us with the question of whether different means of new-gene acquisition by temperate phages—illegitimate recombination versus specialized transduction in a traditional sense—can distinguish morons from non-morons. In this section I suggest that they cannot.

14.2.1 Morons as Products of Illegitimate Recombination

The two main mechanisms of nonviral DNA incorporation into a phage genome, other than imprecise prophage excision (below), is via what are known as homologous recombination on the one hand and illegitimate or non-homologous recombination on the other (Chap. 13). These two mechanisms often are presented as opposites, but really what they mean are molecular recombination events that are based on recognizably sufficient DNA homology to allow base pairing between two strands of DNA versus molecular recombination events that are not based on recognizably sufficient DNA homology but which still might be driven by some degree of DNA homology. While homologous recombination is a relatively common occurrence, potentially resulting in orthologous replacement at a genetic locus of one allele with a different allele, illegitimate recombination is considered to be rare. Nevertheless, illegitimate recombination substantially broadens what DNA may be incorporated into a phage genome, and also will tend to give rise to insertion events rather than swapping of one sequence of DNA for another. In short, whether a phage is temperate or not, during infection of a bacterial cell it may become modified in terms of what genes or alleles it carries due to recombination with bacterial chromosomal DNA, plasmid DNA, or endogenous prophage DNA. The consequence of insertion events resulting from illegitimate recombination can be the formation of phage morons.

14.2.2 Morons as Products of Specialized Transduction

Specialized transduction, as historically defined, involves the transfer, within transducing particles, of a combination of bacterial chromosomal DNA and phage DNA. The bacterial DNA is that which is found adjacent to the previously integrated prophage, and its incorporation into a transducing particle is a consequence of imprecise prophage excision. As such, what genes are carried during specialized transduction tend to be limited in terms of type, i.e., they are 'special'. Unlike with generalized transduction, with specialized transduction it is possible for a transducing particle to also be a viable phage that happens to be carrying some non-phage DNA. As a result, not only may this non-phage DNA be passed from donor bacterium to recipient bacteria, in principle that DNA, given its carriage by a viable virion, may be carried beyond the initial recipient bacterium as well. Nevertheless, for a recipient bacterium to be successfully transduced, the 'transducing particle' cannot give rise directly to either a lytic cycle or an abortive infection. As a consequence, with specialized transduction there is a strong association with lysogeny, both in terms of transducing DNA acquisition from a donor bacterium and in terms of depositing transduced DNA into a recipient bacterium.

14.2.3 Broadening the Definition of Specialized Transduction to Include All Morons

Weisberg (1996), published 4 years before Hendrix et al. (2000), is a chapter specifically on specialized transduction. There he notes that (p. 2442),

> The limited set of host genes that λ naturally transduces can be stably incorporated into its own chromosome, where these genes are then replicated and packaged as efficiently as if they were native phage genes. By contrast, generalized transducing phages usually do not incorporate the genes they transduce into their own chromosomes but occasionally package a segment of host DNA in place of the phage chromosome. Because of this difference, use of the term "specialized transduction" has broadened to encompass any high-efficiency virus-mediated replication and packaging of a nonviral gene no matter how the nonviral gene is incorporated into a virus chromosome. In this broader sense, probably all viruses are capable of specialized transduction.

In this broader sense, morons thus may be described as products of specialized transduction, and specialized transduced genes, particularly as associated with viable phages, may be described as morons. The concept of moron, that is, does not necessarily describe only extra DNA that is acquired via illegitimate recombination, but perhaps should include extra genes acquired by imprecise prophage excision as well. Indeed, due to the nature of prophage excision, i.e., as seen with the phage λ model, the genes acquired via imprecise prophage excision (specialized transduction) end up in the middle of the phage genome upon packaging into virion particles (Chap. 13, Fig. 13.3).

A special case of moron acquisition appears to be of eukaryotic-gene origin. These have been identified in the genomes of the temperate *Wolbachia* phage WO, which have life cycles that take place within eukaryotic cells and eukaryotic bodies (Bordenstein and Bordenstein 2016). Still, other than where these genes appear to have originated, there is no reason to *not* consider them to be products of specialized transduction in this broader sense, i.e., as a "nonviral gene is incorporated into a virus chromosome" (Weisberg, 1996, p. 2442).

14.2.4 Diversity Even with Specialized Transduction Sensu Stricto?

The suite of genes that may be acquired by illegitimate recombination, and where those genes may end up in the phage genome, is exceptionally broad (any gene may be acquired and, at least in terms of intergenic regions, can end up almost anywhere). By contrast, what genes may be acquired by imprecise prophage excision, and where they end up in the phage genome, is exceptionally narrow (a few genes and adjacent to the ends of inserted prophage chromosomes). Nevertheless, across many phage types and across many locations that prophages may insert into bacterial chromosomes, including in terms of secondary prophage insertion sites (Chap. 5), the suite of genes potentially associated with specialized transduction should be quite a bit broader than just the traditional, phage λ model of specialized transduction would seem to suggest. It is possible also that gene acquisition by specialized transduction (*sensu stricto*) can compensate for limitations on diversity to some extent by being more likely than illegitimate recombination events, though alternatively specialized transduction events, again *sensu stricto*, may simply be more easily detected.

The distinction between morons as acquired by illegitimate recombination and genes acquired by temperate phages in the course of specialized transduction (as imprecise prophage excision events) thus may not be as substantial as at first blush they might seem, and this may be particularly so with regard to contribution of these genes to bacterial evolution. I concede, nevertheless, that specialized transduction, as traditionally defined, still must be more constrained than illegitimate recombination in terms of what genes may be acquired and precisely where in phage genomes those genes may at first appear.

14.3 Limitations on Phage Acquisition of Additional Genes

> . . .anyone who is alive in the world of the living has some hope. . .—Ecclesiastes 9:4.
>
> Only the dead are hopeless.—attributed to Telesphorus of Rhodes
>
> Where there's life there's hope. . .—J.R.R. Tolkien

In terms of phages as drivers of evolution—at least in terms of genetic migration of phage-carried genes to bacteria—what we are particularly interested in is temperate phage acquisition of genes and then carriage of those genes that could be useful to bacteria. If those genes are both newly acquired and expressed by successful phages, i.e., are not simply in the process of being lost from populations, then many or perhaps most such genes may be described as morons, as narrowly defined. That is, to the extent that they are expressed from prophages, then these genes have some potential of modifying bacteria phenotypically, possibly toward greater lysogen fitness in comparison to the fitness of the parental, prophage-less bacterium. If phage genes are not expressed from prophages, and do not impact bacteria phenotypically even if they are expressed, then these genes are less likely to have a meaningful, positive impact on bacterial fitness, though they might still be described as morons should they persist within phage or lysogen populations. There are, however, at least three limitations on new-gene acquisition by phages.

14.3.1 Limitations on Encapsidation by Phage Genome Sizes

The first of these limitations is that phage capsids, especially for tailed phages, tend to be limited in terms of the amount of DNA that can be packaged. Once genomes are larger than the amount that is normally carried, then the efficiency of packaging declines, e.g., (Feiss et al. 1977). The result is selection against ever larger phage genomes and therefore against the incorporation of ever more additional phage genes (different forces, by contrast, likely lead to limitations on genome sizes in prokaryotes; Sela et al. 2016). This encapsidation issue places limitations on the sizes of DNA fragments that may be propagated by temperate phages once acquired (Fig. 14.1).

This issue is always a concern with specialized transduction, *sensu stricto*, i.e., as due to imprecise prophage excision, where incorporation of too much new DNA results in infection-incompetent phages, though not necessarily also transduction-incompetent phage particles. Thus, a phage may be able to acquire one or at most a few additional genes via illegitimate recombination and remain replication competent. At some point, however, in terms of amounts of new DNA, in order to package

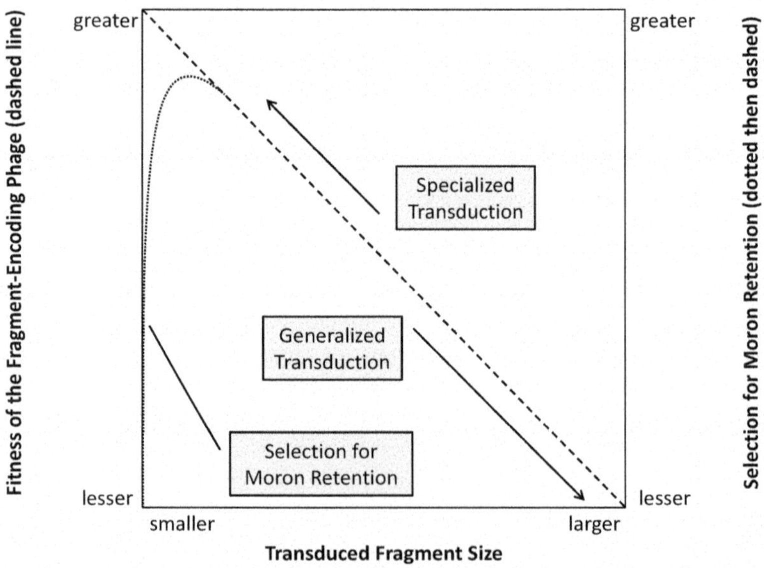

Fig. 14.1 Moron size and likelihood of successful acquisition by bacteria. Independent of the sequence of the acquired DNA, or its potential for functionality, simply smaller fragments are less likely to negatively impact the fitness of carrying phages (*y* axis to the left; dashed line) and thereby the ability of those phages to propagate. At some point in terms of decreasing size, however, a potential for functionality for a phage-carried DNA fragment is likely to be lost—whether as a phage gene or instead if acquired as a bacterial gene—and as a result selection for retention should be lost (*y* axis to the right; dotted then dashed line, going from left to right). Specialized transduction in the figure is defined in the original sense as based on imprecise prophage excision. Generalized transduction is an extreme in terms of transduced fragment size

fully functional phage genomes then either older phage genes must be lost to offset the amount of new DNA gained or phage capsids must be mutationally expanded to accommodate new DNA. This process of accommodation, i.e., compensating losses of genetic material, is not impossible to envisage occurring, however, e.g., as discussed under the heading of moron accretion (Chap. 11, Fig. 11.2) and below.

First, phages often have regions which are deletable, at least in terms of phage population growth in the laboratory, thus implying that a deletion step toward offsetting increases in genome sizes may in principle be possible. Second, acquisition and deletion need not occur simultaneously, at least so long as some packaging remains possible following DNA acquisition, even if this packaging is less efficient. Three, the processes of acquisition and deletion in principle can be an iterative process, that is, occurring over multiple rounds as new genes are added and other genes are lost from a phage's genome. Nevertheless, there are limitations on these processes as imposed by the size of phage capsids. Especially as envisaged processes become more complex or the amount of new DNA acquired becomes larger, the likelihood of such accommodations occurring will be lower.

14.3.2 The Problem of Insertional Mutagenesis

The second limitation is the problem of insertional mutagenesis. The process outlined in the previous paragraph essentially is one of natural selection, that is, selection against those phage genomes that have become too large. Other mechanisms of natural selection against new-gene acquisition by phages exist as well, and these too both can and will place limitations on gene acquisition by phages. First, the process of gene acquisition itself, by illegitimate recombination, will not necessarily result in 'clean' gene incorporation, that is, where previously existing phage DNA is still found in some not-disrupted manner. Thus, it should be assumed that most illegitimate recombination events will give rise to phage genomes that are inherently no longer functional, at least at the new DNA insertion point, and this can be especially problematic if it is essential phage genes that are interfered with by the insertion of new genetic material. Literally this is insertional mutagenesis, though occurring in phage genomes rather than in bacterial genomes. Insertional mutagenesis should not be issue, however, given new-gene acquisition by temperate phages via specialized transduction, *sensu stricto*.

Just as with excessively large phage genomes, these disrupted phage genomes can be rapidly removed by natural selection from the phage gene pool. Though less extreme, even if phage viability is retained following new gene acquisition, this does not mean that phage fitness will not have been negatively affected. Nevertheless, just as with inefficient but not impossible phage genome packaging, so long as some capacity is retained by a phage to reproduce, then iterative evolutionary processes are possible. That is, natural selection can act on additional genetic variation such that the fitness of phages carrying additional DNA may be improved over time, resulting in retention, with higher likelihood, of that DNA within the phage gene pool. This fitness improvement may be a consequence of compensating mutations, i.e., that fully or partially repair the defect or defects introduced in the phage genome upon new DNA acquisition. These mutations, however, would not be ones that necessarily result also in deletion of compensating amounts of phage genetic material, i.e., as can be the case where moron accretion results in excessively large phage genomes (Chap. 11, Fig. 11.2, and previous subsection).

14.3.3 Lack of a Selective Benefit

Continuing this theme of natural selection acting on phages which have acquired new DNA, though now getting closer to the idea of phages as drivers of bacterial evolution, the presence of that new DNA may not be selected for and indeed may be selected against. That is to say, we can have issues with phage fitness due to phage genomes simply being too large or having been disrupted so that previously existing phage functions (besides or in addition to packaging) are less effective, but also because the new DNA itself is detrimental to the carrying phage, or at least not beneficial. A lack of benefit would be expected to be seen especially if newly acquired genes do not possess the necessary sequences to allow for their expression during phage infections, or if they are expressed but have little or no phenotypic impact on either the phage or the hosting lysogen.

In the case of detrimental or neutral fitness effects (the latter being neither detrimental nor beneficial), then natural selection by definition no longer can give rise to increases in the prevalence of these genes. Instead, it is only via random processes, that is, genetic drift, that a newly acquired neural or detrimental alleles may increase in frequency. It is likely, however, that most genes that are acquired by illegitimate recombination in fact will not be beneficial to their newly carrying phages (Hendrix et al. 2003) and thus will be lost from the phage population. Indeed, newly acquired genes that are only slightly beneficial also may lost from populations with relatively high likelihood via random processes.

On the other hand, a possibility exists that new alleles that have an only neutral impact on phage fitness might be retained sufficiently long within populations that additional mutations could be acquired that allow these new alleles to provide a selective benefit, again with evolution toward new gene usefulness being an iterative process. Such iterative processes, however, are more likely if newly acquired DNA has an immediately beneficial impact as therefore this new DNA is likely to be retained for longer within phage populations, as well as increase sufficiently in absolute numbers such that acquisition of mutations may occur with some likelihood. Detrimental impacts of gene acquisition on phage fitness by contrast will make it likely that genes will be lost from populations before evolution via acquisition of compensating mutations can provide improvements in phage fitness. The result is that very few gene-acquisition events are expected to survive over the long term.

We are still here only referring to phage fitness costs and benefits, which represent phages experiencing rather than driving evolution. Thus, we need to transition to considering morons explicitly in terms of how they may impact bacterial fitness rather than only phage fitness. Note, however, that though we might expect that all genes that increase a phage's fitness during lysogenic cycles will also increase a harboring lysogen's fitness, it is possible nevertheless that changes to lysogen phenotypes in at least some cases could be more beneficial to the prophage than to the bacterium (Abedon and LeJeune 2005; Chap. 15).

14.4 Morons as Lysogenic Converting Genes

A natural progression of thought seems to lead to a question: Just what is the difference between a moron and a lysogenic converting gene? One answer to that question is that if we define morons sufficiently broadly, e.g., as we did when considering the moron accretion hypothesis (Chap. 11), then a moron need not be a component even of a temperate phage genome. Alternatively, for those morons that are components of temperate phages, and are expressed during lysogenic cycles, i.e., 'moron' as narrowly defined, then the distinction between a moron and a lysogenic converting gene may be tiny indeed (Brüssow et al. 2004, Taylor et al. 2019).

In the following chapter (Chap. 15) we will dispense entirely with consideration of morons—as newly acquired lysogenic converting genes—and focus instead on lysogenic converting genes especially as phenotypic phenomena, though still from an evolutionary ecological perspective. Morons, however and as noted, might also be important as a route toward gene movement from one bacterium to another, with morons as newly acquired phage genes serving as intermediates in that process, i.e., toward phage gene domestication, as considered in Chap. 17.

References

Abedon ST (2022) Evolution of bacteriophage latent period length. In: Dickins TE, Dickens BJA (eds) Evolutionary biology: new perspectives on its development. Springer, Cham

Abedon ST, LeJeune JT (2005) Why bacteriophage encode exotoxins and other virulence factors. Evol Bioinform Online 1:97–110

Bordenstein SR, Bordenstein SR (2016) Eukaryotic association module in phage WO genomes from *Wolbachia*. Nat Commun 7:13155

Brüssow H, Canchaya C, Hardt WD (2004) Phages and the evolution of bacterial pathogens: from genomic rearrangements to lysogenic conversion. Microbiol Mol Biol Rev 68:560–602

Campbell A (1961) Conditions for the existence of bacteriophages. Evolution 15:153–165

Feiss M, Fisher RA, Crayton MA, Egner C (1977) Packaging of the bacteriophage λ chromosome: effect of chromosome length. Virology 77:281–293

Hendrix RW (2005) Bacteriophage evolution and the role of phages in host evolution. In: Waldor MK, Friedman DI, Adhya SL (eds) Phages: their role in bacterial pathogenesis and biotechnology. ASM Press, Washington DC, pp 55–65

Hendrix RW, Casjens SR (2008) The role of bacteriophages in the generation and spread of bacterial pathogens. In: Hensel M, Schmidt H (eds) Horizontal gene transfer in the evolution of pathogenesis. Cambridge University Press, Cambridge, pp 79–112
Hendrix RW, Hatfull GF, Smith MCM (2003) Bacteriophages with tails: chasing their origins and evolution. Res Microbiol 154:253–257
Hendrix RW, Lawrence JG, Hatfull GF, Casjens S (2000) The origins and ongoing evolution of viruses. Trends Microbiol 8:504–508
Juhala RJ, Ford ME, Duda RL, Youlton A, Hatfull GF, Hendrix RW (2000) Genomic sequences of bacteriophages HK97 and HK022: Pervasive genetic mosaicism in the lambdoid bacteriophages. J Mol Biol 299:27–51
Sela I, Wolf YI, Koonin EV (2016) Theory of prokaryotic genome evolution. Proc Natl Acad Sci U S A 113(41):11399–11407
Taylor VL, Fitzpatrick AD, Islam Z, Maxwell KL (2019) The diverse impacts of phage morons on bacterial fitness and virulence. Adv Virus Res 103:1–31
Weisberg RA (1996) Specialized transduction. In: Neidhardt FC, Curtiss R III, Ingraham JL, Lin ECC, Low KB, Magasanik B, Reznikoff WS, Riley M, Schaechter M, Umbarger HE (eds) *Escherichia coli* and *Salmonella* cellular and molecular biology. ASM Press, Washington, D.C., pp 2442–2448

Chapter 15
Why Lysogenic Conversion?

Contents

Abstract The ability of non-essential prophage genes to modify bacterial phenotype is called lysogenic conversion. Though most notoriously, as well as most recognizably, lysogenic converting genes encode bacterial virulence factors, such as cholera toxin, Shiga toxin, and vibrio toxin, in fact a diversity of other bacterial accessory genes also can be lysogenic converting genes. Generally these otherwise non-essential or not consistently essential genes are thought to provide bacteria with selective benefits under at least some specific circumstances. The primary issue addressed in this chapter, however, is what is the benefit to having these genes encoded by prophages? Specifically, what are the benefits from prophage encoding over especially having these genes encoded more directly by bacterial chromosomes? Does the gene itself benefit in some way from being prophage encoded? Is it especially the phage or prophage that in some manner benefits? Is the bacterial host instead the primary beneficiary? Or is it some combination of these? The answer is not certain, though a number of different possible scenarios are discussed. Reiterating, it's not why or that these genes are useful to bacteria, but instead what is the utility of having these genes encoded especially by prophages?

> . . .it would be to the advantage of the phage to carry genes that enhance the growth rate of lysogenic cells.—Bruce R. Levin and Richard E. Lenski (1983, p. 110)

Temperate phages during lysogenic cycles are able to provide their bacterial hosts with potentially useful phenotypes. This is a process known as lysogenic conversion (Little 2005, Hendrix and Casjens 2008, Łoś et al. 2010, Bondy-Denomy and

S. T. Abedon, *Bacteriophages as Drivers of Evolution*,
https://doi.org/10.1007/978-3-030-94309-7_15

Davidson 2014, Obeng et al. 2016, Harrison and Brockhurst 2017), though some prefer to describe these genes simply as morons (Taylor et al. 2019). Most notably medically, lysogenic converting genes include those that express phage-encoded bacterial virulence factors (Brüssow et al. 2004, Christie et al. 2012, Kuhl et al. 2012, Casas and Maloy 2018, Villa et al. 2019), but there also are prophage genes for example that regulate bacterial gene expression (Fortier and Sekulovic 2013, Feiner et al. 2015, Argov et al. 2017, Altuvia et al. 2018, Liu et al. 2020).

This process of lysogenic conversion we can define with a bit more precision as: expression of prophage *genes* that act to modify lysogen phenotypes, doing so independently of normal phage lysogenic cycle functions as well as independently of various actions of subsequently produced extracellular virions. Thus, strictly speaking, we can exclude functions that (1) are solely associated with productive infections as examples of lysogenic conversion as well as contributions of virion release by temperate phages (2) to biofilm formation (Secor et al. 2020), (3) to antibiotic resistance (Tarafder et al. 2020), (4) to mucosal colonization (Bille et al. 2017), (5) to possibly combatting Muller's ratchet (Chaps. 9 and 12), or instead (6) to lysogen allelopathy (the latter as covered in Chap. 24). Alternatively, though ambiguously, the production of phage-encoded Shiga toxin by *E. coli* O157:H7 typically is included as an example of lysogenic conversion even though it is produced and released in conjunction with lytic cycles rather than lysogenic cycles (Mauro and Koudelka 2011). This is in part, though, because it is not strictly phage release but instead phage-induced bacterial lysis that is key to its functioning as a toxin, and also because Shiga toxin is not always prophage associated, i.e., as found in *Shigella dysenteriae*. We also can exclude superinfection immunity from strict considerations of lysogenic conversion (Łoś et al. 2010), since this function is a consequence of expression of the prophage repressor genes, that is, which are essential for physiologically maintaining the lysogenic cycle . Superinfection immunity therefore is instead discussed in the following chapter (Chap. 16) along with other phage-encoded anti-phage mechanisms besides superinfection immunity, with many of these other phage-encoded anti-phage mechanisms indeed examples of lysogenic conversion. Thus, generally, we can speak of lysogenic *converting genes*, and these do not include either prophage repressor genes or 'converting' virions.

In many cases, converting genes represent morons and vice versa. Morons, though, are more of a 'genotypically' defined phenomenon—newly acquired phage genes particularly as discovered via comparative genomics and additional bioinformatic characterization (Chap. 14)—whereas converting genes are more 'phenotypically' defined (additional phage genes, besides the prophage repressor gene, that are expressed during lysogenic cycles). Note also that phage-inducible chromosomal islands, as satellite phages, may be viewed as an extreme form of lysogenic converting 'phages', where most of the phage functions have been gotten rid of (or never actually existed) in favor instead of maximizing the number of deliverable 'converting' genes (Chap. 13). The primary issue considered by this chapter, however, is that of what might be utility of having lysogenic converting-

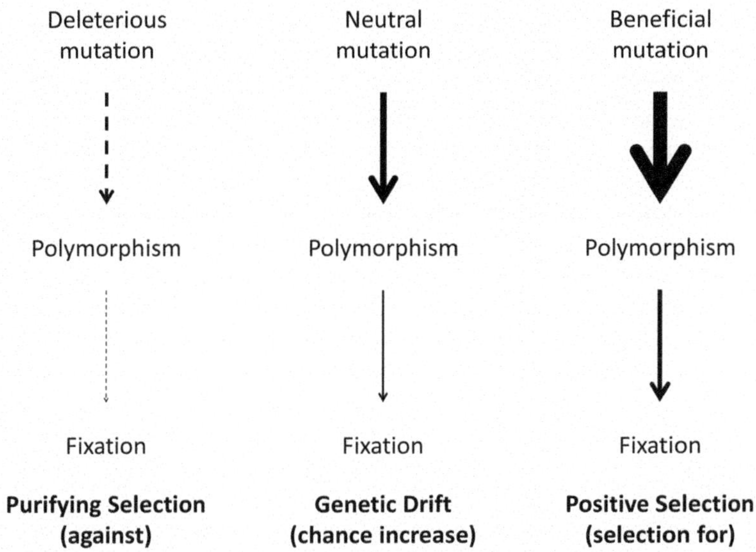

Fig. 15.1 Potential of allele survival (contribution to polymorphism) and fixation as a function of allele utility. Thickness of lines reflects likelihood. Dashed lines further emphasize lowness of likelihood. For both deleterious and neutral alleles, these likelihoods occur as a consequence solely of drift and for deleterious alleles, selection actively serves to reduce the likelihood of their survival, evolutionarily, relative to the evolutionary survival of neutral alleles

type genes encoded by prophages in particular, e.g., rather than by plasmids or rather than being non-prophage chromosomally encoded.

Why, that is, are these genes *lysogenic* converting genes rather than simply *bacterial* genes? Especially, what are the specifics of the presumptive positive selection (Fig. 15.1) presumably supporting the existence of lysogenic conversion? A somewhat broader look at these ideas can be found in Koonin et al. (2020), who explored the question of why certain bacterial genes but not others should tend to be encoded by mobile genetic elements.

15.1 More than Lytic Cycle Buttressing of Accessory Gene Encoding?

> "The game being played by the phage involves weighing the chance of extinction in lysogeny (the bacterium and its descendants may die before the phage escapes it by prophage induction) versus the chance of extinction if it goes lytic (for example, due to a toxic substance that kills phage before they find other bacteria to infect)."—Mikkel Avlund et al. (2009, p. 11417)

Lysogenic converting genes exist as part of a larger category of bacterial genes known as accessory genes, e.g., (Bobay and Ochman 2017). These are genes that are considered to be dispensable depending on circumstances and many are carried by mobile genetic elements, such as by plasmids or prophages. Phage capsids, however, typically impose strict limitations on how much accessory DNA can be carried (Chap. 14), and this has the effect of limiting the number of bacterial accessory genes, i.e., lysogenic converting genes, that might be carried per temperate phage. Temperate phages probably can make up for this deficit in accessory gene coding capacity—which presumably limits the breadth of environments in which what few converting genes that they can carry may be useful—by being able to exist not just as a platform of carriage of accessory genes but also as virions (Levin and Bergstrom 2000, Iranzo and Koonin 2018). That is, lytic temperate phages possess two distinct life cycles from the perspective of converting genes, cycles in which converting genes may be relevant to bacteria (lysogenic cycles) and cycles in which converting genes are less or not relevant to bacteria (lytic cycles). It is possible, therefore, that the population growth that temperate phages can achieve via their lytic cycles helps to make up for possible deficits in the breadth of converting gene utilities during lysogenic cycles. In other words, if a temperate phage at a given moment in time cannot be a highly effective prophage, e.g., as stemming from their display of lysogenic conversion, then at least a temperate phage may be an effective non-prophage instead, with the converse also true (Avlund et al. 2009). Equivalently, from Abedon and LeJuene (2005), p. 100, "It is conceivable that [converting genes] could even be detrimental to the fitness of harboring bacterial lysogens especially if phage-virion mediated horizontal transmission can make up for... fitness costs associated with prophage-mediated vertical transmission."

That conclusion, that lysogenic converting phages can persist because they are phages, despite their lysogenic converting phenotypes not necessarily always being useful to hosts of their lysogenic cycles, sidesteps the broader question, however: Why do prophages encode converting genes at all? The answer to that question turns out to be more complicated, at least in my opinion, than simply that expression of these genes can be helpful to the bacteria whose phenotypes are being modified, and those complications are what I focus on in this chapter. Much of this musing is based on an article published by myself and Jeffrey LeJeune in 2005, "Why bacteriophage encode exotoxins and other virulence factors", which was my contribution to a collaboration we had looking at Shiga toxin in *E. coli* O157:H7 (LeJeune et al. 2004). Admittedly, the Abedon and LeJeune study was narrower in scope than the question posed above, as virulence factor genes are only one subset of converting genes. Nevertheless, this more-focused question can lead to more-general questions of just how the selective benefits stemming from lysogenic converting genes being prophage genes may be manifest, which is what is considered in this chapter.

15.2 A Plethora of Possible Selective Mechanisms

> "In general, it appears that the prophages often (always?) bring genes with them that give the bacterial host some (short-term?) advantage over non-lysogens."—Roger W. Hendrix and Sherwood R. Casjens (2008, p. 99)

Ultimately, we (Abedon and LeJeune 2005) were able to propose ten conceivable selective mechanisms for retention of lysogenic converting genes by prophages. Five of these benefitted primarily the converting gene (1 through 5, below). Two benefited primarily the gene-expressing prophage (9 and 10, below). Another two proposed that the benefit was experienced by the bacterial population, and thus were at least potentially and problematically group selectionist (7 and 8, below). Lastly, one could benefit the carrying bacterial lysogen and/or the larger bacterial population consisting of those lysogens (number 6, below). That is, only one of our answers to the question of why prophages might carry bacterial virulence factor genes, and by extension lysogenic converting genes, proposed that individual bacteria could be the primary beneficiary, but even there this was muddled by the proposed benefit being, at least in some cases, more spread out to the affected bacterial population as a whole.

This is not to say that bacteria cannot benefit from the expression of accessory genes. Indeed, such benefits to bacteria presumably are mostly why accessory genes exist, at least for accessory genes that don't serve in some manner toward the parasitism of prophage-infected bacteria. Rather, the key question is why an accessory gene should necessarily be *encoded by a prophage* and this is rather than considering that a bacterium or a bacterial population might benefit from accessory gene expression. Why then is it beneficial for a given gene, a gene that perhaps has little to do with phage functioning, to be carried by a prophage rather than being chromosomally encoded? And, what specific entity actually benefits from prophage carriage (Fig. 15.2)? That is, is it the gene, the bacterium, the bacterial population, the phage, or some combination of the above?

The benefits specifically to the lysogenic converting gene that might come from being prophage carried are (1) that of rapidity of mobility which could serve as a counter to selection against lysogens carrying that gene; (2) greater potential to

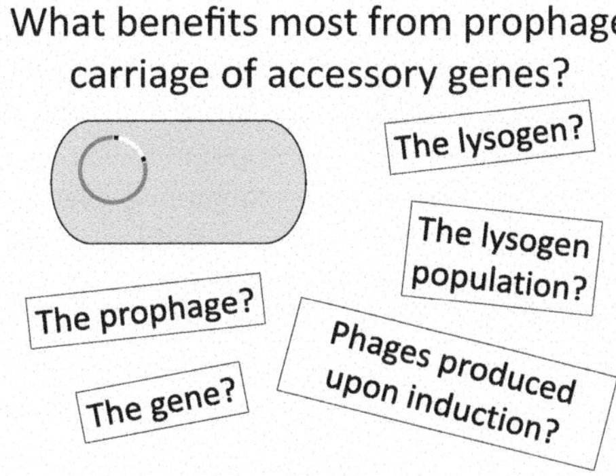

Fig. 15.2 The question is not whether accessory genes are advantageous to bacteria but instead what is the advantage of those genes being encoded by prophages, and which unit of selection benefits?

become associated with new bacterial lineages than a gene that is bacterial chromosome encoded, or instead plasmid encoded, these being new bacterial lineages that could happen to be more successful than a current gene-carrying host (a genetic hitchhiking argument); (3) an ability to reduce vertebrate immune surveillance against the gene due to a potential to become associated with new, immunogenically distinct bacteria or bacterial species (thereby potentially extending its utility as a virulence factor gene rather than just hitchhiking on a different genotype); (4) a greater potential for gene survival due to an ability by the gene to exist at least part time as a virion (virions under some circumstances have a greater survival potential than lysogens); and/or (5) a potential for the converting gene to evolve more rapidly as a phage gene than as bacterial gene (as due to higher mutation rates during productive cycles but also in terms of natural selection across a diversity of bacterial lineages; see Chap. 11, Sect. 11.4).

The benefit to individual bacteria of having a gene encoded by a prophage could be due to (6) there being a phenotypic utility to having the gene so encoded. This, though, would be other than the converting gene's product requiring phage-induced lysis of the carrying bacterial lysogen, as that lysis of course would kill the producing bacterium, thereby eliminating any direct benefit to that same bacterium. It nonetheless is true in the case of Shiga toxin as produced by *E. coli* O157:H7 that this lysis occurs in conjunction with induced lytic cycles (Nawrocki et al. 2020), though Shiga toxin, as noted, also is produced by *S. dysenteriae* strains without being explicitly prophage carried (Lampel et al. 2018). The rest of the same bacterial *population* still should benefit from this Shiga toxin release, though also (7) by lytic cycles allowing for further amplification of the production of a converting gene's product following new phage infections of co-located non-lysogenic bacteria, e.g., (Gamage et al. 2006), or instead (8) from the killing of closely related competitor bacteria (via 'lysogen allelopathy'; Chap. 24), including at the same time that the converting gene is being expressed (again, via new lytic cycles during infection of non-lysogenic bacteria).

Especially benefit (7) requires that the converting gene product be produced during lytic cycles, i.e., as indeed one sees with Shiga toxin. Also, and as suggested, so long as gene expression and gene product release are tied to lytic cycles, then a converting gene's utility cannot be advantageous to the producing bacterial lysogen as they are killed in the course of toxin release, thereby making benefit (6) for Shiga toxin production only a population-level utility rather than something that could be maintained by selection acting on individual bacteria. The counter argument, though, is that if a bacterial population is clonal, then the distinction between what is an individual and what is a population can be vague, as could be relevant to the evolution of gene carriage by temperate phages based on all three of those potential benefits, (6) through (8). This implied kin selection is considered in a different context in Chap. 20, especially Sect. 20.1.4.

Next is a suggestion (9) that the prophage itself could benefit because the bacterial lysogen it is a part of benefits directly from expression of the gene (Sect. 15.4). This presumably is what most researchers default to as an assumption as to why these genes would be prophage encoded, e.g., (Hendrix et al. 2000). That is, 'happy' lysogens should make for 'happy' prophages. Alternatively, (10) the phage may benefit not just because the bacterium's phenotype is being modified but because the bacterium's local environment is being modified as well, what we dubbed as "Indirect enhancement of phage fitness" (Abedon and LeJeune 2005), also as considered further immediately below (Sect. 15.3).

Overall, then, there could be a number of benefits, realized by various units of selection—gene, phage, bacterium, bacterial population (Fig. 15.2)—to prophages encoding bacterial accessory genes, i.e., as toward lysogenic conversion. Particularly, these benefits are in addition to simply that expression of the converting genes may be beneficial to carrying lysogens. That is, and yet again, the issue is not whether converting genes are beneficial to carrying bacteria, but instead why those genes should be carried by prophages. The following sections (Sects. 15.3 and 15.4 as well as 15.5), address further the question of what utility there is for these genes to be carried by temperate phages.

15.3 Indirect Selection for Lysogenic Conversion

By "Indirect" (10, above) what we meant (Abedon and LeJeune 2005) is that converting-gene expression could result in ecosystem modification—as equivalent to the actions of especially soluble virulence factors on bodies—and that ecosystem modification in turn could come back to benefit the converting phage. This is the kind of gene action that Dawkins (1982) might describe as giving rise to an "Extended phenotype", as it represents a 'phenotypic' change that is found outside of the boundaries of an organism. For example is the local breakdown of body tissues following bacterial release of an exotoxin. This might supply the lysogen with additional nutrients or instead might increase the potential for the lysogen to disseminate through a less spatially structured environment. Increased nutrient presence might also make subsequent lytic infections more productive (e.g., larger phage burst sizes) or might help to increase the number of locally available bacteria that could be subsequently infected by released virions (potentially more phage-sensitive bacteria to infect). Thus, benefits may be seen in association not or not just with lysogenic cycles but instead with subsequent lytic cycles.

We concluded by suggesting that such "Indirect" benefits to converting phages displaying subsequent virion-productive infections would be more likely when converting-gene expression occurs within relatively complex bacterial communities rather than when bacterial lysogens had effectively cut themselves off from the outside world. Microbiologically simpler environments one sees, for example, with highly invasive bacterial infections where phage opportunities to disseminate and acquire new bacteria to infect may be limited. Microbiologically more complex environments can be seen when bacterial infections are associated instead, for example, with mucous membranes, where opportunities for virion dissemination and acquisition of new, phage-sensitive bacteria to infect might be more robust. Thus, though lysogenic cycles might benefit from, e.g., increased nutrient availability no matter what additional bacteria are found in their midst, these lysogens may not be the only bacteria to benefit from increased nutrient availability, among which may be new bacteria for phage virions to infect. Due to superinfection immunity by these same lysogens (Chaps. 4, 8, and 16, Sects. 4.4, 8.1.3, and 16.1), however, such new hosts for released virions to infect will be present only within environments that contain additional bacterial strains, that is, besides the converting-gene expressing lysogens.

15.4 Direct selection for Lysogenic Conversion

…if they survive loss when rare.—Lindi M. Wahl and Tyler Pattenden (2017, p. 408)

As repeatedly noted above, we are assuming as a default that lysogenic converting genes can provide a selective benefit to the carrying lysogens. This benefit may not always be present, however, but instead might only be situational. Particularly when the benefit is not present, then whether or not the converting gene persists will be a function solely of avoiding the negative impacts of genetic drift. The power of genetic drift over the converting gene furthermore will increase as an inverse function of their absolute numbers. Thus, the fewer instances of a given allele that exist, then the more likely that the allele will go extinct.

These same ideas can be applied to the prophage as well. If the prophage supplies a benefit to the harboring lysogen, such as by expressing useful converting genes (9, above), then the prophage will tend to increase its representation within the bacterial population. Here, that is, the prophage is the 'plus' allele, a lack of the prophage is the 'minus' allele, and we are assuming that natural selection is acting on the prophage as a whole. As with individual converting genes, if a prophage is not supplying its carrying lysogen with a net benefit, then the prophage will be acted upon especially by genetic drift. Furthermore, if the prophage is present within a bacterial population in low numbers, even if it is being acted upon positively by natural selection, then genetic drift too can strongly impact the frequency of that prophage.

From those musings, should the converting phage (1) be present as a prophage in high frequency within a bacterial population, and especially high absolute numbers, and (2) be beneficial to the carrying lysogen, then we of course will expect that the prophage would persist ('I' in Fig. 15.3). But is this specifically why prophages encode converting genes? Alternatively, if the prophage is present in high absolute numbers, and is selectively neutral in terms of its impact on the fitness of its host, then we would *not* expect the prophage to be lost with high likelihood due to genetic drift, simply because genetic drift under those circumstances would be weak ('II' in Fig. 15.3). On the other hand, what if absolute prophage numbers instead are small? Then we would expect genetic drift to be robust ('III' or 'IV' in Fig. 15.3). It is explicitly when prophage numbers are small that I suggest that converting genes should be most beneficial to converting phages ('III' in Fig. 15.3).

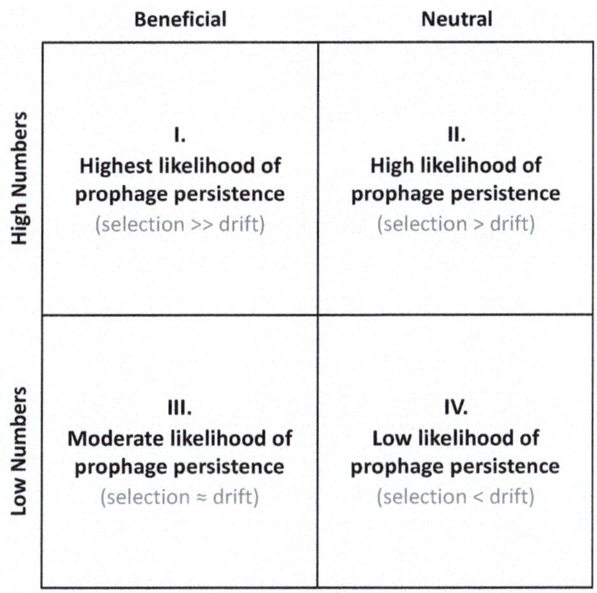

Fig. 15.3 Consequence of fitness impact (top) and prevalence (left) on the likelihood of prophage retention within a bacterial population. Reductions in likelihoods of persistence are a consequence of genetic drift, as suggested in the parentheticals. Note that these considerations are true for any allele

Consider a bacterial population consisting of nothing but non-lysogens, and that this is then followed by the exposure of that population to a converting, temperate phage. To the extent that many or most of the resulting phage infections will be lytic (Chap. 2), then our expectation is that relatively few of those bacteria will be converted to lysogens. Let us assume further that most of the surviving bacteria from this population are not lysogens, including bacteria that either are phage resistant or alternatively that are phage sensitive but phage virions had failed to reach them to infect, e.g., such as due to low starting phage numbers in combination with little phage amplification because the bacteria they are infecting exist within numerical refuges (Chap. 7, Sect. 7.3.2).

If the newly created lysogens are consequently few in number, then unless the bacterial population is experiencing both high resource availability and little competition with other bacteria (Blazanin and Turner 2021), our expectation is that over time the few-in-number new lysogens may be lost. That is, they will go locally extinct with relatively high likelihood as a consequence of genetic drift even if the presence of the prophage is not detrimental to the carrying bacterial lysogen. Alternatively, if the temperate phage in question is highly successful lytically, then the need for lysogenic conversion during lysogenic cycles, at least over the short term and especially if those lysogens are resultingly generated in relatively high numbers, should be lower.

Lysogenic conversion thus could be viewed as a temperate-phage strategy that is used especially under lower-success circumstances, i.e., as particularly useful when neither lytic cycles nor lysogenic cycles are numerically prominent. In other words, sure lysogenic conversion may be useful to temperate phages when they are already successful, but potentially even more useful when temperate phages are not yet successful. Exceptions to that assertion likely exist, however, such as if converting genes are only useful if the encoding lysogens are already present at high densities. It is important to keep in mind, though, that in sufficiently small micro-environments, even a single cell can represent a high density (Redfield 2002).

15.5 Lysogenic Conversion and Ecotypes

An ecotype is a population that has two defining properties. The first is that it exploits a different ecological niche from other ecotypes making up the same species. The second is that periodic selection events, i.e., selective sweeps, will tend to be confined to within an ecotype. This is because clonally related bacteria making up a single, e.g., bacterial ecotype, should be more susceptible to competition especially by selectively fitter variants of those same populations (Cohan 2001). Although above I have argued that converting genes may be especially useful to prophages soon after the formation of a new lysogen, i.e., toward combating genetic drift (Fig. 15.3), over longer time frames the benefits of lysogenic conversion (again, 9, above) may be viewed instead in terms of both ecotypes and periodic selection. I start with periodic selection.

The first way this benefit may be manifest is by the newly formed lysogen selectively sweeping across an ecotype. In other words, the lysogen persists within the same niche as its parental bacterium, but because of a new lysogenic converting gene, the newly formed lysogen now exploits that niche more effectively. Thus, explicitly, the lysogen drives its parental non-lysogen co-ecotypes to extinction (i.e., periodic selection). This idea is equivalent to stating that the prophage supplies to its parental bacterium a selective benefit, but in this case where the selective benefit does not involve a change in the ecological niche being exploited. Thus, for example, a bacterial pathogen might become better at evading its host's immune system, but with the pathogen still otherwise participating in the same infectious life cycle as its non-lysogenic parental bacterium.

The second scenario is where the newly formed lysogen is able to invade a different niche from its parental bacterium. This could be exploiting only that new niche (a strict change in ecotype) or instead remaining as well exploiting the previous niche (a facultative change in ecotype). Such an impact may be viewed as equivalent to differences in genomic island content potentially serving to distinguish ecotypes (Coleman et al. 2006). Whether or not we decide that this new ecotype represents a new bacterial species—a possibility which, frankly, I don't consider to be terribly productive to think further about from a perspective of phages as drivers of evolution (e.g., see Chap. 11, Sect. 11.1)—by generating a new ecotype the lysogen will have freed itself from a likelihood of extinction that could result from selective sweeps of non-lysogens or of different lysogens through the parental ecotype's population.

In the first scenario the lysogen by expressing converting genes provides the selective sweep through the parental ecotype whereas in this second scenario the lysogen by expressing converting genes, and thereby founding a different ecotype, is avoiding selective sweeps that still can occur through the parental ecotype. The new ecotype, that is, would differ in part from the parental ecotype specifically by not being susceptible to the same selective sweeps.

15.6 Phage-Encoded Phage Resistance

In the following chapter (Chap. 16) we consider a special kind of potentially beneficial lysogenic conversion phenotype, that of prophage-encoded phage resistance. Included among those phage resistance mechanisms is superinfection immunity, which is special for three reasons. First, it is a ubiquitous feature of lysogenic cycles and therefore generally present at the same time that lysogenic conversion occurs. Second, at least as defined here, it technically is not an example of lysogenic conversion. Third, it generally is more effective against homospecific phages than it is against heterospecific phages, whereas non-superinfection immunity mechanisms of prophage-encoded phage resistance in many cases should be particularly effective against divergent phages rather than necessarily also effective against additional examples of the converting phage.

References

Abedon ST, LeJeune JT (2005) Why bacteriophage encode exotoxins and other virulence factors. Evol Bioinformatics Online 1:97–110

Altuvia S, Storz G, Papenfort K (2018) Cross-regulation between bacteria and phages at a post-transcriptional level. Microbiol Spectr 6:1–14

Argov T, Azulay G, Pasechnek A, Stadnyuk O, Ran-Sapir S, Borovok I, Sigal N, Herskovits AA (2017) Temperate bacteriophages as regulators of host behavior. Curr Opin Microbiol 38:81–87

Avlund M, Dodd IB, Semsey S, Sneppen K, Krishna S (2009) Why do phage play dice? J Virol 83: 11416–11420

Bille E, Meyer J, Jamet A, Euphrasie D, Barnier JP, Brissac T, Larsen A, Pelissier P, Nassif X (2017) A virulence-associated filamentous bacteriophage of *Neisseria meningitidis* increases host-cell colonisation. PLoS Pathog 13:e1006495

Blazanin M, Turner PE (2021) Community context matters for bacteria-phage ecology and evolution. ISME J 15:3119–3128

Bobay LM, Ochman H (2017) The evolution of bacterial genome architecture. Front Genet 8:72

Bondy-Denomy J, Davidson AR (2014) When a virus is not a parasite: the beneficial effects of prophages on bacterial fitness. J Microbiol 52:235–242

Brüssow H, Canchaya C, Hardt WD (2004) Phages and the evolution of bacterial pathogens: from genomic rearrangements to lysogenic conversion. Microbiol Mol Biol Rev 68:560–602

Casas V, Maloy S (2018) The role of phage in the adaptation of bacteria to new environmental niches. In: Rampelotto PH (ed) Molecular mechanisms of microbial evolution. Springer, Cham, pp 267–306

Christie GE, Allison HA, Kuzio J, McShan M, Waldor MK, Kropinski AM (2012) Prophage-induced changes in cellular cytochemistry and virulence. In: Hyman P, Abedon ST (eds) Bacteriophages in health and disease. CABI Press, Wallingford, pp 33–60

Cohan FM (2001) Bacterial species and speciation. Syst Biol 50:513–524
Coleman ML, Sullivan MB, Martiny AC, Steglich C, Barry K, DeLong EF, Chisholm SW (2006) Genomic islands and the ecology and evolution of *Prochlorococcus*. Science (New York, N Y) 311:1768–1770
Dawkins R (1982) The extended phenotype. Oxford University Press, Oxford
Feiner R, Argov T, Rabinovich L, Sigal N, Borovok I, Herskovits AA (2015) A new perspective on lysogeny: prophages as active regulatory switches of bacteria. Nat Rev Microbiol 13:641–650
Fortier LC, Sekulovic O (2013) Importance of prophages to evolution and virulence of bacterial pathogens. Virulence 4:354–365
Gamage SD, Patton AK, Strasser JE, Chalk CL, Weiss AA (2006) Commensal bacteria influence *Escherichia coli* O157:H7 persistence and Shiga toxin production in the mouse intestine. Infect Immun 74:1977–1983
Harrison E, Brockhurst MA (2017) Ecological and evolutionary benefits of temperate phage: what does or doesn't kill you makes you stronger. Bioessays 39:1700112
Hendrix RW, Casjens SR (2008) The role of bacteriophages in the generation and spread of bacterial pathogens. In: Hensel M, Schmidt H (eds) Horizontal gene transfer in the evolution of pathogenesis. Cambridge University Press, Cambridge, pp 79–112
Hendrix RW, Lawrence JG, Hatfull GF, Casjens S (2000) The origins and ongoing evolution of viruses. Trends Microbiol 8:504–508
Iranzo J, Koonin EV (2018) How genetic parasites persist despite the purge of natural selection. EPL (Europhysics Letters) 122:58001
Koonin EV, Makarova KS, Wolf YI, Krupovic M (2020) Evolutionary entanglement of mobile genetic elements and host defence systems: guns for hire. Nat Rev Genet 21:119–131
Kuhl S, Hyman P, Abedon ST (2012) Diseases caused by phages. In: Hyman P, Abedon ST (eds) Bacteriophages in health and disease. CABI Press, Wallingford, pp 21–32
Lampel KA, Formal SB, Maurelli AT (2018) A brief history of *Shigella*. EcoSal Plus https://doi.org/10.1128/ecosalplus.ESP-0006-2017
LeJeune JT, Abedon ST, Takemua K, Christie NP, Sreevatsan S (2004) Human *Escherichia coli* O157:H7 genetic marker in isolates of bovine origin. Emerg Infect Dis 10:1482–1485
Levin BR, Bergstrom CT (2000) Bacteria are different: observations, interpretations, speculations, and opinions about the mechanisms of adaptive evolution in prokaryotes. Proc Natl Acad Sci U S A 97:6981–6985
Levin BR, Lenski RE (1983) Coevolution in bacteria and their viruses and plasmids. In: Futuyma DJ, Slatkin M (eds) Coevolution. Sinauer Associates, Inc., Sunderland, pp 99–127
Little JW (2005) Lysogeny, prophage induction, and lysogenic conversion. In: Waldor MK, Friedman DI, Adhya SL (eds) Phages: their role in bacterial pathogenesis and biotechnology. ASM Press, Washington DC, pp 37–54
Liu Z, Deng Y, Ji M, Sun W, Fan X (2020) Prophages domesticated by bacteria promote the adaptability of bacterial cells. Biocell 44:157
Łoś M, Kuzio J, McConnell MR, Kropinski AM, Wegrzyn G, Christie GE (2010) Lysogenic conversion in bacteria of importance to the food industry. In: Sabour PM, Griffiths MW (eds) Bacteriophages in the control of food- and waterborne pathogens. ASM Press, Washington, DC, pp 157–198
Mauro SA, Koudelka GB (2011) Shiga toxin: expression, distribution, and its role in the environment. Toxins (Basel) 3:608–625
Nawrocki EM, Mosso HM, Dudley EG (2020) A toxic environment: a growing understanding of how microbial communities affect Shiga toxin expression by *E. coli* O157:H7. Appl Environ Microbiol 86(24):e00509–e00520
Obeng N, Pratama AA, Elsas JD (2016) The significance of mutualistic phages for bacterial ecology and evolution. Trends Microbiol 24:440–449
Redfield RJ (2002) Is quorum sensing a side effect of diffusion sensing? Trends Microbiol 10:365–370

Secor PR, Burgener EB, Kinnersley M, Jennings LK, Roman-Cruz V, Popescu M, Van Belleghem JD, Haddock N, Copeland C, Michaels LA, de Vries CR, Chen Q, Pourtois J, Wheeler TJ, Milla CE, Bollyky PL (2020) Pf bacteriophage and their impact on *Pseudomonas* virulence, mammalian immunity, and chronic infections. Front Immunol 11:244

Tarafder AK, von Kügelgen A, Mellul AJ, Schulze U, Dgal A, Tam B (2020) Phage liquid crystalline droplets form occlusive sheaths that encapsulate and protect infectious rod-shaped bacteria. Proc Natl Acad Sci U S A 117:4724–4731

Taylor VL, Fitzpatrick AD, Islam Z, Maxwell KL (2019) The diverse impacts of phage morons on bacterial fitness and virulence. Adv Virus Res 103:1–31

Villa TG, Feijoo-Siota L, Rama JR, Sánchez-Pérez A, Viñas M (2019) Horizontal gene transfer between bacteriophages and bacteria: antibiotic resistances and toxin production. In: Horizontal gene transfer. Springer, Cham, pp 97–142

Wahl LM, Pattenden T (2017) Prophage provide a safe haven for adaptive exploration in temperate viruses. Genetics 206:407–416

Chapter 16
Prophages Preventing Phage Superinfection

Contents

Abstract One means by which prophages can modify bacterial phenotypes is by protecting lysogens from infections by additional phages. In part this protection is derived from the inherent properties of lysogenic cycles themselves, specifically consisting of in a sense a repurposing of the function prophages employ to continue to exist as prophages rather than these prophages instead transitioning to effecting productive infections. This function involves prophage-encoded repressor proteins and these proteins can repress not only prophage induction to productive cycles but also the infections of superinfecting phages. The result historically has been dubbed as a 'superinfection immunity', but in terms of providing a utility to lysogens, superinfection immunity generally is effective against only a relatively narrow range of only temperate phage types. A variety of additional, lysogenic converting, phage-infection resisting functions exist as well. Some prophages, for example, are able to block phage infections by preventing entrance of phage genomes into the lysogen cytoplasm, a process known as superinfection exclusion. This function too can be relatively narrowly effective in terms of what phage types it impacts, however. Lastly is phage encoding of more generally acting phage resistance mechanisms including abortive infection, CRISPR-Cas-mediated, and restriction-modification systems.

> ...when temperate phage impose a cost to cellular growth but confer immunity to lysogens against infection by lytic viruses, then temperate phage can invade in ecological contexts

S. T. Abedon, *Bacteriophages as Drivers of Evolution*,
https://doi.org/10.1007/978-3-030-94309-7_16

> with pre-existing viruses where they would otherwise not be able to invade in a virus-free case.—Guanlin Li et al. (2020, p. 9)

Lysogenic cycles meld together temperate phages and the bacteria they lysogenically infect into a single unit of selection. As such, what is good for the prophage, so long as it remains a prophage, generally is good for the bacterium as well. One thing that prophages do is to prevent the bacteria they are infecting from being subject to successful infections by other phages. In terms of utility, this primarily prevents the lysogen from being killed by these other phages, but also serves to help limit the number of other prophages also carried by that bacterium. Thus, an important category of prophage-associated modifications of bacterial phenotypes (lysogenic conversions; Chap. 15) are those associated with preventing other phages from successfully infecting the same bacterium. These can be viewed as a special case of phage-resistance mechanisms generally, as considered by and large from Chap. 18 through until the end of the book, but in this chapter the primary emphasis is on these phage-resistance mechanisms being prophage encoded and this is rather than to differentiate substantially among the different types of resistance mechanisms being encoded.

Notwithstanding that last statement, we can differentiate prophage-encoded anti-phage mechanisms into those that explicitly are examples of lysogenic conversion—which *qualitatively* are most of these mechanisms—and those that are not examples of lysogenic conversion. The latter consist especially of superinfection immunity (Łoś et al. 2010), which *quantitatively* should be viewed as the most common of prophage-mediated modifications of bacterial phenotype, among those both effecting and not effecting phage resistance. Toward considering superinfection immunity and also providing interesting examples of lysogenic conversion, particularly ones acting at the level of the bacterium rather than as extended phenotypes (Chap. 15), in this chapter we consider various mechanisms of prophage-mediated interference with phage superinfection. See Table 16.1 for definitions of some relevant terms, many of which have already been touched upon in previous chapters and will be considered again in subsequent chapters.

16.1 Superinfection Immunity

Certainly it is reasonable to use the term 'immunity' in a general sense to describe all sorts of bacterial phage resistance mechanisms, or to distinguish this 'immunity' into innate versus adaptive, as effectively all cellular organisms display some form of immunity against potential parasites, including against phages (Abedon 2012). Nevertheless, the concept of immunity, for better or for worse, has a very specific as well as historically robust meaning for phages, and particularly as mediated by resident prophages. Here immunity means a blocking of the progression of infections by secondary phages (superinfections) by means of the same mechanisms that

Table 16.1 Basic Concepts of direct phage-phage interactions and infection

Term	Description
Abortive infection	Destruction of a phage infection, following adsorption and as mediated by some cell-internally acting mechanism, in which the infected bacterium also *does not* survive. This is a form of bacterial 'self-sacrifice' as considered in Sects. 18.4 and 20.1 of Chaps. 18 and 20.
Abortive infection system	Evolved mechanism involving expression of one or more genes that results in the display by a bacterial infection of an abortive infection.
Adsorption	Irreversible attachment of a virion to a bacterium.
Coinfection	Presence within a bacterium of two infecting phages as resulting from two separate virion adsorption events, i.e., as results from secondary infection.
Destructive infection	Infection that ends in loss of phage viability, e.g., such as an abortive infection (a.k.a., bacterial self-sacrifice) but also such as a restrictive infection (a form of phage negation).
Heteroimmunity	Description of the potential for a secondary phage to be subject to superinfection immunity by a prophage: a heteroimmune phage will *not* be subject to superinfection immunity.
Homoimmunity	Description of the potential for a secondary phage to be subject to superinfection immunity by a prophage: a homoimmune phage will be subject to superinfection immunity.
Induction	Transition of a lysogenic cycle to a productive cycle.
Infection	Entrance of a phage genome into a cell's cytoplasm, i.e., as tends to follow adsorption/attachment.
Primary ($1'$)	First phage to infect a bacterium, as tends to follow primary adsorption.
Reductive infection	Lysogenic cycle, or alternatively pseudolysogeny.
Restriction	Destruction of a phage infection, following adsorption and as mediated by some cell-internally acting mechanism, in which the infected bacterium *does* survive. This is a form of bacterial 'negation' as considered in Sects. 18.3 and 19.2, of Chaps. 18 and 19.
Secondary ($2'$)	Adsorption ($2'$ adsorption) or infection ($2'$ infection) of an already phage-infected bacterium. Secondary infection results in coinfection.
Successful infection	Infection that does *not* end in loss of phage viability. This explicitly is not a phage-destructive infection.
Superinfection	Equivalent to secondary infection, i.e., infection of an already ($1'$) phage-infected bacterium.
Superinfection exclusion	Mechanisms resulting in blocking of the initiation of a phage infection but not a blocking of the initiation of phage adsorption. Superinfection exclusion explicitly is a block on phage genome uptake into an adsorbed bacterium and is not equivalent to superinfection immunity.
Superinfection immunity	Following successful secondary phage genome uptake, prevention of a successful secondary infection, particularly as mediated by the action of prophage repressor proteins.

these prophages employ to retain their own lysogenic cycles (Blasdel and Abedon 2017; Mavrich and Hatfull 2019; Domingo-Calap et al. 2020).

Specifically, prophages encode proteins known as repressors which bind to temperate phage DNA. In so binding, these proteins prevent prophage induction. When a phage of the same type as a resident prophage superinfects a lysogen, those

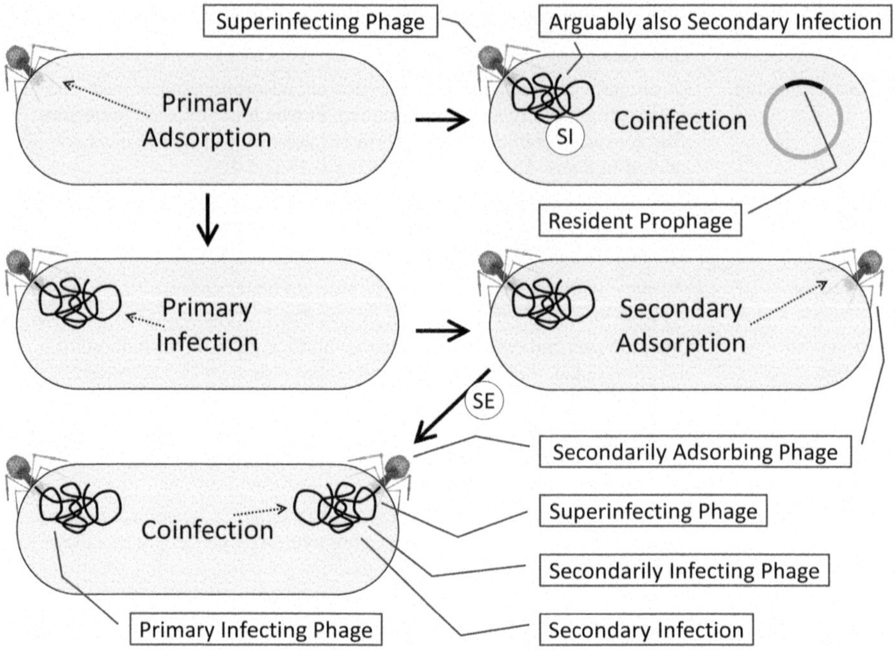

Fig. 16.1 Primary adsorption, primary infection, secondary adsorption, secondary infection, superinfection, and coinfection. Here secondary infection is being used synonymously with superinfection and both result in coinfections. Coinfections also occur given primary infections of bacterial lysogens (upper-right). Note that superinfection exclusion is a block at the step indicated as "SE", as found between secondary adsorption and secondary infection in the figure (center, below the middle point). Thus, not all secondary adsorptions lead to secondary infections, particularly if 'infection' is defined as starting after phage genome translocation into a cell. Superinfection immunity by contrast is a block at the step following secondary infection of a lysogen so is indicated in the upper-right of the figure. Some prophages also are capable of effecting superinfection exclusion, though this is not shown in the figure

phage chromosomes too are exposed to these repressor proteins. The net effect is a sufficient blocking of superinfecting phage gene expression that continuation of these superinfections is blocked (see "SI", upper-right corner, Fig. 16.1). Technically, as defined here, this would be a form of 'negation' (Chap. 18) of these superinfecting phages, as those secondary phages are essentially inactivated but the bacterium as a lysogen is not lost. Superinfection immunity is a particularly important mechanism of phage 'negation' as it is more likely to be acquired in a bacterial population exposed to temperate phages than other forms of phage resistance (Levin and Lenski 1983).

As noted previously (above and Chap. 15), Łoś et al. (2010) prefer to avoid equating superinfection immunity with lysogenic conversion, and this is a perspective that I agree with. Łoś et al's reasoning is that superinfection immunity is such a basic aspect of lysogeny that it cannot be categorized as an accessory function, while generally lysogenic conversion functions instead *are* accessory functions. My reasoning goes further than that, however. Specifically, I ask whether repressor genes have been subject to natural selection specifically for their superinfection immunity function. If not, then repressor genes presumably exist perhaps exclusively for repressing prophage productive cycles rather than for preventing infection by subsequently infecting phages, even if they nevertheless are able to effect superinfection immunity.

16.1.1 Selection for Diversification in Phage Immunity Types

The problem that a prophage would have with superinfecting phages is subsequent display by those superinfecting phages of reductive (Table 16.1) or especially lytic productive infections. Reductive or chronically productive infections by superinfecting phages would be problematic to prophages because of the resources diverted away from the bacterial lysogen, and therefore away from the previously existing prophage, while lytic productive infections of course ultimately would lyse and kill these same bacteria. From the perspective of a superinfecting phage, however, display of a productive or reductive infection must be far superior to instead their being subject to destruction, i.e., as mediated by superinfection immunity. Thus, we would expect that selection acting on superinfecting phages would be to in some manner evade superinfection immunity effected by resident prophages (Campbell and Botstein 1983).

In addition to the terms 'immunity' and 'superinfection immunity', also associated with this phenomenon are 'homoimmunity' and 'heteroimmunity' (Chap. 4). A resident prophage that is homoimmune to a superinfecting phage will subject that superinfecting phage to superinfection immunity. A resident prophage that is not homoimmune to a superinfecting phage, that is, which instead is heteroimmune, will *not* subject that superinfecting phage to superinfection immunity. A heteroimmune superinfecting phage therefore will be able to either productively or reductively infect a lysogen, rather than being inactivated by superinfection immunity. Though all temperate phages that are genetically identical will also be homoimmune, as they will all encode the same repressor as well as possess the same repressor binding sequences on their chromosomes, phages that are not genetically identical may or may not be homoimmune (Mavrich and Hatfull 2019). Indeed, stabilizing frequency-dependent selection (Chap. 24) should select in temperate phages for a diversification of immunity types or, from Campbell and Botstein (1983), p. 371, "divergence rather than conservation of functional specificities." That is, natural selection should favor heteroimmunity over homoimmunity among different temperate phage types, and this is so that especially otherwise divergent superinfecting phages will be more likely to display successful infections of bacterial lysogens.

16.1.2 Superinfection Immunity as a Benefit to Bacteria

What does all of this have to do with the impact of bacteriophages on evolution? One answer is that clearly stabilizing frequency-dependent selection acting on phage immunity is a mechanism of phage-motivated phage evolution (again, Chap. 24). But is superinfection immunity also driving bacterial evolution? Clearly as well the answer here is yes, at least to the extent that any modification of bacterial genotype, such as by prophage acquisition, represents a form of bacterial evolution. But can we expect natural selection to always favor display of superinfection immunity by bacterial lysogens? I present three selective scenarios in answering that question.

First, to the extent that prophages are homoimmune to a given superinfecting phage, then both the prophage and its lysogen should benefit from the prophage's potential to display superinfection immunity (Bossi et al. 2003). However, the more effective diversifying selection is in acting on immunity, that is, the fewer potential superinfecting phages in the environment that are homoimmune, then the less likely that a superinfecting phage will be homoimmune and therefore the less relevant superinfection immunity might be in terms of preventing successful phage secondary infections of lysogens. Nevertheless, when the lysogenizing phage and the superinfecting phage are clonally related, then clearly superinfection immunity as a means of blocking productive or additional reductive secondary infections must be useful to the bacterial lysogen. Clearly, in particular, preventing its own lysis has to be beneficial to a bacterium, at least in most circumstances.

Nonetheless and second, one could argue that to the extent that homoimmunity rather than heteroimmunity is rare, then there may not be that much utility to displaying superinfection immunity toward preventing lytic infections, i.e., as qualitatively most lytic infections would not be prevented. Even if different temperate phage types are more likely to be heteroimmune than homoimmune, however, a lysogen probably is still more likely to encounter a homoimmune phage than a heteroimmune one. This is because to extent that sister lysogens will occupy the same micro-localities, then these localities likely will be enriched in homoimmune temperate phages as released from these same lysogens upon induction. If those released phages were able to lytically infect these sister lysogens, then that clearly would not be supportive of ongoing bacterium/lysogen persistence within those environments. Thus, even though it may be simply fortuitous that prophages display superinfection immunity, rather than this strictly being an evolved phenotype (above), these mechanisms likely are beneficial to both prophage and bacterium/lysogen as it means that single prophage induction events within a group of clonally related lysogens will not result in the eradication of those lysogens.

As a third answer, clearly there must be limits, in terms of metabolic demands as well as potentials for prophage induction, to the utility of bacterial acquisition of yet more prophages. Whether that means that a bacterium would benefit from preventing the acquisition of any given new prophage, however, is questionable. One circumstance where it would be less questionable, though, would be were the incoming prophage identical, or nearly so, to an existing prophage (Fogg et al. 2007). In that case, what benefits to a bacterial host might there possibly be in acquiring a genetically redundant prophage? Therefore, to the extent that homoimmunity is correlated with genetic similarity, as to some degree it certainly is, then an additional advantage to the carrying bacterium/lysogen of superinfection immunity could be as a block on the acquisition of truly redundant genetic material.

Note that many of these benefits are more than just effects that are "exerted at the population level" (Bossi et al. 2003), p. 6471. This is because, as emphasized in this section, both newly formed and existing lysogens should benefit from their display of superinfection immunity given encounter with homoimmune phages. Alternatively, as discussed in Chap. 24, is lysogen allelopathy, where temperate phages are used as anti-bacterial competitor agents, which indeed is something that is (as continued from Bossi et al.) "exerted at the population level, since the individual bacteria that suffer induction are presumably killed in the process..." That is, benefits are to individual lysogens in the above examples (one, two, and three) even though that would not be case when bacterial lysogens are lytically releasing phages to be used as weapons against susceptible bacterial competitors. Nonetheless, these same benefits are realized also at the individual level even with lysogen allelopathy, as it is only via superinfection immunity that the rest of the lysogen population exerting this allelopathic attack would survive.

Though not often considered in these terms, note that lysogen allelopathy is perhaps most likely to occur when a temperate phage enters a bacterial population which is naïve to that phage. In this case, lytic infections can occur (given that this is a lytic temperate phage) while at the same time lysogens may be established. If bacterial densities exceed those of a numerical refuge (Chap. 7, Sect. 7.3.2), then non-lysogen bacterial numbers may be substantially reduced at the same time that lysogens are being established. Thus, though these lysogens may not be supplying the bulk of the phages establishing lytic infections, nor possibly even any of those phages, nevertheless the impact is the same as what is expected in the case of lysogen allelopathy: selective elimination of non-lysogens that are closely related to co-located lysogenic bacteria. The 'selective' part of that process, just as would be the case for lysogen allelopathy proper, is due to superinfection immunity.

16.2 Superinfection Exclusion

Superinfection exclusion, unlike superinfection immunity (Fig. 16.2), would seem obviously to be a product of natural selection. It is possible, however, that the primary role of superinfection immunity in fact is not always one of resistance to superinfection. Superinfection exclusion, specifically and as I and others have defined it, is a phage-encoded phage-resistance mechanism that operates after phage adsorption and prior to phage infection (Fig. 16.1), e.g., from Labrie et al. (2010), p. 319: "Superinfection exclusion systems are proteins that block the entry of phage DNA into host cells..." Importantly for our discussion here, these proteins are

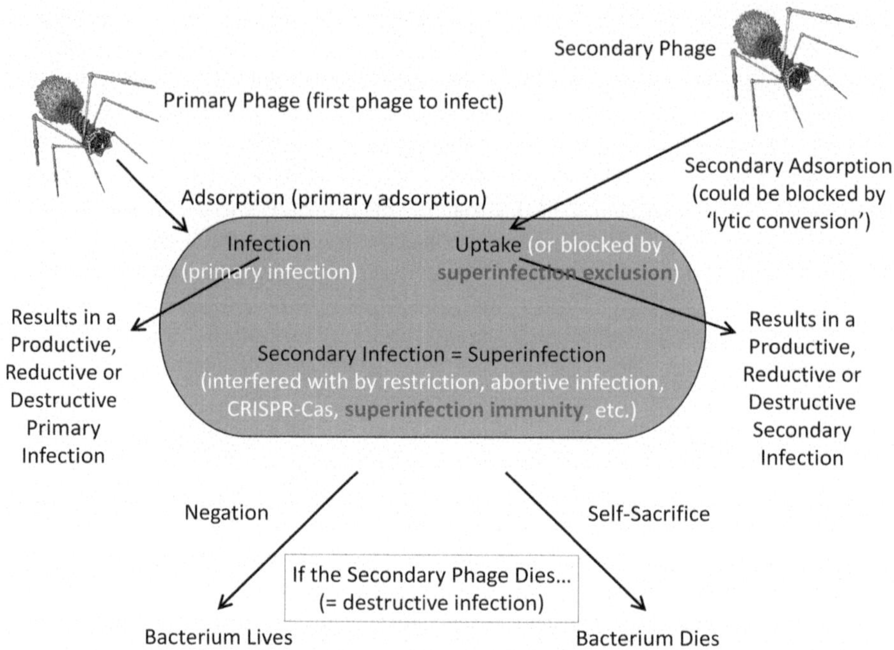

Fig. 16.2 Primary infections can be followed by secondary adsorptions and also secondary infections (superinfections). Superinfections may or may not be successful (productive or reductive) but if not successful then they are destructive of, in this case, the secondary phage. Blocks on superinfection success can be a consequence of number of mechanisms encoded by primary phages, including especially by prophages (parenthetical list found in the figure under "Secondary infection. . ."). See Decker et al. (1994) for description of "Lytic conversion"

often products of prophage genes, e.g., (Kliem and Dreiseikelmann 1989; Bondy-Denomy and Davidson 2014; van Houte et al. 2016; Kirchberger et al. 2021).

16.2.1 Superinfection Exclusion in a Non-temperate Phage

One well studied superinfection exclusion mechanism is that associated with the phage T4 gene *immunity*- and *spackle*-mediated system. Yes, it is unfortunate that the *immunity* gene (or, *imm*) was given that name, since its actions otherwise have nothing to do with superinfection immunity. In fact, however, *imm* was named for neither superinfection immunity nor superinfection exclusion, but instead for resistance to the harmful effects of secondary adsorptions (but called "Superinfections" by the authors) of what are known as phage ghosts (Vallée and Cornett 1972). Ghosts are virions that have been osmotically shocked so that they have lost their genomes but not necessarily also their ability to adsorb to bacteria. These adsorptions by phage T4 ghosts are, for whatever reason, more damaging than adsorptions by unmodified T4 virions, and phage-encoded resistance to them was described as immunity. Phage T4 gene *spackle* (abbreviated, *sp*) in turn was first discovered because in its absence phage T4 infected bacteria were able to lyse their bacterial host without producing their normal endolysin, called lysozyme in phage T4 (Emrich 1968). Gene *spackle* is also involved in effecting a resistance to a phenomenon known as lysis from without (Abedon 2011). In fact, these two genes are involved in a number of phenomena that mostly involve phage infection interactions with secondarily adsorbing phages (Cornett 1974; Abedon 1994, 1999, 2019). Therefore, we can question—just as we can with superinfection immunity—whether the underlying superinfection exclusion as displayed by T4 phage infections exists primarily for the sake of effecting superinfection exclusion, and certainly we can question whether these genes exist solely for the sake of effecting superinfection exclusion.

Another issue with superinfection exclusion as displayed by phage T4, at least for our purposes here, is that phage T4 is a strictly lytic rather than temperate phage. This means that no matter what explains why superinfection exclusion is displayed by phage T4, that display likely does not drive bacterial evolution. Superinfection exclusion, however and as noted above, has also been found to be expressed by various temperate phages, which therefore like superinfection immunity could serve to modify bacteria both genetically (as part of the prophage) and phenotypically. Indeed, superinfection exclusion as expressed by prophages, unlike superinfection immunity, is clearly an example of lysogenic conversion, and indeed potentially expressed by morons (Cumby et al. 2012). There is still a question of what sort of benefits it might supply to the encoding prophage/lysogen, however.

16.2.2 Why Should Temperate Phages Express Superinfection Exclusion?

The first issue here is the same as that for phage T4's *imm* and *sp* genes, and that is the question of whether prophage-encoded superinfection exclusion genes may have been selected for superinfection exclusion versus some other phenomenon associated with primary phage infections interacting with secondarily adsorbing phages (Abedon 2019; Domingo-Calap et al. 2020). The second issue only extends the first, and that is, if prophages already display superinfection immunity, then why would they also need a mechanism of superinfection exclusion? Perhaps some sort redundancy? Perhaps (Wiesmeyer 1966) to keep superinfection immunity systems from being overwhelmed by higher multiplicities of phage adsorption? The third issue is also seen with superinfection immunity and this is the possible breadth of activity of superinfection exclusion systems.

If only very similar phages are affected, as analogous to superinfection immunity affecting only the same immunity types, then the range of phages affected could be slight. If many different phage types are affected, however, then that would be a different story, i.e., with superinfection exclusion serving essentially as a more general phage-resistance mechanism. It is uncertain, though, whether such breadth of activity might generally be the case. Of interest, though, Kirchberger et al. (2021) in particular argue that selection could act specifically toward prophages blocking the infection of dissimilar phages, though in this case all as found within the same phage family, *Microviridae*, which are a kind of not-tailed, ssDNA phage.

Knowles et al. (2016) seem to rely at least in part on an assumption that prophage-mediated phage resistance can act relatively broadly. This is in their piggy-back-the-winner model of favoring of lysogenic infections by phages over lytic infections by phages at higher bacterial densities, i.e., they state: "...increasing resistance via lysogeny. Lysogeny, with its implicit super-infection immunity dynamic" (p. 467). I personally find it difficult to envisage, however, a utility to lysogeny that is based on prophages supplying to their host bacteria general mechanisms of resistance especially to unrelated and particularly to highly divergent phages if relying solely upon either superinfection exclusion or superinfection immunity. Still, expanding on the idea of prophage-encoded immunity to something more broadly defined, the following section (Sect. 16.3) considers additional phage resistance-conferring, prophage-encoded mechanisms. Whether they are sufficiently ubiquitous and broadly acting to support this piggy-back-the-winner model is, however, not certain. See Weitz et al. (2017) for additional discussion of the Knowles et al. (2016) piggy-back-the-winner model.

16.3 Prophage Encoding of Other Phage-Resistance Mechanisms

> The solution to this conundrum, at least in prokaryotes, is mobilization of defence systems via [mobile genetic elements], which creates a constant flux of defence genes, making them available to microorganisms at times of need.—Eugene V. Koonin et al. (2020, p. 127)

As noted, the utilities associated with both superinfection immunity and superinfection exclusion, especially as prophage-encoded mechanisms of phage resistance, may be uncertain. This, however, does not mean that these mechanisms do not provide bacteria with some degree of phage resistance nor that prophages generally do not encode phage resistance mechanisms, far from that. Instead, a number of different prophage-encoded phage-resistance mechanisms are known. These include abortive infection systems, CRISPR-Cas systems, and restriction-modification systems (Susskind et al. 1974; Fang et al. 2012; Bondy-Denomy et al. 2016; Dedrick et al. 2017; Azam and Tanji 2019; Faure et al. 2019; Roux et al. 2019; Taylor et al. 2019; Hampton et al. 2020) (see, e.g., Chaps. 5 and 16, Figs. 5.4 and 16.2). Note, though, that sometimes these anti-phage systems are described as examples of "Superinfection exclusion" but with that terms used broadly to include not just cell-envelope-level blocks on secondary phage infection.

What these systems can provide that superinfection immunity as well as superinfection exclusion mechanisms often or typically fail to provide are broader activities. That is, potentially more or at least additional types of secondarily adsorbing or infecting phage types may be impacted. Thus, like lysogenic conversion genes generally, it is easy to imagine that these mechanisms can provide to carrying lysogens/bacteria some sort of selective benefit, which in this case presumably would be explicitly protection from phage infection, such as by lytically infecting phages. Koonin et al. (2020) in fact make the argument that phage-defense systems should tend to be encoded by mobile genetic elements, whether plasmids or prophages. Indeed, they describe an "entanglement of mobile genetic elements and host defence systems" (p. 119), the evolution of which across multiple, interacting bacterial species appears to be largely driven by phages acting as harmful antibacterial agents.

Continuing the theme of phages are drivers of evolution, the relevant question then should be why do prophages encode these more broadly effective anti-phage defense systems rather than, e.g., plasmids? I think the answer is simply that it is the case that not all anti-phage defense systems are effective against all types of phages. Therefore, even if other anti-phage defense systems already exist in a bacterium, their existence essentially is irrelevant if a bacterium is being impacted by a phage type against which existing defenses are, for whatever reason, not effective. This means that should a bacterium acquire a prophage that encodes a phage-resistance mechanism that is effective against phages which that bacterium is not yet protected from, then that should supply to that bacterium and therefore to that prophage a selective advantage. Furthermore, there should exist hard selection (Chap. 23) for that advantage to the extent that those bacteria which lack that prophage are eliminated by the phage population that the prophage provides protection against. In any case, a lysogen will not realize an advantage from mechanisms of phage resistance to phages until a phage against which those mechanisms are active is present within its immediate environment in great enough numbers.

In the following chapter (Chap. 17) we consider not why prophages may encode converting genes but instead that bacteria often appear to just end up 'stealing' prophage genes for themselves.

References

Abedon ST (1994) Lysis and the interaction between free phages and infected cells. In: Karam JD, Kutter E, Carlson K, Guttman B (eds) The molecular biology of bacteriophage T4. ASM Press, Washington, DC, pp 397–405

Abedon ST (1999) Bacteriophage T4 resistance to lysis-inhibition collapse. Genet Res 74:1–11

Abedon ST (2011) Lysis from without. Bacteriophage 1:46–49

Abedon ST (2012) Bacterial 'Immunity' against bacteriophages. Bacteriophage 2:50–54

Abedon ST (2019) Look who's talking: T-even phage lysis inhibition, the granddaddy of virus-virus intercellular communication research. Viruses 11:951

Azam AH, Tanji Y (2019) Bacteriophage-host arm race: an update on the mechanism of phage resistance in bacteria and revenge of the phage with the perspective for phage therapy. Appl Microbiol Biotechnol 103:2121–2131
Blasdel BG, Abedon ST (2017) Superinfection immunity. In: Reference module in life sciences. Elsevier, Amsterdam
Bondy-Denomy J, Davidson AR (2014) When a virus is not a parasite: the beneficial effects of prophages on bacterial fitness. J Microbiol 52:235–242
Bondy-Denomy J, Qian J, Westra ER, Buckling A, Guttman DS, Davidson AR, Maxwell KL (2016) Prophages mediate defense against phage infection through diverse mechanisms. ISME J 10:2854–2866
Bossi L, Fuentes JA, Mora G, Figuero-Bossi N (2003) Prophage contribution to bacterial population dynamics. J Bacteriol 185:6467–6471
Campbell A, Botstein D (1983) Evolution of the lambdoid phages. In: Hendrix RW, Roberts JW, Stahl FW, Weisberg RA (eds) Lambda II. Cold Spring Harbor Laboratory, Cold Spring Harbor, pp 365–380
Cornett JB (1974) Spackle and immunity functions of bacteriophage T4. J Virol 13:312–321
Cumby N, Edwards AM, Davidson AR, Maxwell KL (2012) The bacteriophage HK97 gp15 moron element encodes a novel superinfection exclusion protein. J Bacteriol 194:5012–5019
Decker K, Krauel V, Meesmann A, Heller KJ (1994) Lytic conversion of *Escherichia coli* by bacteriophage T5: blocking of the FhuA receptor protein by a lipoprotein expressed early during infection. Mol Microbiol 12:321–332
Dedrick RM, Jacobs-Sera D, Bustamante CA, Garlena RA, Mavrich TN, Pope WH, Reyes JC, Russell DA, Adair T, Alvey R, Bonilla JA, Bricker JS, Brown BR, Byrnes D, Cresawn SG, Davis WB, Dickson LA, Edgington NP, Findley AM, Golebiewska U, Grose JH, Hayes CF, Hughes LE, Hutchison KW, Isern S, Johnson AA, Kenna MA, Klyczek KK, Mageeney CM, Michael SF, Molloy SD, Montgomery MT, Neitzel J, Page ST, Pizzorno MC, Poxleitner MK, Rinehart CA, Robinson CJ, Rubin MR, Teyim JN, Vazquez E, Ware VC, Washington J, Hatfull GF (2017) Prophage-mediated defence against viral attack and viral counter-defence. Nat Microbiol 2:16251
Domingo-Calap P, Mora-Quilis L, Sanjuan R (2020) Social bacteriophages. Microorganisms 8:533
Emrich J (1968) Lysis of T4-infected bacteria in the absence of lysozyme. Virology 35:158–165
Fang G, Munera D, Friedman DI, Mandlik A, Chao MC, Banerjee O, Feng Z, Losic B, Mahajan MC, Jabado OJ, Deikus G, Clark TA, Luong K, Murray IA, Davis BM, Keren-Paz A, Chess A, Roberts RJ, Korlach J, Turner SW, Kumar V, Waldor MK, Schadt EE (2012) Genome-wide mapping of methylated adenine residues in pathogenic *Escherichia coli* using single-molecule real-time sequencing. Nat Biotech 30:1232–1239
Faure G, Shmakov SA, Yan WX, Cheng DR, Scott DA, Peters JE, Makarova KS, Koonin EV (2019) CRISPR-Cas in mobile genetic elements: counter-defence and beyond. Nat Rev Microbiol 17:513–525
Fogg PCM, Gossage SM, Smith DL, Saunders JR, McCarthy AJ, Allison HE (2007) Identification of multiple integration sites for Stx-phage ϕ24B in the *Escherichia coli* genome, description of a novel integrase and evidence for a functional anti-repressor. Microbiology 153:4098–4110
Hampton HG, Watson BNJ, Fineran PC (2020) The arms race between bacteria and their phage foes. Nature (London) 577:327–336
van Houte S, Buckling A, Westra ER (2016) Evolutionary ecology of prokaryotic immune mechanisms. Microbiol Mol Biol Rev 80:745–763
Kirchberger PC, Martinez ZA, Luker LJ, Ochman H (2021) Defensive hypervariable regions confer superinfection exclusion in microviruses. Proc Natl Acad Sci U S A 118:e2102786118
Kliem M, Dreiseikelmann B (1989) The superimmunity gene *sim* of bacteriophage P1 causes superinfection exclusion. Virology 171:350–355
Knowles B, Silveira CB, Bailey BA, Barott K, Cantu VA, Cobián-Güemes AG, Coutinho FH, Dinsdale EA, Felts B, Furby KA, George EE, Green KT, Gregoracci GB, Haas AF, Haggerty JM, Hester ER, Hisakawa N, Kelly LW, Lim YW, Little M, Luque A, McDole-Somera T,

McNair K, de Oliveira LS, Quistad SD, Robinett NL, Sala E, Salamon P, Sanchez SE, Sandin S, Silva GG, Smith J, Sullivan C, Thompson C, Vermeij MJ, Youle M, Young C, Zgliczynski B, Brainard R, Edwards RA, Nulton J, Thompson F, Rohwer F (2016) Lytic to temperate switching of viral communities. Nature (London) 531:466–470

Koonin EV, Makarova KS, Wolf YI, Krupovic M (2020) Evolutionary entanglement of mobile genetic elements and host defence systems: guns for hire. Nat Rev Genet 21:119–131

Labrie SJ, Samson JE, Moineau S (2010) Bacteriophage resistance mechanisms. Nat Rev Microbiol 8:317–327

Levin BR, Lenski RE (1983) Coevolution in bacteria and their viruses and plasmids. In: Futuyma DJ, Slatkin M (eds) Coevolution. Sinauer Associates, Inc., Sunderland, pp 99–127

Li G, Cortez MH, Dushoff J, Weitz JS (2020) When to be temperate: on the fitness benefits of lysis vs. lysogeny. Virus Evolution 6:veaa042

Łoś M, Kuzio J, McConnell MR, Kropinski AM, Wegrzyn G, Christie GE (2010) Lysogenic conversion in bacteria of importance to the food industry. In: Sabour PM, Griffiths MW (eds) Bacteriophages in the control of food- and waterborne pathogens. ASM Press, Washington, pp 157–198

Mavrich TN, Hatfull GF (2019) Evolution of superinfection immunity in cluster a mycobacteriophages. MBio 10:e00971–e00919

Roux S, Krupovic M, Daly RA, Borges AL, Nayfach S, Schulz F, Sharrar A, Matheus Carnevali PB, Cheng JF, Ivanova NN, Bondy-Denomy J, Wrighton KC, Woyke T, Visel A, Kyrpides NC, Eloe-Fadrosh EA (2019) Cryptic inoviruses revealed as pervasive in bacteria and archaea across Earth's biomes. Nat Microbiol 4:1895–1906

Susskind MM, Wright A, Botstein D (1974) Superinfection exclusion by P22 prophage in lysogens of *Salmonella typhimurium*. IV. Genetics and physiology of *sie*B exclusion. Virology 62:367–384

Taylor VL, Fitzpatrick AD, Islam Z, Maxwell KL (2019) The diverse impacts of phage morons on bacterial fitness and virulence. Adv Virus Res 103:1–31

Vallée M, Cornett JB (1972) A new gene of bacteriophage T4 determining immunity against superinfecting ghosts and phage in T4-infected *Escherichia coli*. Virology 48:777–784

Weitz JS, Beckett SJ, Brum JR, Cael BB, Dushoff J (2017) Lysis, lysogeny and virus-microbe ratios. Nature (London) 549:E1–E3

Wiesmeyer H (1966) Prophage repression as a model for the study of gene regulation. I. Titration of the lambda repressor. J Bacteriol 91:89–94

Chapter 17
Domestication of Phage Genes

Contents

Abstract Phages can accumulate genes over time, derived either from other phages or instead from non-virus bacterial genetic material. So too genes can go from being phage encoded to instead being directly bacterium encoded. Considered are three scenarios by which this process can take place. These are (1) the loss of other genes associated with integrated prophages (prophage decay) leaving behind genes that provide benefits to the harboring bacteria, (2) recombination into bacterial genomes of genes associated with unsuccessfully infecting phages, and (3) loss of phage functions from plasmid prophages, though with the caveat that the latter results in plasmids depleted of phage functions rather than former phage genes necessarily being encoded by bacterial chromosomes. Pointed out is that it is unlikely that prophage genes would be deleted due to their direct impact on bacterial metabolisms, other than in terms of the requirement that their DNA be replicated, since most of these genes in fact are not expressed during lysogenic cycles. The merging of the genomes of organisms more generally is discussed, where prophages merging with bacteria into a single organism, as a single unit of selection, represents a specific case of such genomic merging.

> . . .a sort of 'moron cycle' by which phages acquire, as morons, novel genes that will be beneficial to the host and contribute them in a multistep process to the bacterial genome. . . . Over the long term, the expectation is that all of the prophage genes will be deleted, with the exception of those that provide a benefit to the host.—Roger W. Hendrix (2005, p. 62).

'Domestication' inherently is an evolutionary process. Specifically, some entity is transformed from a less useful to a more useful (to us) form, and this occurs via

S. T. Abedon, *Bacteriophages as Drivers of Evolution*,
https://doi.org/10.1007/978-3-030-94309-7_17

microevolutionary processes, especially mutation in combination with natural selection, though genetic migration, i.e., the crossing of different varieties, can be important as well. For the plants and animals that human civilizations rely upon, especially as our domesticated food supply, the process of domestication involves what is known as artificial selection, that is, where more desirable varieties (again, to us) are selectively bred by humans, and less desirable ones are discarded or, more likely, are simply eaten. Often these less desirable traits are ones that would be more useful to an animal or plant as a wild rather than domesticated organism, and correspondingly these wild traits can be less useful to humans, e.g., such as an excessive willingness to attack their captors!

Such selection for loss of certain traits may be seen as well with bacteria and prophages. Prophages can be inherently dangerous to bacteria as they literally can have a life of their own, and in fact also a willingness to attack their bacterial 'captors', in this case in the form of prophage induction and resulting phage lytic cycles. Indeed, in Chap. 6 we considered that bacteria may possess mechanisms that result in greater propensities to turn perhaps especially more readily inducible prophages into ones that no longer are able to kill their bacterial hosts. Notwithstanding the potential dangers of prophages, as we have seen, phages also can carry genes that are helpful to carrying bacteria, especially lysogenic converting genes (Chap. 15). The domestication of phage genes by bacteria thus represents a process whereby useful prophage genes are retained by bacteria while less useful or dangerous prophage sequences are discarded.

This is a process, like domestication more generally, that is driven especially by both mutation and natural selection, with the prophage having supplied an initial genetic migration. Here, natural selection is acting on the bacterial population, presumably increasing the frequency of those bacteria that have succeeded in inactivating more dangerous or at least more detrimental prophage genes while retaining prophage genes that are still useful to the bacterial host. As a consequence of these processes, likely with genetic drift playing some roles as well (as considered further, below), within bacterial genomes domesticated prophage genes may be relatively abundant (Bobay et al. 2014). However, to partly paraphrase Lawrence and Hendrickson (2008), p. 12, "It is not clear what fraction of laterally transferred genes currently observed in bacterial genomes once arrived in larger genomic" segments—they said "Islands", but these 'segments' could just as well be "Prophages".

17.1 Merging of Genomes

The process of domestication of phage genes is an example of a broader phenomenon of the merging of genomes of symbiotic and especially endosymbiotic organisms. This is perhaps most familiarly appreciated in terms the domestication, by eukaryotic cells, of the bacteria that would become mitochondria and chloroplasts (Archibald 2015), an area I in fact once reviewed (Abedon et al. 2008). This process involved and indeed involves a combination of what can be described as gene transfers and the loss of redundant functions (Archibald 2006). That is, over evolutionary time, genes that were originally found in endosymbiont genomes migrate—basically via accidental processes—into the host cell nuclei, where through time those genes also accidentally acquire sequences necessary for their expression. When transferred genes create redundancy in function between nuclear and endosymbiont genes, it is the endosymbiont genes that tend to be lost, resulting in more streamlined endosymbiont genomes, but also with formerly bacterial genes having become nuclear genes. The genomes of host and endosymbiont, in other words, merge, while especially the endosymbiont genes, and indeed the endosymbiont itself, becomes domesticated by the larger, eukaryotic cell.

The basic process of genome merger and gene domestication is one of long-term, close physical as well as functional association of genomes in combination with depletion of no longer needed genes. The latter occurs in association with genetic drift but also can be driven by natural selection if superfluous genes are also detrimental (i.e., see Chap. 6). This process of genome merger and gene domestication is also possible with bacteriophages, as illustrated in the following section (Sect. 17.2) by an elegant laboratory experiment.

17.2 Merging of Phage Genomes

> . . .comparative analysis of phage genome sequences imply that a major component of phage evolution is large numbers of intrinsically very improbable events. This view of the manner in which phage evolution occurs becomes plausible only with the realization that there has been ample opportunity for these events to have occurred, despite their improbable nature.—Roger W. Hendrix (2005, p. 56)

Phages can be masters of mosaic evolution, with one phage picking up genes from another in the course of either homologous or illegitimate recombination (Hendrix 2005; 2008); from the slightly older literature (Campbell and Botstein 1983), this can be (p)restated (p. 372) as, "Regions of homology (in the same order and orientation) are interspersed with regions of nonhomology." Though technically that represents a merging of genomes, just as is the case with sexual processes generally, it is not a process that is rooted in symbiosis. Rather, in the case of phages it is simply gene exchange that is occurring. For two phages to truly merge their genomes, in a manner that is equivalent to the forging of the modern eukaryotic cell via the merging of nuclear and endosymbiont genomes, then the two phages first must actually merge without loss of genetic material into a single phage. This could constitute an approximate doubling of phage genetic material, and given limitations on how much DNA a tailed phage can encapsidate (Chap. 14), it would seem to be impossible to achieve. For filamentous phages, i.e., members of phage family *Inoviridae*, equivalent constraints on genome packaging are less present, meaning that indeed in principle two phage genomes can become co-packaged, thereby supplying long-term, close physical association, and upon infection close functional association as well.

Sachs and Bull (2005) set up just such a situation using the inoviruses f1 and Ike. Because these phages infect chronically, that is, rather than lytically (Chap. 2), they can establish symbiotic rather than predatory relationships with their host bacteria. Consequently, antibiotic resistance genes cloned into these phages' genomes can bestow long-term antibiotic resistance to their hosts. What Sachs and Bull did was to clone antibiotic resistance genes targeting two different antibiotics, one into each of these phages, and then exposed the phage-infected bacteria to both antibiotics. The result of this manipulation was to select for bacteria which were experiencing coinfection by the two different phages, and with all other bacteria being negatively impacted by one or both of the supplied antibiotics. Thus, close physical association between the genomes of the two phages was established within doubly antibiotic-resistant, doubly phage-infected bacteria, and especially in terms of antibiotic resistance there was close functional association as well between the two phages. The difficulty, though, is in efficiently propagating this now phage-phage symbiotic relationship horizontally to new bacterial hosts.

To propagate the phages, Sachs and Bull simply collected extracellular virions, with infected host bacteria removed, and then added these phages to uninfected bacteria. Since coinfection by the two phages is required to achieve full antibiotic resistance in this experiment, only those bacteria receiving at least one of each phage type would survive the subsequent round of antibiotic selection. Due to the statistics of phage adsorption (Abedon 2017; Abedon and Katsaounis 2018), such coinfection is somewhat likely at higher phage concentrations (high multiplicities) but somewhat less likely at lower phage concentrations (low multiplicities). At this point, note that this is still more or less just a standard phage experiment, though involving two rather than just one type of propagating phage.

Since low multiplicity virion propagation inherently in this system results in low phage as well as low bacterial survival, this could provide a selective advantage to the two types of phages if they were able to link their propagation together. In fact, this is just what Sachs and Bull observed, as was achieved by co-packaging of the two phages into a single capsid. This allowed for a more long-term, close physical association of the two phage's genomes, i.e., over multiple generations of phage propagation, as well as close functional association during infection of the same cell. In terms of endosymbioses, this step would be equivalent to the initial partnership between a proto-mitochondrion, or proto-cyanobacterium in terms of chloroplasts, and a eukaryotic or proto-eukaryotic cell.

A genetic streamlining of the system then occurred, i.e., evolved. In effect, the genes of one phage, particularly the key antibiotic resistance gene, were domesticated by the other phage. This involved a loss of genetic material, that is, a deletion of genes that had been associated with the now domesticated phage. Those genes in fact were explicitly redundant to the virion morphogenesis genes found in the other phage, since genome co-packaging was occurring, and we can speculate that these genes might even have been detrimental to the system by interfering with this co-packaging. Thus, a long-term interaction was established between two phages that was associated with a combination of genetic linkage of genes (as due to co-packaging of genomes) and this led to depletion of redundant, otherwise unneeded, and/or detrimental genes, but retention in this case of at least one known mutually beneficial gene, that encoding antibiotic resistance. The antibiotic resistance gene encoded by one phage, in other words, was domesticated by a second phage.

17.3 Merging of Phage and Bacterial Genomes

> A... defective phage could arise through deletion of phage specific DNA or the scrambling of the genetic information of selected portions of the phage genome.—Anthony J. Garro and Julius Marmur (1970, p. 260)

Prophages upon establishment of lysogenic cycles essentially by definition merge their genomes with those of their bacterial hosts. These phage infections also are long lasting and involve, given especially integration, a close physical association between genomes. Particularly in terms of lysogenic conversion, there also is a close functional association between the phage and the bacterium. In short, in the course of lysogenic cycles, prophages as "Genetic symbionts" (Obeng et al. 2016) provide genes to bacteria (as hosts) that are beneficial to the bacterium. Prophages are not entirely beneficial or benign entities, however, as they represent additional genetic material that a bacterium must replicate and prophages, as potentially independent units of selection, also retain a potential to outright kill their bacterial hosts (above and Chap. 6). Harrison and Brockhurst (2017) also list as a possible cost of prophage carriage regulatory interference by prophage genes with bacterial homeostasis. Khan et al. (2020) suggest from bioinformatic analysis that it is especially those prophage genes that are involved in "replication, packaging, and release" that tend to be preferentially lost as prophage's decay. In Chap. 6 we discussed how bacteria may display greater than expected propensities toward acquiring deletion mutations. Thus, while prophages can provide to bacteria beneficial genes, the rest of a prophage's genome may be useful to bacteria to eliminate (Fig. 17.1), and bacteria may possess enhanced propensities to do this eliminating by excessively acquiring deletion mutations. Hendrix and Casjens (2008) referred to this process as one of "Moron assimilation" (p. 100).

The net effect of these processes is that already merged phage and bacterial genomes can be streamlined by eliminating genes that are not needed by the bacterium, along with elimination of phage genes that are potentially detrimental to the bacterial host, particularly as resulting in an accumulation of defective prophages or their remnants (Bobay et al. 2014). Indeed, the overall process parallels that of streamlining of endosymbionts as well as the streamlining seen with Sachs and Bull's (2005) filamentous phages: Two genomes become physically associated, initially or ultimately with one serving as the host genome and the other as the symbiont genome. In the course of time, mutations accumulate that result in the symbiont losing its ability to replicate independently of its host. Selection therefore

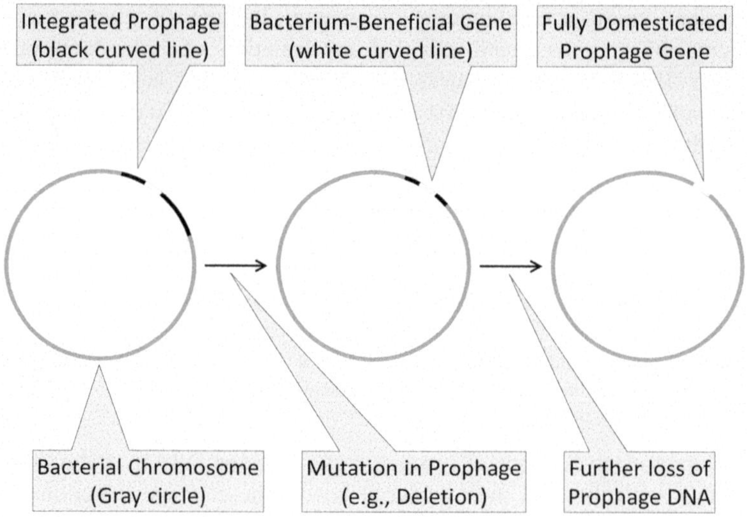

Fig. 17.1 Conversion of prophage gene to a 'domesticated' bacterial gene via loss of prophage DNA. Fixation of defective prophages within bacterial populations could be a consequence of either natural selection or genetic drift, while *retention* of the 'domesticated' gene is likely a consequence of natural selection acting on the bacterial population, i.e., as due to some benefit provided by the gene to the carrier

no longer favors the retention of symbiont genes that had contributed to its independence from its host, and ultimately only those symbiont genes that provide benefits to the host may be retained. For prophages, it is once they have lost an ability to display productive cycles, that is, an ability to produce and release functional virions, that they have transitioned to being a collection of bacterial genes rather than potentially independent prophage genes. At that point, it is natural selection acting on the bacterium as the unit of selection rather than the phage as a unit of selection that drives the further adaptive evolution of these now defective prophages.

Keep in mind with this previous statement, though, that whether or not a prophage is still functional as a potentially virion-productive phage genome has little impact on its evolutionary fate within the bacteria they are infecting. First, prophage induction by definition has not yet occurred for lysogens possessing not-yet-induced prophages. Second, the selective advantage of being able to display a productive infection has not been realized for a prophage that also has not been induced. It is certainly true that within a bacterial population there could be selection against the retention of a non-defective prophage as these prophages are weeded out in the course of their induction, and for these phages it certainly should remain selectively beneficial to retain a potential produce virions, else, of course, why be a phage? Nevertheless, the susceptibility of a prophage gene to mutational inactivation towards eventual domestication of associated phage genes should not vary as a function of whether or not a prophage is defective, except that with non-defective prophages there should be fewer phage genes left to inactivate given that some already have been. In other words, don't assume that the genes of intact prophages as units of selection are any less prone to domestication than the genes of defective prophages not as units of selection, unless there are more of latter than there are of the former within a bacterial population. Intact prophages will lose more ecological and evolutionary potential if their genes are inactivated than will a defective prophage, but the processes of mutagenesis that occur within bacterial cells do not care!

In any case, in the course of loss of prophage sequence, formerly lysogenic converting genes can become bacterial genes. More generally, which genes are being selectively retained by bacteria from within otherwise decaying prophages can be inferred from non-synonymous over synonymous substitution rates (Bobay et al. 2014). This is the ratio of fixed mutations that result in a change in amino acid (non-synonymous substitution) relative to those that don't result in a change in amino acid (synonymous substitution), where excesses of synonymous substitutions suggest that natural selection for retention of gene function is operating. Complicating things, however, an appearance of excessive synonymous substitution also can occur given orthologous replacement of prophage genes by homologous genes from functional, superinfecting phages, that is, restoring through recombination a more synonymous approximation of the original prophage gene. As further suggested by Bobay et al. (2014), even bacteria-useful and otherwise selectively retained domesticated phage genes may become functionally redundant within bacteria and thereby subsequently may be lost, that is if bacteria acquire yet additional prophages that happen to encode the same bacteria-useful functions as a decaying prophage.

Note as an historical aside that the concept of defective prophage has 'evolved' through the years. In the pre-sequencing era a defective prophage often could not be easily detected, as defective, unless it were inducible or, if not inducible, at least able to supply functions to defective superinfecting phages. Nevertheless, defective prophages could still be described as having (p. 260) a "Widespread distribution" (Garro and Marmur 1970).

17.4 Phage Gene Domestication without Prophage Integration

Contrasting phage gene domestication in the course of loss of integrated prophage sequence (this loss described in this section as 'prophage depletion'), it is also possible for phage genes to become domesticated—and to become domesticated in a sense instantaneously—via illegitimate recombination events. For resulting gene insertions to give rise to evolutionary change in bacteria, then the phages supplying the genes cannot have been infecting lytically. The latter concern is easily assuaged, however, if the infected bacterium possesses especially 'negating' phage-resistance mechanisms (Chaps. 18 and 19), i.e., in which the progression of a phage infection is blocked without loss of the viability of the host bacteria, e.g., as effected by bacterial restriction-modification systems. That is, explicitly, not all phage genes that become domesticated by bacteria necessarily are derived from temperate phages, nor necessarily from decaying prophages. Unfortunately, though, the degree to which negated phage DNA is available for recombination is not well understood. Arber (2000) for example argued that DNA cut by restriction endonucleases can be recombinogenic. Restriction endonucleases nevertheless still might interfere with horizontal gene transfer (Cohan 2001). A reasonable conclusion, though, seems to be that otherwise inert phage DNA that makes its way into bacterial cytoplasms is not fully excluded from becoming incorporated into the genomes of these bacteria (Touchon et al. 2017).

Unfortunately, it is not necessarily always bioinformatically trivial to distinguish between illegitimate recombination gain of a new locus and domestication of an equivalent locus by prophage depletion. Some clues can be present, however. For instance, if the gene is found adjacent to a likely prophage insertion point, such as a tRNA, then that may be indicative of a depleted prophage origin, though so too possible genomic island insertion instead (Juhas et al. 2009). Similarly, if the prophage sequence is present but is found in what otherwise would be noncontiguous parts due to deletions, versus uninterrupted sequence found adjacent to the gene in question, then that too could be suggestive of depleted prophage origin. Gaps in prophage sequence, and indeed prophage sequence generally, would however still need to be identifiable as having been derived from a prophage. Alternatively, in the absence of prophage sequence, including domesticated gene-adjacent prophage sequence, ascribing prophage origin to a gene can be challenging, beyond evidence stemming from sequence homology by the inserted gene to known temperate phage genes. Again, however, gains of phage genes by bacteria need not occur solely in the course prophage depletion, nor phage genes domesticated by bacteria even necessarily derived from temperate phages. Indeed, just because a gene has been identified in a temperate phage's genome does not mean that it cannot also be found in non-temperate phage genomes.

I provide that preamble to this section in response to an article by Randich et al. (2019), who show that a common *Caulobacterales* gene, *spmX*, likely evolved from a gene that is of phage origin: an endolysin. Endolysins (lysins) are proteins that lytic phages use to digest the cell wall of infected bacteria toward effecting bacterial lysis (Young and Wang 2006; Shen et al. 2012; Danis-Wlodarczyk et al. 2021; Dennehy and Abedon 2021; Heselpoth et al. 2021). Lysins themselves are produced inside of bacteria, i.e., within the cytoplasm, and consequently are harmless to bacteria unless they are able to reach the bacterial cell wall in an active form. Generally endolysins are unable reach cell walls in an active form, however, without the help of other phage proteins, particularly phage proteins called holins (Young 1992). I belabor this point—that endolysins on their own actually are not toxic—because that is not how the authors describe endolysins ("A toxic phage gene"), and without toxicity it is certainly possible for a typical endolysin gene to be illegitimately inserted by itself into a bacterial chromosome in a fully expressible form, that is, not otherwise constrained in its expression by other prophage sequence. Truly toxic-as-expressed phage genes, by contrast, may find it more difficult to become domesticated.

The authors also suggest that this endolysin was of prophage origin. Without other clues, though, it is not possible to distinguish a prophage endolysin from an endolysin encoded by a strictly lytic phages, nor bacterial acquisition of an endolysin gene via an illegitimate recombination event versus a prophage depletion event. Notwithstanding these quibbles, the authors in fact suggest (p. 11) that "very few examples of novel gene emergence from prophage exist in the literature," that is, versus domesticated phage genes retaining some approximation of their phage-associated function. Bobay et al. (2014) meanwhile point out (p. 12131), as consistent with the Randich et al. conclusions, that "Occasionally prophage-derived elements may evolve toward a new highly specialized function. This will lead to their enduring domestication." Thus, Randich et al. (2019) provide an interesting observation of a likely phage gene (encoding an endolysin) that has evolved toward a completely different function (that of encoding the *Caulobacterales* SpmX protein) in the course of having been domesticated by a bacterial lineage, but precisely how that domestication took place in my opinion is uncertain.

Overall, then, there are at least two direct routes toward prophage gene domestication. One is as equivalent to the development of endosymbioses and involves depletion of those aspects of the genetic complement of the symbiont that are not useful to the host. The other is simply horizontal gene transfer involving illegitimate recombination and gene insertion events. A third route exists as well, and that is bacteriophage gene domestication via homologous recombination, that is, as resulting in orthologous replacement. In the case of evolutionarily successful orthologous replacement of single bacterial genes, though, there should be little or no phage sequence other than that associated with the gene itself, and it would not be expected that such adjacent phage sequence was ever present following the recombination event. That is, orthologous replacement is not a mechanism involving prophage depletion, nor illegitimate recombination.

17.5 Getting Rid of Plasmid Prophage Genes?

One converse to the loss of prophage genes while previously prophage-encoded but bacteria-useful genes remain is the loss of bacterial accessory genes while their carrying mobile genetic element is retained, for example as encoded by plasmids (Zhang et al. 2019). As certain temperate phages exist as plasmids (Chap. 2, Fig. 2.4), this mechanism presumably could occur for prophages as well, but rather in the opposite direction. That is, the plasmid and accessory genes might be retained but, in a sense, the prophage gotten rid of in the course of loss of genes other than those required for plasmid retention. A scheme at least for loss of accessory genes from plasmids is as follows:

The first step in the proposed process is a reduction in the copy number of accessory gene-expressing plasmids as they are partially replaced by descendant plasmids. These descendant plasmids, dubbed satellite plasmids are described, perhaps arguably, as molecular parasites, having lost an ability to express these same accessory genes. This would be useful to the extent that per gene-copy expression of the now partially numerically lost accessory genes is costly. The proposed mechanism of subsequent accessory gene loss, but retention of the carrying mobile genetic element, thus is thought to represent a combination of natural selection and genetic drift: (1) natural selection acting on bacterial populations for the less costly plasmid propagation and then (2) genetic drift within the plasmid-carrying bacterium resulting in an accidental curing of those plasmids that contain accessory genes while their accessory gene-less descendent plasmids by chance are retained. The latter then may be followed (3) by natural selection for those bacteria not expressing the accessory gene or plasmid genes at all. At each step, natural selection acting on bacteria for accessory gene retention presumably would be relatively weak. When accessory genes are useful to bacteria, by contrast, then bacteria which have accidentally lost those genes would presumably be selected against.

One difference between a plasmid-prophage scenario and one involving the loss of expressed accessory genes from plasmids is that presumably most of the plasmid's prophage genes would not be expressed during lysogenic cycles, thereby resulting perhaps in somewhat less selection for their loss. Indeed, for integrated prophages as well, to the extent that most genes are not expressed also during lysogenic cycles, except for repressor and lysogenic converting genes, then our expectation as well should be at best weak selection for their individual loss, even as those prophage genes become pseudogenes (Chap. 6). Contrasting that possibility, the rest of this monograph will emphasize especially the role of natural selection as a motivator of phage-driven evolution.

References

Abedon ST (2017) Multiplicity of Infection. In: Reference module in life sciences. Elsevier, Amsterdam

Abedon ST, Breitenberger C, Roden EE, Williams JB (2008) Ecological processes: respiration. In: Jørgensen SE, Fath BD (eds) Encyclopedia of ecology. Elsevier, Oxford, pp 3010–3020

Abedon ST, Katsaounis TI (2018) Basic phage mathematics. Meth Mol Biol 1681:3–30

Arber W (2000) Genetic variation: molecular mechanisms and impact on microbial evolution. FEMS Microbiol Rev. 24:1–7

Archibald JM (2006) Algal genomics: exploring the imprint of endosymbiosis. Curr Biol 16: R1033–R1035

Archibald JM (2015) Endosymbiosis and eukaryotic cell evolution. Curr Biol 25:R911–R921

Bobay LM, Touchon M, Rocha EP (2014) Pervasive domestication of defective prophages by bacteria. Proc Natl Acad Sci USA 111:12127–12132

Campbell A, Botstein D (1983) Evolution of the lambdoid phages. In: Hendrix RW, Roberts JW, Stahl FW, Weisberg RA (eds) Lambda II. Cold Spring Harbor Laboratory, Cold Spring Harbor, NY, pp 365–380

Cohan FM (2001) Bacterial species and speciation. Systems Biol 50:513–524

Danis-Wlodarczyk KM, Wozniak DJ, Abedon ST (2021) Treating bacterial infections with bacteriophage-based enzybiotics: in vitro, in vivo and clinical application. Antibiotics 10:1497

Dennehy JJ, Abedon ST (2021) Phage infection and lysis. In: Harper D, Abedon ST, Burrowes BH, McConville M (eds) Bacteriophages: biology, technology, therapy. Springer, New York, pp 341–383

Garro AJ, Marmur J (1970) Defective bacteriophages. J Cell Physiol 76:253–264

Harrison E, Brockhurst MA (2017) Ecological and evolutionary benefits of temperate phage: what does or doesn't kill you makes you stronger. Bioessays 39:1700112

Hendrix RW (2005) Bacteriophage evolution and the role of phages in host evolution. In: Waldor MK, Friedman DI, Adhya SL (eds) Phages: their role in bacterial pathogenesis and biotechnology. ASM Press, Washington DC, pp 55–65

Hendrix RW (2008) Phage evolution. In: Abedon ST (ed) Bacteriophage ecology. Cambridge University Press, Cambridge, pp 177–194

Hendrix RW, Casjens SR (2008) The role of bacteriophages in the generation and spread of bacterial pathogens. In: Hensel M, Schmidt H (eds) Horizontal gene transfer in the evolution of pathogenesis. Cambridge University Press, Cambridge, pp 79–112

Heselpoth RD, Swift SM, Linden SB, Mitchell MS, Nelson DC (2021) Enzybiotics: endolysins and bacteriocins. In: Harper DR, Abedon ST, Burrowes BH, McConville M (eds) Bacteriophages: biology, technology, therapy. Springer International Publishing AG, Cham, pp 989–1030

Juhas M, van der Meer JR, Gaillard M, Harding RM, Hood DW, Crook DW (2009) Genomic islands: tools of bacterial horizontal gene transfer and evolution. FEMS Microbiol Rev. 33:376–393

Khan A, Burmeister AR, Wahl LM (2020) Evolution along the parasitism-mutualism continuum determines the genetic repertoire of prophages. PLoS Comput Biol 16:e1008482

Lawrence JG, Hendrickson H (2008) Genomes in motion: gene transfer as a catalyst for genome change. In: Schmidt H, Hensel M (eds) Horizontal gene transfer in the evolution of pathogenesis. Cambridge University Press, Cambridge, pp 3–22

Obeng N, Pratama AA, Elsas JD (2016) The significance of mutualistic phages for bacterial ecology and evolution. Trends Microbiol 24:440–449

Randich AM, Kysela DT, Morlot C, Brun YV (2019) Origin of a core bacterial gene via co-option and detoxification of a phage lysin. Curr Biol 29:1634–1646

Sachs JL, Bull JJ (2005) Experimental evolution of conflict mediation between genomes. Proc Natl Acad Sci USA 102:390–395

Shen Y, Mitchell MS, Donovan DM, Nelson DC (2012) Phage-based enzybiotics. In: Hyman P, Abedon ST (eds) Bacteriophages in health and disease. CABI Press, Wallingford, pp 217–239

Touchon M, de Sousa JAM, Rocha EP (2017) Embracing the enemy: the diversification of microbial gene repertoires by phage-mediated horizontal gene transfer. Curr Opin Mirobiol 38:66–73

Young R (1992) Bacteriophage lysis: mechanisms and regulation. Microbiol Rev. 56:430–481

Young R, Wang I-N (2006) Phage lysis. In: Calendar R, Abedon ST (eds) The bacteriophages. Oxford University Press, Oxford, pp 104–125

Zhang X, Deatherage DE, Zheng H, Georgoulis SJ, Barrick JE (2019) Evolution of satellite plasmids can prolong the maintenance of newly acquired accessory genes in bacteria. Nat Commun 10:5809

Part V
Phage Resistance

Chapter 18
Resistance to Phages, Part I: Overview

Contents

Abstract Phages can decrease the fitness of infected bacteria, as is most obviously seen with phage lytic cycles. This as a consequence can result in selection acting on bacteria to acquire phage-resistance mechanisms. This chapter provides an overview of phage resistance mechanisms as viewed from evolutionary ecological perspectives. Especially as based on at what stage phage infection activity is interfered with, we can differentiate these bacterial phage-resistance mechanisms into four categories as associated with (1) no possibility of phage gene expression, (2) early blocks on phage gene expression, (3) later blocks on the success of especially lytic phage infections, and (4) incomplete inhibition of productive phage infections. These I describe in this chapter as phage avoidance, phage negation, bacterial self-sacrifice, and phage delay. Phage infections and thereby bacterial death are prevented with phage avoidance; phage infection processes are interfered with at a sufficiently early stage given phage negation that this also results in survival of phage-exposed bacteria; phage infection success is suicidally blocked by bacteria with self-sacrifice; and phage population growth is slowed rather than eliminated by phage delay. Examples include adsorption resistance or superinfection exclusion; superinfection immunity or CRISPR-Cas systems; numerous abortive infection systems; and bacteria-mediated reductions in phage burst sizes, respectively. The evolutionary ecology of these mechanisms are considered in greater detail in subsequent chapters.

> From a virus γ active on strain B of *Escherichia coli* a spontaneous virus mutant γ' has now been isolated that attacks a mutant bacterial strain A resistant to γ.—Salvador E. Luria (1944, p. 416)

S. T. Abedon, *Bacteriophages as Drivers of Evolution*,
https://doi.org/10.1007/978-3-030-94309-7_18

A common concern when considering the utility of antibacterial agents, such as antibiotics, is the potential for bacteria to acquire resistance, that is, to *evolve* resistance. Otherwise-sensitive bacteria similarly can acquire resistance to phages, as then can be selected for as a direct consequence of phages acting as predators of bacteria. This resistance acquisition can occur by mutation such that the bacterial target of an antibacterial agent becomes modified, as can be seen with resistance to antibiotics or phages. Alternatively, targets may be outright mutationally deleted, such as is often seen with resistance to phages (Markwitz et al. 2021). Such modifications of phage or antibiotic targets typically involve changes to genes located on the bacterial chromosome, i.e., as can be described as resulting in chromosomal resistance. Alternatively, resistance can be acquired through the gain of new genes, and these new genes typically are encoded extrachromosomally rather than chromosomally, that is, by plasmids. A fair number of general reviews of phage resistance mechanisms have been published, e.g., (Hyman and Abedon 2010; Labrie et al. 2010; Seed 2015; van Houte et al. 2016; McCallin and Oechslin 2019; Rostøl and Marraffini 2019; Hampton et al. 2020).

Resistance mechanisms such as to phages can vary in terms of their costs to expressing bacteria (van Houte et al. 2016). These costs can be present, for example, during use. This is seen especially with abortive infection mechanisms, as resulting in 'self-sacrifice' as I am also referring to these here. Costs instead could be associated simply with gene or allele carriage and expression (Westra et al. 2015; van Houte et al. 2016; Iranzo et al. 2017). The latter includes costs of full or partial loss of gene functions as is often the case with chromosomal resistance (Chap. 22), costs associated with protection from "autoimmunity" against the expressing bacterium (Koonin et al. 2020), or instead costs of carriage by bacteria of plasmids. Bacterial mechanisms of phage resistance typically will vary as well in terms of both which and how many phages are affected (e.g., Chap. 16). Acquisition of resistance mechanisms by bacteria furthermore may be more or less likely depending on phage, host, and circumstances, with some bacteria having a lower ability to mutationally acquire resistance to certain phages (Hyman 2017; Dickey and Perrot 2019). In short, the subject of phage resistance to bacteria is complicated, including from an evolutionary perspective, though it also is one of great interest especially regarding the development of phages as antibacterial agents (Dąbrowska and Abedon 2019; McCallin and Oechslin 2019; Abedon et al. 2021).

I've broken up the initial discussion of phage resistance into three parts, as individual chapters. The first, this chapter, is an overview of various ideas and the second and third (Chaps. 19 and 20) consider those ideas in greater detail. Specifically, in this chapter I discuss the four new terms that I am using as categories to describe general mechanism of phage resistance (as introduced above), while in the companion chapters I consider further evolutionary ecological implications associated with each of these categories. These latter chapters are distinguished particularly in terms of whether the bacterium survives (Chap. 19) or instead does not (Chap. 20), the latter despite the phage infection to some extent being resisted. For a complementary perspective on the evolutionary ecology of phage resistance, see van Houte et al. (2016). The subsequent two chapters in this unit, Chaps. 21 and 22, involve further considerations of a subset of these mechanisms, particularly as involving mutations to phage resistance. Bacterial resistance to phages I then consider further in the final unit of this monograph (Chaps. 23, 24, and 25).

18.1 Categorizing Phage Resistance by Outcomes

> Traditionally in lactic acid bacteria, natural phage defense systems are divided in four main groups, namely, the inhibition of phage adsorption [avoidance], the inhibition of DNA ejection [avoidance], restriction-modification systems [negation], and abortive infection (Abi) systems [self-sacrifice but also phage delay].—Hélène Deveau et al. (2008, p. 1390, with bracketed text added)

We can consider four general mechanisms that bacteria can employ toward resisting phages. These I describe here using the above-noted shorthand designations: 'avoidance', 'negation', 'self-sacrifice', and 'delay' (Fig. 18.1). These differ in terms of (1) whether or not either phages or bacteria survive following phage infection and (2) the means by which natural selection may support their evolution. I provide summaries of these mechanisms in this chapter and then, as noted, discuss them with

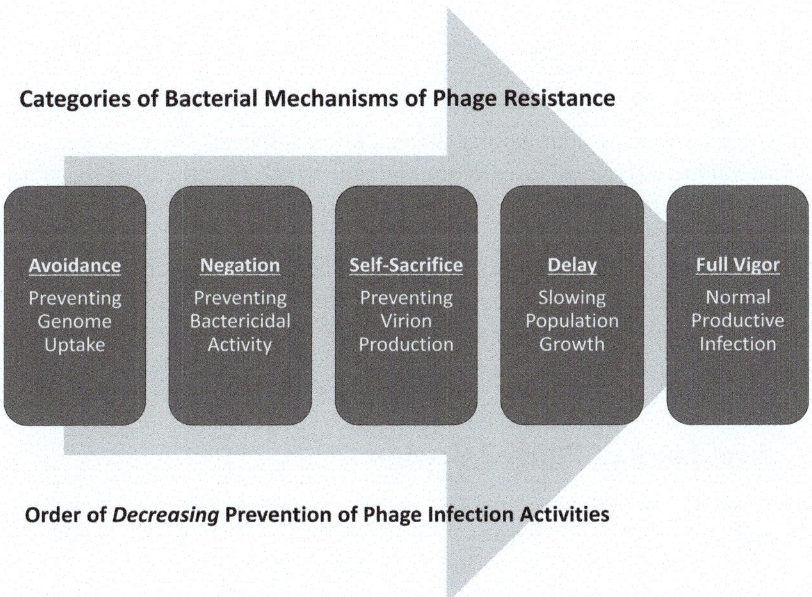

Fig. 18.1 Multiple categories of prevention of phage infection activity, ranging from preventing infection (Avoidance) to preventing infection progression to the point of killing the bacterial host (Negation) to preventing phage virion production (Self-sacrifice) to allowing but otherwise reducing rates or amounts of virion production (Delay). Full vigor represents a normal phage infection not subject to a bacterial mechanism of phage resistance

Table 18.1 Avoidance, negation, self-sacrifice, and delay

General mechanism	Also described as...	Bacteria survive?	Phages survive?	Description
Avoidance	Infection prevention	Yes	Yes/no	Prevention of phage genome uptake; phages survive if they don't adsorb (as due to encounter blocks or adsorption resistance) but fail to survive if they do adsorb (as due to superinfection exclusion)
Negation	Restriction	Yes	No	Blocking phage infections sufficiently early that phage genes effecting lethality are not expressed
Self-sacrifice	Abortive infection	No	No	Either blocking phage infections sufficiently late that phage genes effecting lethality are expressed or actively causing cell lethality so as to prevent progression of phage infections to the point of virion release
Delay	Reduced infection vigor	No	Yes	Slowing down of phage population growth by reducing phage burst sizes or extending phage latent periods, but also slowing rates of phage adsorption

greater elaboration in Chapters 19 and 20. Note that an alternative way to generalize bacterial resistance mechanisms is into inducible versus constitutive (Westra et al. 2015; van Houte et al. 2016), though that approach will not be considered here.

The need for these general categories—as summarized in Table 18.1—is toward not losing track of the evolutionary ecological implications of the different phenomena. Most importantly, this is whether the individual bacteria as well as whole bacterial populations survive the phage-bacterium encounter. As always, I define phage 'infections' as beginning with phage genome entrance into the bacterial cytoplasm (Dennehy and Abedon 2021b), and this is rather than equating phage infection with phage virion adsorption and attachment to a bacterium (Dennehy and Abedon 2021a). As follows I introduce further the individual categories.

18.2 Avoidance of Phage Infection

'Avoidance' corresponds to blocks on phage initiation of infections, where 'infection', as I have noted, I define as entry of phage nucleic acid into the bacterial cytoplasm. The bacterium thus 'avoids' phage infection. These blockages consist in part of encounter blocks or adsorption resistance (Table 18.1). They either can be innate blocks on phage infection, as seems to often be the case with encounter blocks, or instead are associated with mutation of bacterial chromosome-encoded surface receptors, as is the case with adsorption resistance. In addition is superinfection exclusion (Chap. 16), which acts following adsorption but still avoids the initiation of infection since the phage genome never physically reaches the bacterial cytoplasm. Levin and Lenski (1983) alternatively have called some or all of these 'avoidance' mechanisms, "Exclusion", i.e., p. 120: "Preventing novel DNAs from entering cells".

With encounter blocks and adsorption resistance, both phage and the encountered bacterium survive. The phage's survival, however, can be provisional as the phage, if failing to adsorb, still must successful infect another bacterium to continue its life cycle. Or, if subject to superinfection exclusion—as well as for many or most other mechanisms of bacteria-encoded phage resistance (below)—the adsorbing phage simply does not survive. Because the bacterium survives, avoidance should be subject to positive selection acting on those bacteria, selection as effected by the presence of phages to which bacteria lacking avoidance mechanisms are sensitive.

18.3 Negation of Phage Infections

'Negation' is a block on a lytic phage's ability to carry through its infection once that infection has been initiated by the phage's genome entering the bacterium's cytoplasm. Mechanisms of phage negation particularly act by blocking the progression of phage infections prior to a point where the phage infection becomes capable of killing the infected bacterium. Thus, the infection, once begun, is negated rather than being avoided altogether. More broadly stated: Given such phage-infection negation, then whatever the mechanism that serves to block progression of a phage infection beyond a certain point, that point is prior to the phage itself fatally harming the infected bacterium, and expression of the phage-resistance mechanism itself also must not substantially harm the phage-infected bacterium (thus contrasting self-sacrifice, below). In addition, though not directly preventing bacteria lethality, negation mechanisms of bacterial resistance to phages also can interfere with the establishment of lysogenic cycles.

Mechanisms equivalent to negation I have called phage 'restriction' in the past as too have Levin and Lenski (1983), meaning (p. 120), "Destroying novel DNAs that do enter" cells. This is because historically, at least after bacterial restriction endonuclease enzymes were discovered, restriction has been associated especially with bacterial restriction-modification systems. Restriction-modification systems kill phages by cutting up their DNA genomes, which at least in principle may be sufficient to interfere with phage expression of bacteria-lethal genes. Negation is also seen with superinfection immunity, and indeed that is included under their heading of "Restriction" by Levin and Lenski (1983), and more recently negation has been seen with BREX (a.k.a., bacteriophage exclusion, as discussed in Chap. 19) as well as CRISPR-Cas systems (ditto). These latter mechanisms, however, do not necessarily always effect phage infection negation, that is, in terms of killing phages but allowing phage-infected bacteria to survive.

Negation, like avoidance, should be subject to positive selection in bacteria, though unlike avoidance, negation both allows the initiation of phage infections and always gives rise to phage killing. Note, though, that arguably Levin and Lenski's (1983) definition of restriction also applies to what here I am referring to as 'self-sacrifice' (next section, Sect. 18.4), and indeed the concept of restriction, as in 'restricting' a phage infection from producing phage progeny, could be used generally, as I discuss in Chap. 19, to describe especially 'restrictions' on a phage's host range. I don't believe, therefore, that it would be terribly useful to continue to use the term 'restriction' in this more general manner—despite my past preferences to do so—and, as indicated, I have turned here instead to 'negation' as an alternative term that lacks the ambiguity associated with 'restriction'.

18.4 Bacterial Self-Sacrifice upon Phage Infection

'Self-sacrifice'—by bacteria upon phage infection—blocks the progression of phage infections (infecting phages are killed) though at the expense of the survival of the infected bacterium (infected bacteria also are killed). That is, progression of the phage infection is blocked but not prior to expression of bacteria-lethal phage genes or, instead, the blocking mechanism itself causes not just the phage infection but the phage-infected bacterium as well to die. Key here as a bacterial mechanism of phage resistance is that the infecting phages fail to produce new phage progeny, and key in terms of bacterial self-sacrifice is the noted loss of infected-bacterium viability. Contrast self-sacrifice with both avoidance and negation of phage infections (above) in which the bacterium instead survives, in the latter case despite becoming phage infected.

Self-sacrifice by bacteria as a mechanism of phage resistance is represented most notably by a diversity of abortive infection systems. In principle, though, bacteria might also display chromosomal mutations that interfere with phage progeny production without otherwise stopping the bactericidal progression of phage infections. Note though that not all mechanisms, especially as resulting in low phage efficiency of plating in association with phage-mediated bacteria killing, necessarily fully result in a lack of production of new phage virions (see instead 'delay', Sect. 18.5). Here, however, I am defining both bacterial self-sacrifice and abortive infections as indeed preventing phage infection success, at least when functioning properly, whether that success is defined in terms of production of new virions or instead in terms of phage establishment of a lysogenic cycle.

Self-sacrifice as abortive infection mechanisms is another example, like negation/restriction, of both post-adsorption and post-infection phage killing. Because self-sacrifice mechanisms also involve bacteria killing, however, evolution of abortive infection systems presumably requires kin selection rather than positive selection (Chap. 20). That is, these mechanisms, unlike negation/restriction, cannot be selected for by directly plating out bacteria in the presence of excesses of phages. Instead, one generally identifies a population of bacteria which are resistant to a given phage and only later identify that resistance as being associated with an abortive infection mechanism.

18.5 Delay of Phage Propagation

'Delay' refers to mechanisms of phage resistance by bacteria where the infected bacterium does not survive but production of phage virions nonetheless is not completely blocked (strictly speaking, therefore, these are not truly 'aborted' phage infections). Phage delay instead refers to rates of phage population growth being interfered with, e.g., as due to phage-resistance mechanisms that act to reduce infection burst sizes, extend infection latent periods (collectively, 'reduced infection vigor'), or instead delay but not completely prevent phage virion encounter with not yet phage-infected bacteria. These mechanisms, that is, are ones that do not directly result in the death of bacteria-infecting phages, but nonetheless still interfere with phage infections relative to when there is an absence of such mechanisms. Arguably, Levin and Lenski call this "Incompatibility", meaning (p. 120), "Preventing foreign DNA from replicating at a rate sufficient for persistence", though it seems that their emphasis there was more on intracellular survival of phages and plasmids rather than in terms of rates of phage population growth involving new bacterial hosts.

Mechanisms of phage delay at least hypothetically can be used especially by clonal bacterial microcolonies to slow phage lytic progression, thereby increasing the potential for successful bacterial dissemination from these microcolonies to other locations (Abedon 2017). As mechanisms of phage delay both do not fully prevent phage replication or virion release, nor prevent phages from killing bacteria, they are not easily discovered, requiring instead relatively time consuming and subtle analyses of phage growth parameters such as using single-step growth experiments or virion adsorption-rate determinations (Chap. 20). As with abortive infection systems, evolution of bacterial mechanisms that impose delays on phage population growth likely also require kin selection. Such kin selection acting on bacteria presumably can be readily supplied given the clonal nature of bacterial microcolonies, however.

18.6 Summary

Specific bacterial mechanisms of phage resistance can be differentiated into ones of avoidance, negation, self-destruction, and delay, and also in terms of whether the bacterium lives or dies and the phage also lives or dies upon phage-bacterium encounter. See Fig. 18.2 for summary and the next two chapters (Chaps. 19 and 20) for additional discussion of these various mechanisms of bacterial resistance to phages.

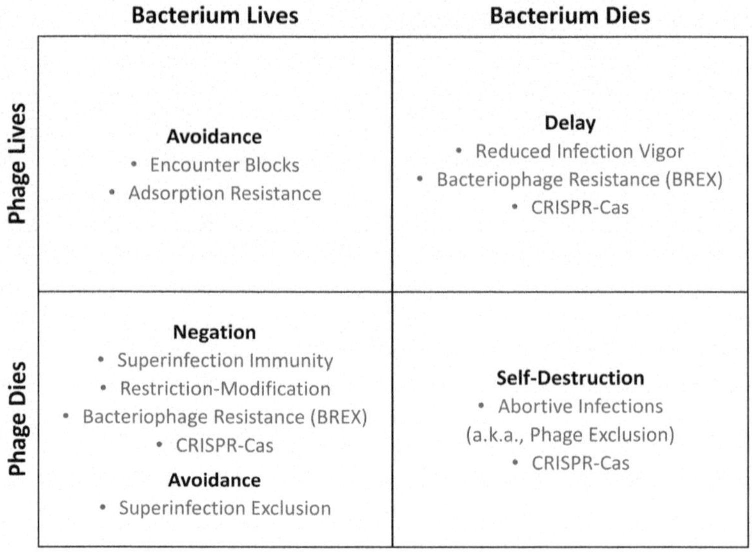

Fig. 18.2 Basic evolutionary ecology and general examples of bacterial phage-resistance mechanisms

References

Abedon ST (2017) Phage "delay" towards enhancing bacterial escape from biofilms: a more comprehensive way of viewing resistance to bacteriophages. AIMS Microbiol 3:186–226

Abedon ST, Danis-Wlodarczyk KM, Wozniak DJ (2021) Phage cocktail development for bacteriophage therapy: toward improving spectrum of activity breadth and depth. Pharmaceuticals (Basel) 14:1019

Dąbrowska K, Abedon ST (2019) Pharmacologically aware phage therapy: pharmacodynamic and pharmacokinetic obstacles to phage antibacterial action in animal and human bodies. Microbiol Mol Biol Rev 83:e00012–e00019

Dennehy JJ, Abedon ST (2021a) Adsorption: phage acquisition of bacteria. In: Harper D, Abedon ST, Burrowes BH, McConville M (eds) Bacteriophages: biology, technology, therapy. Springer, New York City, pp 93–117

Dennehy JJ, Abedon ST (2021b) Phage infection and lysis. In: Harper D, Abedon ST, Burrowes BH, McConville M (eds) Bacteriophages: biology, technology, therapy. Springer, New York City, pp 341–383

Deveau H, Barrangou R, Garneau JE, Labonté J, Fremaux C, Boyaval P, Romero DA, Horvath P, Moineau S (2008) Phage response to CRISPR-encoded resistance in *Streptococcus thermophilus*. J Bacteriol 190:1390–1400

Dickey J, Perrot V (2019) Adjunct phage treatment enhances the effectiveness of low antibiotic concentration against *Staphylococcus aureus* biofilms *in vitro*. PLoS One 14:e0209390

Hampton HG, Watson BNJ, Fineran PC (2020) The arms race between bacteria and their phage foes. Nature (London) 577:327–336

van Houte S, Buckling A, Westra ER (2016) Evolutionary ecology of prokaryotic immune mechanisms. Microbiol Mol Biol Rev 80:745–763

Hyman P (2017) Phage receptor. In: Reference module in life sciences. Elsevier, Amsterdam

Hyman P, Abedon ST (2010) Bacteriophage host range and bacterial resistance. Adv Appl Microbiol 70:217–248

Iranzo J, Cuesta JA, Manrubia S, Katsnelson MI, Koonin EV (2017) Disentangling the effects of selection and loss bias on gene dynamics. Proc Natl Acad Sci USA 114:E5616–E5624

Koonin EV, Makarova KS, Wolf YI, Krupovic M (2020) Evolutionary entanglement of mobile genetic elements and host defence systems: guns for hire. Nat Rev Genet 21:119–131

Labrie SJ, Samson JE, Moineau S (2010) Bacteriophage resistance mechanisms. Nat Rev Microbiol 8:317–327

Levin BR, Lenski RE (1983) Coevolution in bacteria and their viruses and plasmids. In: Futuyma DJ, Slatkin M (eds) Coevolution. Sinauer Associates, Inc., Sunderland, pp 99–127

Luria SE (1944) Mutations of bacterial viruses affecting the host-range, and their relation to bacterial mutations. J Bacteriol 47:416–417

Markwitz P, Lood C, Olszak T, van Noort V, Lavigne R, Drulis-Kawa Z (2021) Genome-driven elucidation of phage-host interplay and impact of phage resistance evolution on bacterial fitness. ISME J 16:533–542

McCallin S, Oechslin F (2019) Bacterial resistance to phage and its impact on clinical therapy. In: Górski A, Miedzybrodzki R, Borysowski J (eds) Phage therapy: a practical approach. Springer, Cham, pp 59–88

Rostøl JT, Marraffini L (2019) (Ph)ighting phages: how bacteria resist their parasites. Cell Host Microbe 25:184–194

Seed KD (2015) Battling phages: how bacteria defend against viral attack. PLoS Path 11:e1004847

Westra ER, van Houte S, Oyesiku-Blakemore S, Makin B, Broniewski JM, Best A, Bondy-Denomy J, Davidson A, Boots M, Buckling A (2015) Parasite exposure drives selective evolution of constitutive versus inducible defense. Curr Biol 25:1043–1049

Chapter 19
Resistance to Phages, Part II: Bacteria Live!

Contents

Abstract Acquired resistance to phage infections that retain bacterial survival can be differentiated into those acting prior to the delivery of phage DNA into bacterial cytoplasms and those acting after that delivery. These I describe here as phage avoidance versus phage negation, respectively. Phage avoidance blocks even the possibility of phage gene expression and involves either preventing phage adsorption or preventing the delivery of phage genomes into the bacterial cytoplasm given phage adsorption. The latter can be accomplished by prophage-expressed superinfection exclusion functions. To achieve the negation of a phage infection, phage gene expression must be blocked sufficiently early that bacteria-lethal gene products are not produced. Classically, this phage-resistance approach has been associated with bacterial expression of restriction-modification systems, while more recently this has been seen with CRISPR-Cas and BREX systems. Both phage avoidance and phage negation mechanisms in bacteria can be positively selected, e.g., by plating bacteria in the presence of large numbers of phages, phages to which those bacteria are otherwise susceptible. Phage avoidance, particularly in terms of the prevention of phage adsorption, appears to be more easily achieved mutationally, however, whereas phage negation tends to be associated with phage-resistance systems acquired by bacteria via horizontal gene transfer.

S. T. Abedon, *Bacteriophages as Drivers of Evolution*,
https://doi.org/10.1007/978-3-030-94309-7_19

The principal function of the [restriction-modification] system is thought to be protecting the cell against invading DNA, including viruses.—Simon J. Labrie et al. (2010, p. 320)

In this chapter, I go into more detail discussing avoidance and negation, both categories of bacterial anti-phage strategies with a defining common characteristic that the expressing bacterium both survives phage encounter and prevents phage propagation. For the sake of keeping track of how I am using these terms, along with other ways that one can describe them and how to think about them in terms of bacterial evolution, see Table 19.1, which serves as an elaboration on Table 18.1 from the previous chapter (Chap. 18). Though included in Table 19.1, bacterial self-sacrifice and phage delay are both considered in more detail in the following chapter (Chap. 20) rather than this one.

19.1 Avoidance of Phage Genome Uptake

The consistent aspect of mechanisms of phage avoidance by bacteria is that the genomes of encountered phages fail to reach the bacterial cytoplasm. Therefore, phage infection of the bacterial cell is avoided in the sense that there is no possibility that a virus can express its genes from a location that is outside of the cell's plasma membrane. The types of mechanisms that are found in this category of phage resistance are illustrated in the top half of Fig. 19.1.

19.1.1 Blocking Adsorption

The simplest means of phage avoidance is to prevent phage adsorption in the first place. This can be especially simple evolutionarily since preventing adsorption can involve loss of gene function by a bacterium rather than gain of gene function. Especially, mutational knock out of production of a conformationally intact bacterial cell surface-located adsorption receptor molecule is a common means of bacterial evolution of phage resistance (Gurney et al. 2019), though this does not necessarily occur without cost to the bacterium (Chap. 22). Alternatively, but more complicated—as it should involve gain rather than loss of function—is acquisition of mechanisms of control of expression of bacterial surface molecules. It can be possible, for example, (1) to temporarily not display phage adsorption-receptor molecules, (2) to display those molecules in very low numbers, or (3) to vary what molecules, as possessing equivalent functions, are displayed at a given time, the latter, i.e., as called phase variation (Bikard and Marraffini 2012; Bull et al. 2014; Gencay et al. 2018); see also (Sekulovic et al. 2015).

Other bacterial molecules that phages interact with also in principle can mutate such that phage infections once initiated are blocked. As these are not bacterial surface molecules that are mutated, however, then phage genome access to the

Table 19.1 Avoidance, negation, self-sacrifice, and delay as bacterial mechanisms of phage resistance

Category	Resistance mechanism	Bacteria survive?[a]	Natural selection[b]	Phages survive?[c]	Description
Avoidance	Encounter blocks	Yes	Positive	Yes	Prevention of phages from reaching bacterial surfaces as due to requirements for diffusion though extracellular polymeric substances, such as bacterial capsules
Avoidance	Adsorption resistance, a.k.a., envelope-level resistance	Yes	Positive	Yes	Bacteria either lacking in phage surface-receptor molecules, thereby preventing successful virion attachment, or in which those receptors are either mutationally modified or masked by specific bacterial products
Avoidance	Superinfection exclusion	Yes	Positive	No	Mechanisms that prevents phage genome uptake into the cytoplasm of an adsorbed bacterium; generally these mechanisms are phage encoded, including by prophages
Negation	Superinfection immunity	Yes	Positive	No	Blocking of the progression of infections by homoimmune temperate phages due to the action of prophage-encoded repressor proteins, which are proteins that function primarily to maintain lysogenic cycles in expressing phages
Negation	Restriction-modification	Yes	Positive	No	Intracellular cutting of unmodified phage DNA at fixed cutting sites, terminating the progression of infections unless phages survive at low frequency due to their post-infection epigenetic modificaiton
Negation but also delay	Bacteriophage exclusion (BREX)	Yes	Positive	No or Yes	Blocking of the replication of unmodified phage DNA but without DNA cutting
Negation but also self-sacrifice and delay	CRISPR-Cas systems	Yes	Positive	No or Yes	Intracellular cutting of unmodified phage DNA at acquired cutting sites, terminating the progression of infections

(continued)

Table 19.1 (continued)

Category	Resistance mechanism	Bacteria survive?[a]	Natural selection[b]	Phages survive?[c]	Description
Self-sacrifice but potentially instead delay	Abortive infections, a.k.a., phage exclusion	No	Negative	No or Yes	Inhibition of progression of phage infections but in association with loss of the viability of the phage-infected bacterium
Self-sacrifice or delay	Satellite phages	No	Negative	No or Yes	By parasitizing productive phage infections, satellite phages reduce the total number of virions produced that carry the genome of the productively infecting phage, in some cases to zero
Delay	Reduced infection vigor	No	Negative	Yes	Host-mediated reductions in rates of phage population expansion, but with phage population growth rates not reduced to zero nor virion production completely blocked

[a]This explicitly is survival of a substantial fraction of bacteria following phage encounter with a bacterium

[b]This is natural selection acting on the expressing, phage-encountering bacterium. Positive refers to increases in the prevalence of cells expressing the resistance mechanism given encounter with phages. Negative refers to decreases in the prevalence of cells expressing the resistance mechanism given encounter with phages. Generally with positive selection the trait can be directly selected for by the presence of the selecting agent, in this case phages. With negative selection, mechanisms for increasing the representation of alleles conferring phage resistance are more complicated than with positive selection. These ideas are discussed further in this chapter and the next (Chap. 20)

[c]This is explicitly whether the resistance mechanism can directly result in loss of phage viability. Lack of loss of viability, i.e., survival, is indicated with a "Yes" while loss of viability is indicated with a "No". "No or Yes" implies phages may survive with relatively high likelihood despite display of the resistance mechanism, e.g., as resulting in delay rather than negation or self-sacrifice

bacterial cytoplasm may not actually be avoided. Loss of the receptor molecules involved in phage adsorption, that is, particularly have the utility of allowing bacteria to avoid the initiation of phage infections, and there is no question of whether a bacterium will still survive its encounter with a phage (it will) if that phage cannot adsorb to it.

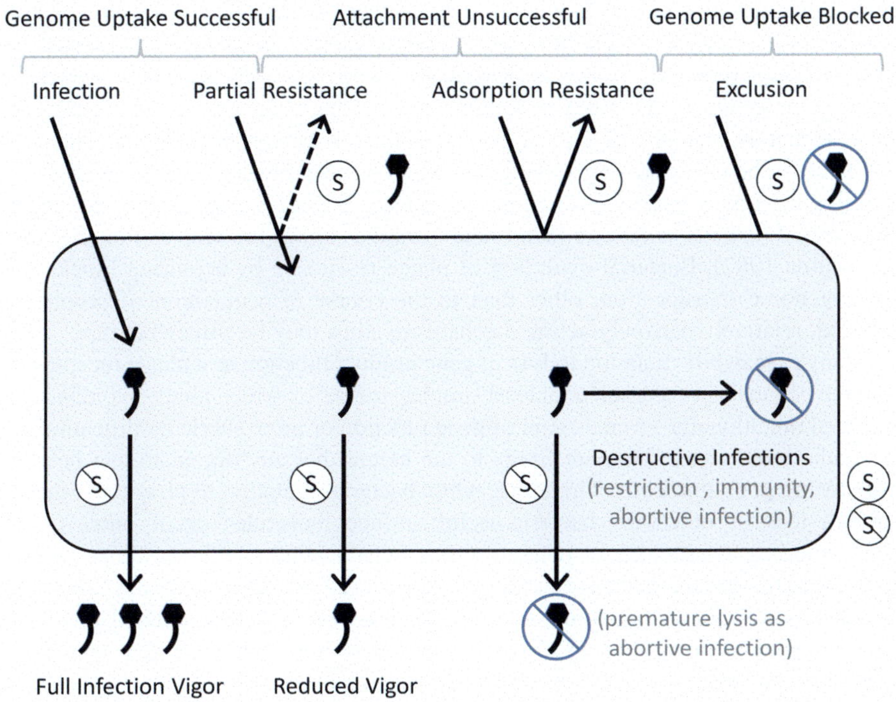

Fig. 19.1 Impacts of bacterial resistance on phages and their infections, and on affected bacterium survival. A circled S indicates bacterium survival while a slash through these circles indicates instead a lack of bacterium survival. Similarly, a depiction of a tailed phage indicates phage survival and replication, whereas a circle with a slash indicates a lack of phage survival following bacterium encounter. The canonical perspective of unhindered phage adsorption leading to unhindered phage infection, as resulting in a robust phage burst size (infection followed by full infection vigor), is shown to the lower, left. Cell-envelope-level mechanisms of bacterial resistance are shown across the top. Partial resistance (top, towards the left) indicates that the likelihood of virion attachment given encounter with a bacterium is reduced but not to zero. Reduced vigor (bottom, towards the left) is explicitly indicated in terms of a reduced burst size (i.e., fewer virions produced) but also can be a consequence of extended latent periods or even reductions in virion survival potential. Phage-destructive infections are shown to the lower, right. Abortive infections (right-bottom) are differentiated into those involving simply phage and bacterial death versus those involving instead premature bacterial lysis

19.1.2 Blocking Phage Genome Uptake

In principle, it should be possible to interfere with phage genome translocation after phage adsorption has occurred, also toward preventing phage infection if not phage adsorption as well. This mechanism if encoded by a prophage would be described as superinfection exclusion (Chap. 16). Such phage-encoded mechanisms, however, presumably involve gains of gene function rather than loss of gene function. Superinfection exclusion also can be narrow in terms of what phages it is effective against. It is possible to interfere with phage genome translocation following phage attachment to a bacterium also through mutation, such as for phage λ as impacting the mannose permease proteins associated with *E. coli* plasma membrane (Burmeister et al. 2021). Like mutational loss of bacterial surface molecules as phage receptors, mutational loss, in this latter case of inner membrane receptors, is also likely to be only narrowly effective in terms of what phages are prevented from infecting the now adsorbed bacterium, e.g., phage T4 instead appears to recognize lipids found in the *E. coli* inner membrane toward a similar function (Furukawa and Mizushima 1982). Bacterial evolution of phage resistance by acquiring blocks on phage-genome translocation, other than in the course of acquisition of prophage encoded, relatively narrowly acting mechanisms, thus may be somewhat rare.

In any case, while mutation to loss of gene product function as a phage receptor is likely to be the most common or at least simplest means by which phage avoidance is achieved mutationally—versus gain of gene function or more subtle modification of molecular structures—there are limits to the extent that any organism can benefit from losing gene functions. Therefore, while bacterial mutation to phage resistance through loss of otherwise bacteria-useful surface molecules or of other functions including as required for phage genome translocation may be easily acquired, over the long term a strategy of evolution of bacterial resistance to phages through repeated loss of gene functions, in terms of a different genes mutated to resist different phages (Wright et al. 2018), must be of only limited utility (see also Chaps. 22 and 25).

Though presumably quite rare today as an acquired, broadly acting gain-of-function phage resistance mechanism, Prangishvili (2013) suggests that in the past bacteria may have evolved a general resistance to viral infections that is explicitly based on interference with phage genome translocation. The idea here is that the peptidoglycan layer that makes up the bacterial cell wall serves as a difficult-to-cross barrier, and thereby provides a profound limitation on what sorts of virion morphology types are able to infect bacteria. Specifically, the virion morphology types that are able to infect members of domain *Archaea* and domain *Eukarya* in many cases differ from those that can infect bacteria. Particularly with regard to archaeal viruses, Prangishvili suggests that their high virion morphological diversity may be possible due to a lack of need by archaeal viruses to breach a peptidoglycan-like layer in the course of their adsorption. Thus, while phage resistance mediated via blocks on phage genome translocation may be relatively rare and/or today narrowly effective as a mechanism of acquired phage resistance, at the same time bacteria possession of a peptidoglycan cell wall may be both profoundly and broadly acting as a mechanism of virus resistance.

19.1.3 Blocking Virion Encounter with the Bacterial Surface

Besides resisting adsorption, or resisting phage genome translocation into the bacterial cytoplasm given adsorption, it is also possible to interfere with the ability of phage virions to even encounter a bacterium's surface. Without surface encounter, that is, there cannot be an initiation of phage adsorption. This mechanism is most commonly considered to be associated with the production by bacteria of extracellular polymeric substances, e.g., such as bacterial capsules or as make up the matrix of biofilms (Abedon 2011; Danis-Wlodarczyk et al. 2021). The idea is that virion diffusion is slowed or even stopped by these substances, interfering with the ability of phages to reach bacterial surfaces and therefore to interact with surface-located adsorption receptors.

A number of studies have suggested that bacterial extracellular polymers can interfere with phage adsorption, e.g., (Ohshima et al. 1988; Scholl et al. 2005; Scanlan and Buckling 2012; Lin et al. 2017; Chaudhry et al. 2020). To what extent extracellular polymers can serve broadly as absolute blocks on phage encounter with bacteria, however, is uncertain, i.e., how many types of phages may be affected by the polymers produced by a given bacterium or biofilm? Indeed, to what extent has the production of extracellular polymers evolved as a mechanism of phage resistance also is uncertain, particularly as phage resistance likely is a pleiotropic aspect of the evolution of these polymers toward solving different problems encountered by bacteria, such as contributing to bacterial virulence (Cross 1990; Scanlan and Buckling 2012; Lin et al. 2018). Furthermore, regarding breadth of resistance, certain phages carry enzymes on their virions which can degrade bacterial extracellular polymers (Pires et al. 2016), perhaps even allowing these polymers to be used as receptors toward phage adsorption, i.e., "[extracellular polymers] and capsules might 'invite' viral infections" (Weinbauer 2004), p. 147; see also, e.g., (Smith and Huggins 1982; Scholl et al. 2005; Abedon 2011; Altamirano et al. 2021).

19.2 Negating Phage Infections Soon after Initiation

Once a phage has entered a bacterium's cytoplasm, then it can either proceed to successfully infect, whether productively or reductively, or instead infection success can be prevented. Still, on a flowchart of possible outcomes of phage infections (Fig. 19.2), we can distinguish lack of or at least reduced infection success into three different categories: 'negation', 'self-sacrifice', and 'delay'. Negation of the infecting phage is associated with little or no impact on the infected bacterium, and is what is considered in this section. Included among mechanisms of phage infection negation are prophage-expressed mechanisms of superinfection immunity as covered in Chap. 16.

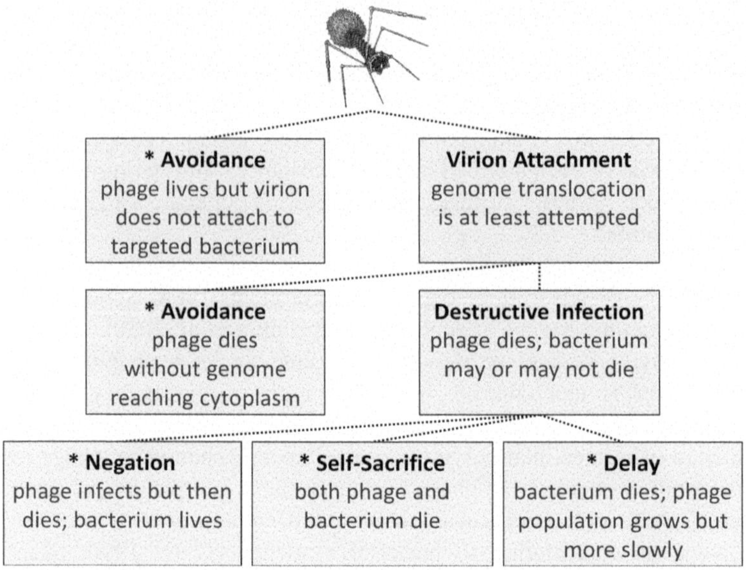

Fig. 19.2 Alternatives to phages successfully infecting a bacterium following encounter. These are, as indicated with the asterisks, avoidance of phage genome uptake, negation of a phage infection if genome uptake has occurred, self-sacrifice of the bacterium in order to prevent virion production by a phage infection, and delay, which is partial interference with phage population growth but without outright preventing virion production by a phage-infected bacterium. For example, for the latter, this can involve reducing a phage's burst size, but not to zero

19.2.1 Gains and Losses of Gene Function, and Positive Selection

It is conceivable that once a phage has already entered a bacterium's cytoplasm, that phage's negation could occur as a consequence of a bacterium lacking a crucial gene function necessary for continuation of the phage infection. For that to happen, however, then this interference with the progression of a phage's infection cycle would need to occur both sufficiently early and sufficiently completely, that is, before enough phage gene expression occurs that bacteria do not survive the infection despite the phage infection having been prevented from fully proceeding. As a consequence of the limitations of such 'passive' interference with phage infection progression, negation should be more readily accomplished instead through more 'active' mechanisms, that is, bacteria gaining in gene function. While loss of gene function typically can be easily achieved mutationally, gain in gene function, especially gain in complex gene functions and over shorter time frames, typically will require horizontal gene transfer. Mechanisms of such active phage negation are thought to include the above-noted restriction-modification, BREX, and CRISPR-Cas systems, though also superinfection immunity.

Like mechanisms leading to avoidance of phage infection, negation of infecting phages has the utility of being subjectable to positive selection. Specifically, the same bacterium that expresses the resistance mechanism also directly benefits as an individual from that expression. This may seem like an obvious property, and indeed it *is* an obvious property, but it also is not a property of all phage-resistance mechanisms, i.e., self-sacrifice and delay as considered in the following chapter (Chap. 20) do not possess this property. To the extent that negation is not easily attainable by a bacterium mutationally, however, then such positive selection simply will tend to distinguish either between those bacteria that already possess a more active resistance mechanism from those that do not, or instead distinguish wild-type bacteria from phage-avoidance mutants. In addition, efficiency of plating assays (Kutter 2009), as they may determine phage viability but do not determine the viability of adsorbed bacteria, also will not distinguish negation mechanisms from either avoidance or self-sacrifice. Still, the fact that expressing bacteria are protected from phage-mediated killing means that phage-negation mutations, to the extent that they do arise, should be viewed as selectively more powerful than bacterial mutations resulting in self-sacrifice or phage delay, though not necessarily any more selectively powerful than bacterial mutations resulting in phage avoidance.

19.2.2 Restriction as a Concept: A Bit of History

As mentioned in Chap. 18, phage-resistance mechanisms that are able effect phage negation might also be considered to be ones of phage restriction, a usage that I have often employed. This particularly has been as a contrast to abortive infections (Chap. 20) when distinguishing among post-phage-infection bacterial phage-resistance mechanisms, e.g., (Abedon 2008). The concept of 'restriction', however, historically was derived from a more general meaning. Adams (1959) for instance writes of "restricted host range" (p. 292), "restricted growth on certain hosts" (ditto), or "restricted phage" (e.g., p. 296). Getting closer to our more modern understanding of the concept of phage restriction in a narrow sense, i.e., as equivalent to phage negation rather than bacterial self-sacrifice, is this quotation from Stent (1963), p. 377: "The instances of host-controlled modifications that have been recognized thus far all concern either a restriction or an extension of the host range of bacteriophages..." Indeed, the following passage should be very recognizable to those familiar with the functioning of restriction-modification systems, also from Stent (1963), p. 377, emphasis his:

> ...all but a small minority of the individuals in a stock of a phage P grown on a bacterial strain, A, are unable to grow on a second bacterial strain, B. Hence P phages grown on strain A, designated as P·A, are *restricted* in their host range. Those rare P·A phages that *are* able to grow on strain B give rise in their first cycle of growth on strain B to a population of *unrestricted* phage progeny of type P·B that can now grow with full efficiency on both strains A and B.

Thus, it seems clear that restriction-modification systems (Arber and Linn 1969), at least in terms of the 'restriction' part, likely were named because they served to *restrict* phage host ranges. This is rather than such restriction necessarily always consisting of mechanisms that prevent phage-infected bacteria from also dying, i.e., not strictly representing mechanisms of negation rather than self-sacrifice, nor even necessarily rather than avoidance. 'Restriction', in other words, is a term that was used to describe the impact of bacteria on phages rather than the impact of phages on bacteria, whereas 'negation' is being used here to describe an impact both on phages (they die) and on bacteria (they live).

19.2.3 Bacteriophage Exclusion (BREX) as a Mechanism of Negation

In addition to restriction-modification systems, the more recently discovered bacteriophage exclusion (BREX) system (Goldfarb et al. 2015) also appears to result in infected phage negation. This is a six-gene system discovered in *B. subtilis* that blocks the replication of phage DNA. As with restriction-modification systems, DNA is methylated to presumably protect host DNA. That BREX truly represents a mechanism of negation rather than self-sacrifice is seen in an experiment presented by Goldfarb et al. in which phages were added to bacteria at a multiplicity of infection of 5 but this did not appear to directly interfere with ongoing bacterial population growth, though adsorption does not appear to be affected.

19.2.4 CRISPR-Cas, Not Always Resulting in Phage-Infection Negation?

The other key mechanism of phage negation is provided by CRISPR-Cas systems. CRISPR-Cas systems can be more specific in terms of their breadth of activity than other mechanisms of negation of phage infections, but also are trainable, allowing for targeted increases in breadth of activity. See Chap. 5 and Fig. 5.4 for an introduction to CRISPR-Cas systems.

The earliest claim that I am aware of that expression of CRISPR-Cas systems can result in phage negation rather than bacterial self-sacrifice is provided by Deveau et al. (2008). While many studies rely solely on efficiency of plating measurements to determine CRISPR-Cas impact on bacteria, e.g., including earlier studies (Barrangou et al. 2007; Brouns et al. 2008), Deveau et al. also presented cell survival data. They found that when CRISPR-Cas was effectively expressed in *Streptococcus thermophilus* against phage 2972, reductions in efficiency of plating of over 10,000 fold were associated also with survival of bacteria in the range of 70%. This is indicative that in roughly 7 in 10 infected bacteria the phage infection was negated. It is also indicative, however, that in roughly 3 in 10 infected the phage infection instead was not negated. This relatively high failure of CRISPR-Cas to negate phage infections possibly, though, is a consequence of the relatively high multiplicity of infection, 5, that was used in bacteria survival experiments, where perhaps the presence of multiple phage genomes promotes bypass of CRISPR-Cas control. Note that CRISPR-Cas presence otherwise was shown to have little impact on phage adsorption in this study.

Further evidence of incomplete control of phage infection by CRISPR-Cas in this system is seen with reported burst sizes despite CRISPR-Cas presence, which were on average about 7% of that seen without CRISPR-Cas. If we consider just those bacteria that did not survive (on average 32%), then this burst size translates instead to roughly 20% of that seen without CRISPR-Cas control. Alternatively, if we consider only that fraction which likely produced phages (on average 16% when infected with a multiplicity of 5), then burst sizes when virions are produced instead would be roughly 40% with CRISPR-Cas present than without. Together, these data serve as evidence that depending on circumstances, CRISPR-Cas systems can mediate phage infection negation (~70% of infections in this data set), instead can achieve only a blocking of phage production via bacterial self-sacrifice (about 16% in this data set, i.e., 32% minus 16%), or instead can allow phage production though with reduced burst sizes (also perhaps 16% and indicating phage delay).

Under lower multiplicity conditions it is conceivable that the percentage of infections that are negated could be somewhat higher. The reported burst size determinations, though, were done with a multiplicity of infection of 0.2 rather than 5, suggesting that phage progeny production despite CRISPR-Cas presence in fact can occur even when phages are infecting bacteria singly. It is fairly striking, therefore, that efficiency of plating nevertheless is so substantially reduced, likely suggestive of an inadequacy of efficiency of plating assays as measures of CRISPR-Cas anti-phage effectiveness. Ecologically, presumably such inefficiencies in CRISPR-Cas activity could serve to narrow the circumstances under which selection might favor CRISPR-mediated resistance relative to other anti-phage resistance mechanisms displayed by bacteria (Levin 2010).

Though it is possible that CRISPR-Cas systems do not always completely protect bacteria from phage infection, perhaps especially if phage multiplicities of infection are greater than 1, there are two circumstances in which CRISPR-Cas systems are not protective of either phage infected bacteria or bacterial populations. The first circumstance is prior to CRISPR adaptation. That is, if bacteria do not yet possess the correct spacer sequences for a given infecting phage, then CRISPR protection will not occur (Hynes et al. 2014). The second has to do with existence in phages of mechanisms of resistance to CRISPR, as reviewed in (Malone et al. 2020). It is the circumstances where phages are completely sensitive to CRISPR-Cas but still seem to produce virion progeny or kill bacteria that has prompted me to list CRISPR-Cas systems under not only negation but also under both self-sacrifice and phage delay in Fig. 18.2 (Chap. 18) and Table 19.1.

The following chapter (Chap. 20) considers other routes toward both self-sacrifice and delay, along with difficulties in discovering such mechanisms in comparison to ones of avoidance and negation.

References

Abedon ST (2008) Phages, ecology, evolution. In: Abedon ST (ed) Bacteriophage ecology. Cambridge University Press, Cambridge, pp 1–28

Abedon ST (2011) Bacteriophages and biofilms: ecology, phage therapy. Nova Science Publishers, Hauppauge, New York, Plaques

Adams MH (1959) Bacteriophages. InterScience, New York

Altamirano FG, Forsyth JH, Patwa R, Kostoulias X, Trim M, Subedi D, Archer SK, Morris FC, Oliveira C, Kielty L, Korneev D, O'Bryan MK, Lithgow TJ, Peleg AY, Barr JJ (2021) Bacteriophage-resistant *Acinetobacter baumannii* are resensitized to antimicrobials. Nat Microbiol 6:157–161

Arber W, Linn S (1969) DNA modification and restriction. Annu Rev Biochem 38:467–500

Barrangou R, Fremaux C, Deveau H, Richards M, Boyaval P, Moineau S, Romero DA, Horvath P (2007) CRISPR provides acquired resistance against viruses in prokaryotes. Science (New York, N Y) 315:1709–1712

Bikard D, Marraffini LA (2012) Innate and adaptive immunity in bacteria: mechanisms of programmed genetic variation to fight bacteriophages. Curr Opin Immunol 24:15–20

Brouns SJ, Jore MM, Lundgren M, Westra ER, Slijkhuis RJ, Snijders AP, Dickman MJ, Makarova KS, Koonin EV, van der Oost J (2008) Small CRISPR RNAs guide antiviral defense in prokaryotes. Science (New York, N Y) 321:960–964

Bull JJ, Vegge CS, Schmerer M, Chaudhry WN, Levin BR (2014) Phenotypic resistance and the dynamics of bacterial escape from phage control. PLoS One 9:e94690

Burmeister AR, Sullivan RM, Gallie J, Lenski RE (2021) Sustained coevolution of phage lambda and *Escherichia coli* involves inner- as well as outer-membrane defences and counter-defences. Microbiology 167:001063

Chaudhry W, Lee E, Worthy A, Weiss Z, Grabowicz M, Vega N, Levin B (2020) Mucoidy, a general mechanism for maintaining lytic phage in populations of bacteria. FEMS Microbiol Ecol 96:fiaa162

Cross AS (1990) The biological significance of bacterial encapsulation. In: Jann K, Jann B (eds) Bacterial capsules. Springer, Berlin, pp 87–96

Danis-Wlodarczyk KM, Wozniak DJ, Abedon ST (2021) Treating bacterial infections with bacteriophage-based enzybiotics: in vitro, in vivo and clinical application. Antibiotics 10:1497

Deveau H, Barrangou R, Garneau JE, Labonté J, Fremaux C, Boyaval P, Romero DA, Horvath P, Moineau S (2008) Phage response to CRISPR-encoded resistance in *Streptococcus thermophilus*. J Bacteriol 190:1390–1400

Furukawa H, Mizushima S (1982) Roles of cell surface components of *Escherichia coli* K-12 in bacteriophage T4 infection: interaction of tail core with phospholipids. J Bacteriol 150:916–924

Gencay YE, Sorensen MCH, Wenzel CQ, Szymanski CM, Brondsted L (2018) Phase variable expression of a single phage receptor in *campylobacter jejuni* NCTC12662 influences sensitivity toward several diverse CPS-dependent phages. Front Microbiol 9:82

Goldfarb T, Sberro H, Weinstock E, Cohen O, Doron S, Charpak-Amikam Y, Afik S, Ofir G, Sorek R (2015) BREX is a novel phage resistance system widespread in microbial genomes. EMBO J 34:169–183

Gurney J, Pleska M, Levin BR (2019) Why put up with immunity when there is resistance: an excursion into the population and evolutionary dynamics of restriction-modification and CRISPR-Cas. Philos Trans R Soc Lond Ser B Biol Sci 374:20180096

Hynes AP, Villion M, Moineau S (2014) Adaptation in bacterial CRISPR-Cas immunity can be driven by defective phages. Nat Commun 5:4399

Kutter E (2009) Phage host range and efficiency of plating. Meth Mol Biol 501:141–149

Labrie SJ, Samson JE, Moineau S (2010) Bacteriophage resistance mechanisms. Nat Rev Microbiol 8:317–327

Levin BR (2010) Nasty viruses, costly plasmids, population dynamics, and the conditions for establishing and maintaining CRISPR-mediated adaptive immunity in bacteria. PLoS Genet 6: e1001171

Lin H, Paff ML, Molineux IJ, Bull JJ (2017) Therapeutic application of phage capsule depolymerases against K1, K5, and K30 capsulated *E. coli* in mice. Front Microbiol 8:2257

Lin H, Paff ML, Molineux IJ, Bull JJ (2018) Antibiotic therapy using phage depolymerases: robustness across a range of conditions. Viruses 10:622

Malone LM, Birkholz N, Fineran PC (2020) Conquering CRISPR: how phages overcome bacterial adaptive immunity. Curr Opin Biotechnol 68:30–36

Ohshima Y, Schumacher-Perdreau F, Peters G, Pulverer G (1988) The role of capsule as a barrier to bacteriophage adsorption in an encapsulated *Staphylococcus simulans* strain. Med Microbiol Immunol 177:229–233

Pires DP, Oliveira H, Melo LD, Sillankorva S, Azeredo J (2016) Bacteriophage-encoded depolymerases: their diversity and biotechnological applications. Appl Microbiol Biotechnol 100:2141–2151

Prangishvili D (2013) The wonderful world of archaeal viruses. Ann Rev Microbiol 67:565–585

Scanlan PD, Buckling A (2012) Co-evolution with lytic phage selects for the mucoid phenotype of *Pseudomonas fluorescens* SBW25. ISME J 6:1148–1158

Scholl D, Adhya S, Merril C (2005) *Escherichia coli* K1's capsule is a barrier to bacteriophage T7. Appl Environ Microbiol 71:4872–4874

Sekulovic O, Ospina BM, Fivian-Hughes AS, Fairweather NF, Fortier LC (2015) The *Clostridium difficile* cell wall protein CwpV confers phase-variable phage resistance. Mol Microbiol 98:329–342

Smith HW, Huggins MB (1982) Successful treatment of experimental *Escherichia coli* infections in mice using phage: its general superiority over antibiotics. J Gen Microbiol 128:307–318

Stent GS (1963) Molecular biology of bacterial viruses. WH Freeman and Co., San Francisco, CA

Weinbauer MG (2004) Ecology of prokaryotic viruses. FEMS Microbiol Rev 28:127–181

Wright RCT, Friman VP, Smith MCM, Brockhurst MA (2018) Cross-resistance is modular in bacteria-phage interactions. PLoS Biol 16:e2006057

Chapter 20
Resistance to Phages, Part III: Bacteria Die. . .

Contents

Abstract Intuitively, bacterial mechanisms of phage resistance that prevent the encountered phage from killing the expressing bacterium would seem to be ideal. In fact, however, a large number of phage-resistance mechanisms do not result in the encountered bacterium surviving. One argument for why such 'non-ideal' mechanisms of phage existence should even exist is that it may simply be easier for bacteria to develop blocks on the progression of phage infections if those blocks don't need to spare the bacterium. In fact, we can describe a progression of phage-resistance mechanisms from prevention of infection (phage avoidance) to prevention of bacterial death (phage negation) to prevention of phage infection only, with the latter dubbed here as bacterial self-sacrifice and as is emphasized in this chapter. A fourth category, though, is an only partial interference with the phage infection, resulting particularly in a slowing of phage population growth rather than elimination of that growth altogether. The latter I've dubbed as phage delay and which is also an emphasis of this chapter. What bacterial self-sacrifice and phage delay particularly have in common is an inability for positive selection acting on individual bacteria to favor these mechanisms, suggesting a role for kin selection instead.

> . . .this interference . . .presumably provides a selective advantage to bacterial populations. . . by reducing phage predation.—José R. Penadés and Gail E. Christie (2015, p. 188)

S. T. Abedon, *Bacteriophages as Drivers of Evolution*,
https://doi.org/10.1007/978-3-030-94309-7_20

In this chapter, I go into more detail discussing bacterial self-sacrifice and phage delay, both categories of bacterial anti-phage strategies with a defining common characteristic: The expressing bacterium does not survive. As above, for the sake of keeping track of (1) how I am using these terms, (2) other ways that one can describe them, and (3) how to think about them in terms of bacterial evolution, see Table 19.1 from Chap. 19.

20.1 The Bacterial Self-Sacrifice of Abortive Infections

What is prevented when would-be productive phage infections are aborted is the formation and release of virions. Prevention of phage progeny production, or prevention of phage infection success generally, is also seen with negation of phage infections. Negation, however, I am defining as blocking those phage infections so completely that the bacterium also survives. With abortive infections, by contrast, both the phage and bacterium do not survive. Because the bacterium does not survive, because we are considering explicitly here mechanisms of bacteria-mediated phage resistance, and also because with abortive infections the benefits of such resistance should be seen particularly by bacteria that are other than those which have been phage infected, I describe such abortive-type infections as being associated with bacterial self-sacrifice.

It is important to note that efficiency of plating determinations alone, as I indicated as well in Chap. 19, can do a poor job of distinguishing among negation, self-sacrifice, and phage delay—and even distinguishing phage viability from phage inviability (Abedon and Yin 2009). A combination of both bacteria killing by infecting phages and low efficiency of plating, on the other hand, would be sufficient to distinguish bacterial self-sacrifice from either phage avoidance or phage negation, as in these latter cases bacteria instead should survive phage exposure. Demonstration of a lack of phage production along with a lack of bacterial survival, rather than just low efficiencies of plating, in turn is required to distinguish bacterial self-sacrifice (no phage production) from phage delay (some phage production). This is why I explicitly define bacterial self-sacrifice—as a mechanism of phage resistance, as well as associated abortive infections—to include only phage infections that generally end in neither phage nor bacterium survival, rather in terms of just low efficiency of plating.

As noted in Chap. 18 (and in Table 19.1 from Chap. 19), as a consequence of bacteria not surviving a phage infection given self-sacrifice, they cannot be isolated based on protocols involving positive selection. That is, despite abortive infections serving as a mechanism of bacterial resistance to phages, bacteria displaying abortive infection systems that even are highly effective against a given phage type will *not* form colonies in the presence of excesses of those specific phages based on the action of those abortive infection systems alone. In this section I consider various additional concepts associated with abortive infections as self-sacrificial mechanisms of bacterial resistance to phages. Subsequently in this chapter I consider what can perhaps be thought of as even less-efficient bacterial anti-phage mechanisms, i.e., as resulting in phage delay rather than explicitly in phage death.

20.1.1 The Need for Making Distinctions

Why am I trying so hard to distinguish between negation, self-sacrifice, and phage delay? As noted, and considered further below, selection acting on bacteria for especially negation differs from that for self-sacrifice or for phage delay. Similarly, ecologically, the circumstances in which negation can be useful to a bacterium certainly will differ from those of self-sacrifice and phage delay, where with phage negation individual bacteria survive their encounter with a phage, but with self-sacrifice or phage delay they do not. Furthermore, while negation and self-sacrifice fully block phage progeny production, phage delay does not. Therefore, phage population growth can continue, even if delayed given only phage delay. This means also, as noted, that simply the means of identifying bacteria displaying these different mechanisms can differ, with negation as well as avoidance most easily discovered via positive selection, self-sacrifice less easily discovered as it requires demonstration of bacterial death in response to phage infection but not phage survival, and phage delay requiring even more effort to identify, as associated mechanisms of this latter form of phage resistance simply give rise to less- but still-effective phage infections and indeed slower phage population growth.

There needs to be, in other words, convenient ways of distinguishing between the various extents to which bacterial mechanisms of phage resistance can impact both phages and bacteria, both from evolutionary and ecological perspectives. At least for me, distinguishing these into categories of avoidance, negation, self-sacrifice, and delay make the most sense (Table 18.1), and particularly so as they reflect qualitatively different extents or timing of blockage of phage gene expression, i.e., complete, early, potentially later, and perhaps only subtly, respectively.

20.1.2 Self-Sacrifice Acquisition

In the previous chapter (Chap. 19), I mentioned the difficulty of blocking phage infections via mutational loss of gene function toward phage infection negation. That is, while it may be possible to interfere with a phage's replication by denying them some bacterial gene function, it is a taller order to do so while at the same time also preventing that phage from killing the infected bacterium. With bacterial self-sacrifice, that concern of the phage still killing the bacterium is eliminated, not by some fancy means of preventing the phage from killing the bacterium, but instead because by definition, here, with bacterial self-sacrifice and associated abortive infections the bacterium *is* killed. Therefore, and unlike mechanisms of phage negation, it is easier to envisage a mechanism involving bacterial self-sacrifice being acquired as a consequence of mutational loss of bacterial gene function than it is for a mechanism of phage infection negation to be acquired by a similar route.

Though in principle self-sacrifice may be potentially easier to mutationally acquire as a phage-resistance mechanism than negation, in fact most bacterial mutations to phage resistance seem to result instead in avoidance, particularly through bacteria mutationally either losing or modifying phage adsorption receptors. Rather than self-sacrifice being acquired mutationally, instead self-sacrifice mechanisms usually are associated with abortive infections systems that are acquirable via horizontal gene transfer, e.g., as constituents of plasmids (Chopin et al. 2005) but also as can be encoded by prophages (Chap. 16). Perhaps consistent with the idea of abortive infection mechanisms potentially being more easily achieved than phage infection negation/restriction, a fairly large diversity of categories of phage-resistance mechanisms that result in bacterial death when expressed have been identified in bacteria, more it seems than categories of phage-resistance mechanisms not resulting in bacterial death when expressed against an infecting phage, e.g., (Labrie et al. 2010; McCallin and Oechslin 2019; Rostøl and Marraffini 2019; Hampton et al. 2020). Again, this apparently greater diversity of mechanisms makes some sense since it should be easier mechanistically to block a phage infection *without* also keeping the infected bacterium alive, than to block a phage infection while keeping the infected bacterium alive.

20.1.3 Negative Selection for Bacterial Self-Sacrifice

Contrasting positive selection, there is of course negative selection. Positive selection is typified, e.g., by selection for antibiotic resistant mutants in bacteria. This involves simply plating in the presence of an antibiotic, where the antibiotic-resistant bacteria are able to form colonies whereas the antibiotic-sensitive bacteria are not (ignoring here issues of clinical sensitivity versus clinical resistance). Similarly, in terms of bacterial phage-resistance mechanisms, bacteria displaying mechanisms of avoidance or negation of phage infections can be successfully plated in the presence of excesses of the free phages to which they are resistant. Contrast bacteria displaying abortive infection systems, or for that matter bacterial auxotrophic mutants. There the process of selection for so-displaying bacteria is somewhat more complicated, i.e., as involving negative selection.

In microbiology, negative selection typically is associated with replica plating. In replica plating, a bacterium, for example, is simultaneously plated using two different conditions, one permissive to colony formation and one not necessarily permissive. Further, the location of colonies on the two different plates is maintained. Thereby, the status of a bacterial colony growing under permissive conditions can be directly (and easily!) compared to its status under non-permissive conditions. An example could be the noted auxotrophic mutants. If a bacterium is unable to synthesize a specific, essential organic factor, such as an amino acid, then it may be able to form colonies in the presence of that factor but certainly not in its absence.

Spontaneously forming auxotrophic mutants thus would be able form colonies under permissive conditions such as when a specific amino acid is present but not under non-permissive conditions such as on minimal media. (There are 'tricks' for isolating auxotrophic mutations involving positive selection, e.g., such as by employing penicillin to select against replicating bacteria, but those here are beside the point.)

With negative selection for bacterial phage-resistance mechanisms, permissive conditions are phage free and the non-permissive conditions are in the presence of excesses of phages. If a bacterium is able to replicate in the absence of phages but not in their presence, but these bacteria otherwise are known to be resistant by some means to the phage in question, e.g., as determined by phage efficiency of plating, then those bacteria might be displaying an abortive infection system. This is because (yes, as repeatedly noted) with the action of abortive infection systems, both bacterium and infecting phage are killed. Thus, even though the bacterium can inactivate the infecting phage, thereby lowering that phage's efficiency of plating, it cannot survive doing so and so therefore cannot form a colony in the presence of excesses of that phage. In a sense, though, by inactivating the infecting phage, it sets up the environment for 'negative selection' in a microbiology laboratory sense, that is, by potentially creating conditions where its bacterial clone mates are able to survive phage free.

20.1.4 Kin Selection for Bacterial Self-Sacrifice

> This cellular self-immolation succeeds in aborting the infection… thereby presumably benefits the cell's siblings in the population by preventing the production of phages that could infect them.—Roger W. Hendrix and Sherwood R. Casjens (2008, pp. 88–89)

What good is a system of phage resistance in which the resisting bacterium dies anyway? We can put this into human terms. What good would be an immune response that brings a pathogen under control but in which you die anyway? Of course, the answer is no good at all, at least on a personal level, but from the perspective of herd immunity, a different conclusion may reached. The concept of herd immunity within modern human societies, however, is clearly group selectionist, i.e., in which people for example are vaccinated at least in part for the altruistic reason of protecting potentially complete strangers who have not been vaccinated against the same pathogen, perhaps because those strangers are too young or immunocompromised to also be successfully vaccinated. (This is a concept that, as I re-read these words in the Fall of 2021, I realize that most everyone who obtains their news from mainstream sources must now be familiar, i.e., as within a context of the current at the time COVID-19 pandemic.)

This latter perspective on herd immunity, with vaccination thereby potentially an altruistic act, ignores the concepts of both spatial structure and kin. That is, who are you most likely to protect by being vaccinated? These are the people you most often come into contact with, and you certainly don't come into contact with every possible person with equal likelihood, e.g., coming into contact with someone living in Los Angeles if you live in Madrid. Indeed, in many cases the people you come into contact with most are members of your own family. Thus, herd immunity is less altruistic as well as less group selectionist if it is accomplished to protect those who are most personally a part of your world and perhaps especially your kin.

As you might imagine with that primer, the same general idea may explain the evolution of abortive infection mechanisms in bacteria (Abedon 2012; Fukuyo et al. 2012; Berngruber et al. 2013; Iranzo et al. 2015; van Houte et al. 2016; Abedon 2017). The spatial structure would be seen in bacterial microcolonies, biofilms, or cellular arrangements, and the kinship would be the clonal relatedness of the bacteria making up especially bacterial microcolonies. Specifically—at least in principle—a single phage infecting a susceptible bacterial microbiology might kill off all of the bacteria making up that microcolony (Abedon 2012, 2017). If the initially adsorbed bacterium instead displays an abortive infection system, however, then that phage's population growth in the vicinity of bacterial kin (clone mates) will literally be aborted, thereby protecting the microcolony, though not the individual phage-infected bacterium.

Of course, the same result could be achieved via display instead of a negation/restriction mechanism. However, just as with our own immune systems, which possess many complementary mechanisms of anti-pathogen activity, so too can a single bacterium possess more than one mechanism of phage resistance, all with different spectra of anti-phage activities. That is, just because a bacterium may possess mechanisms of phage infection negation does not mean that it cannot also encode abortive infection systems. Furthermore, and as noted, given that abortive infection systems do not have to protect the actually infected bacterium, it is a reasonable assumption that self-sacrifice may be easier to evolve than mechanisms of negation. But then again, only incompletely preventing phage production might be easier still to achieve.

20.2 Delay: Slowing Down Phage Population Growth

As a slightly different perspective from above, phage resistance by bacteria can involve (1) both phages and bacteria surviving (a consequence of 'avoidance', particularly bacteria preventing phage adsorption), (2) bacteria surviving but phages not surviving ('negation' but also blocks on phage genome uptake), or (3) both the infected bacterium and the infecting phage not surviving ('self-sacrifice'; Table 18.1, Chap. 18). This leaves, however, a potential fourth category in which bacteria die but phages do not, with phages instead experiencing bacterium-forced but still incomplete impediments to their population growth. This combination of features could still be helpful to bacteria, and in fact could very well represent an even easier resistance mechanism to evolve than bacterial self-sacrifice since not only do phage-infected bacteria not need to stay alive, but phage infections need not be completely stopped from being productive. Such a mechanism I have described, as noted, as 'delay', or 'phage delay', since by slowing down phage population growth they can delay though not completely eliminate the occurrence of subsequent phage infections of bacteria (Abedon 2017).

20.2.1 Temporary Protection of Microcolonies

Though potentially easier for bacteria to evolve, the usefulness of mechanisms of phage delay to bacteria likely is lower as well. Specifically, what good is not even killing the infecting phage? The answer to this might be a bit subtle. First, we have to assume a utility to delaying phage propagation versus not delaying phage propagation (versus fully blocking phage propagation). If such a utility exists, then reducing phage burst sizes, extending phage latent periods, and also slowing the rate at which phages encounter bacteria to adsorb (versus preventing encounters altogether) could have utility to the bacterium. I suggest that this utility again may be found within a context of bacterial microcolonies and kin selection.

Bacterial microcolonies, like biofilms, display life cycles that consist of an initiation step, a growth step, and then a dissemination step (e.g., attachment to a surface, binary fission without substantial spatial separation of daughter cells, and then formation and release of motile bacteria; Fig. 20.1). The ultimate goal of these life cycles, both ecologically and evolutionarily (the latter, i.e., in terms of bacterial fitness) is formation of disseminating bacteria that are able to disperse to found new microcolonies. Disseminating bacteria, however, are not produced immediately after the microcolony initiation step. Instead, bacterial dissemination tends to be deferred to some extent while microcolonies grow and mature, presumably toward increasing the number of disseminating bacteria that may be released. Were a phage to gain access to a microcolony, even if that phage could still propagate in association with that microcolony, but was slowed in that propagation, then the potential for the microcolony to still disseminate bacteria progeny may be temporarily preserved; hence the title of the article (Abedon 2017): "Phage 'delay' towards enhancing bacterial escape from biofilms: a more comprehensive way of viewing resistance to bacteriophages."

An alternative perspective could be that delaying phage population growth can reduce the size that a microcolony needs to achieve in order for rates of bacterial replication to exceed rates of bacterial losses due to phage infection. Specifically, Eriksen et al. (2018) found that *E. coli* microcolonies could persist in the face of phage attack so long as microcolonies were large enough, i.e., so that surface-to-volume ratios were smaller while the number of interior bacteria were larger in absolute terms. If rates of phage replication are slowed, including perhaps particularly in terms of rates of virion penetration into biofilms, then rates of bacterial losses should be lower, suggesting that smaller microcolonies could persist in the presence of phages than would be the case were mechanisms of phage delay not present.

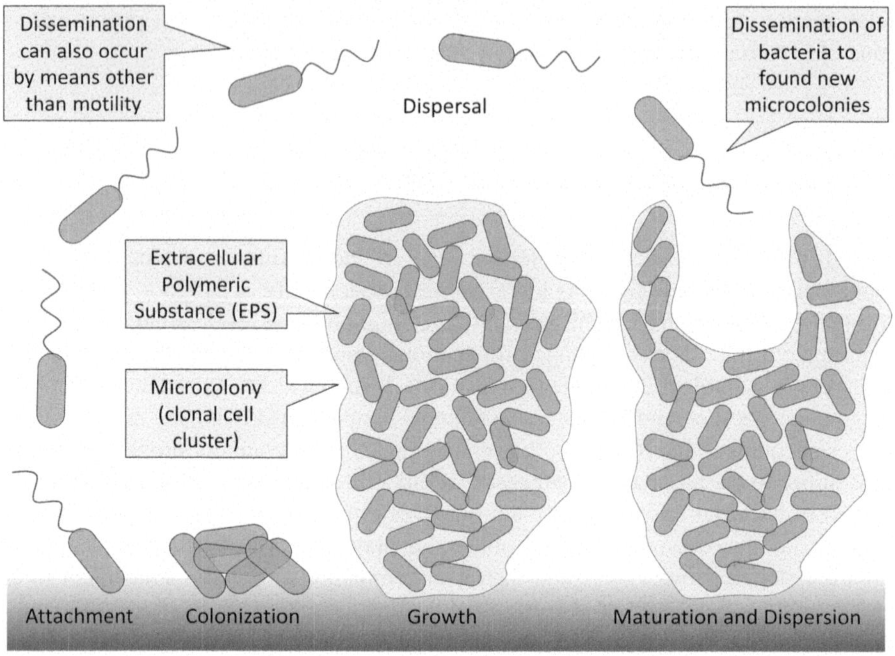

Fig. 20.1 Life cycle of a bacterial microcolony. Equivalent illustrations are often labeled instead as life cycles (or equivalent) of a bacterial biofilm, but there is no difference in terms of the underlying biology

20.2.2 A Discovery Challenge

While bacterial mechanisms of phage delay, as mechanisms of phage resistance, could very well be both easier to evolve and even somewhat common (Abedon 2017, and Sect. 20.2.3), they also likely are much more difficult to detect. Specifically, phage infection avoidance in the form of encounter blocks or adsorption resistance is indicated by bacterial survival along with phage failure to adsorb. Phage infection negation, as well as superinfection exclusion, are indicated by bacterial survival occurring in combination with phage inactivation. And bacterial self-sacrifice is indicated by a combination of both phage and bacterial death. Phage delay, on the other hand, would be detected by looking for bacterial death without necessarily efficiency of plating also being low (though it could be), but more definitively by finding that phage single-step growth characteristics are less 'robust'. That is, one would be looking for the noted reduced burst sizes or extended latent periods, though reduced rates of phage adsorption to bacteria could have a similar impact.

For determining single-step growth characteristics see (Hyman and Abedon 2009; Kropinski 2018). For a higher-throughput alternative to detecting substantial reductions in the robustness of single-step growth characteristics, see Danis-Wlodarczyk et al. (2021).

If these factors that could delay phage population growth tend to be present particularly under the conditions observed in association with infection of bacteria making up microcolonies, then these characteristics may be technically difficult to measure in the laboratory. Similarly, if obvious phage resistance systems or resistance mutations are not present, then it also may be difficult to determine exactly what to compare these phage growth parameters to in deciding whether phage delay is present. In other words, what does it mean when a wild-type phage infecting a wild-type bacterium displays a relatively small burst size? That this is just how things are, or instead that this is an example of phage delay? Not surprisingly, therefore, phage delay clearly is not the first thing that necessarily comes to mind when bacterial resistance mechanisms are being considered. Nonetheless, that does not mean that phage delay does not exist, as discussed in the following subsection (Sect. 20.2.3).

20.2.3 Some Examples of Delay

With bacteriophage resistance (BREX; Chap. 19, Sect. 19.2.3), phage infections are not always fully prevented from forming and releasing virions. The timing of phage release also appears to be delayed, though not all adsorbed bacteria appear to be productively infected (Goldfarb et al. 2015). The implication therefore is that BREX can serve not only as a mechanism of phage negation but also as a mechanism that slows phage population growth, i.e., delay. Thus, another issue of assessing the presence of mechanisms of delay is that they may present—depending on the phage, the overall state of the infected bacterium, or just chance—as mechanisms that we would otherwise tend to classify as examples of negation or self-sacrifice but which simply are not always 100% effective in preventing virion production and release.

CRISPR-Cas systems also may to some degree effect delay rather than necessarily phage infection negation or bacterial self-sacrifice. As also considered in Chap. 19, Sect. 19.2.4, Deveau et al. (2008) observed that some fraction of CRISPR-Cas-positive bacteria were still able to produce virions, though perhaps with reduced burst sizes. In one instance, the latent period was also extended, from about 40 min to about 55. Thus, again, delay could be viewed here as one consequence of a failure to fully effect negation and/or self-sacrifice.

In Abedon (2017), I also suggest that bacteria and bacterial microcolonies might possess numerous mechanisms by which delay is imposed on phages but which do not strictly involve bacterial adaptations. For example is the reduced ability of less rapidly replicating bacteria to support more vigorous phage infections (Hadas et al. 1997) or the potential for phage adsorption to bacteria to be interfered with as a result of 'sorptive scavenging' by already phage-infected bacteria (Abedon 2020). Included among such potentially useful bacterial 'non-adaptations', or at least pleiotropies, may also be biofilm matrix material (Chap. 19, Sect. 19.1.3), i.e., extracellular polymeric substances, that is to the extent that these serve to delay phage virion movement through bacterial microcolonies but otherwise exist for reasons other than interfering with phage propagation.

Alternatively is the case of satellite phages (Chaps. 4 and 13, Sects. 4.5 and 13.3), as the quotation at the start of this chapter is alluding to. These are DNA sequences that parasitize productively infecting helper phages, the latter which are simply normal phages that can help satellite phages reproduce. Satellite phage propagation can impose severe reductions on helper phage infection vigor—including ten- and even 100-fold reductions in burst size—to the point where helper phage plaque formation can be substantially compromised (Lindsay et al. 1998; Ram et al. 2014; Tormo-Más et al. 2010). Indeed, in some cases helper phage replication can be reduced to zero, essentially with the satellite phage thereby serving as an abortive infection system (Barth et al. 2020). The resulting costs also can select for satellite phage-resistant helper phage mutants (Frigols et al. 2015). Carriage by bacterial hosts of satellite phages is probably motivated not so much by natural selection toward delay in helper phage population growth, however, as by natural selection acting directly on the satellite phage, that is, toward satellite phage propagation. Nonetheless, the presence of satellite phages, as essentially prophage equivalents found in bacterial genomes (Chap. 15), would seem to serve as a means of delaying helper phage population growth, though mostly without completely blocking helper phage production.

Lastly are bacteria-produced extracellularly excreted substances that have anti-phage virion activities (Kronheim et al. 2018). Depending on their *in situ* concentrations along with the rapidity of their action, these substances might act to reduce a phage's effective burst size (Chan and Abedon 2012; Abedon 2022), that is, the number phages from a burst which succeed in initiating infections of their own. These substances would be effecting a delay on phage population growth to the extent that they are not inactivating extracellular phages altogether. Other relevant issues include whether inactivated phages, depending on the mechanisms or severity of inactivation, are prevented from adsorbing (avoidance), prevented from killing bacteria if they do adsorb (negation), or are able to kill bacteria but not produce virions, thereby resulting in a form of bacterial self-sacrifice. Indeed, it is conceivable that phage virions could also be impacted intracellularly by bacterial anti-virion mechanisms after their maturation but prior to their release.

20.3 Considerations of Costs and Further Considerations of Ecology

In the following chapter (Chap. 21), we return especially to considerations of avoidance, i.e., in the form of modification of bacterial surface molecules that can serve as receptors for phage adsorption. This is because the underlying mutations to phage resistance that are most commonly seen in the laboratory seem to be of this type. Therefore, analyses of reductions in bacterial fitness due to laboratory evolution of phage resistance in most cases are likely observing the impact of bacterial loss of phage receptors. It is that impact on bacterial fitness as can be associated with bacterial mutation to phage resistance that the yet following chapter considers (Chap. 22).

References

Abedon ST (2012) Spatial vulnerability: bacterial arrangements, microcolonies, and biofilms as responses to low rather than high phage densities. Viruses 4:663–687

Abedon ST (2017) Phage "delay" towards enhancing bacterial escape from biofilms: a more comprehensive way of viewing resistance to bacteriophages. AIMS Microbiol 3:186–226

Abedon ST (2020) Phage-phage, phage-bacteria, and phage-environment communication. In: Witzany G (ed) Biocommunication of phages. Springer, Cham, pp 23–70

Abedon ST (2022) Evolution of bacteriophage latent period length. In: Dickins TE, Dickens BJA (eds) Evolutionary biology: new perspectives on its development. Springer, Cham

Abedon ST, Yin J (2009) Bacteriophage plaques: theory and analysis. Meth Mol Biol 501:161–174

Barth ZK, Silvas TV, Angermeyer A, Seed KD (2020) Genome replication dynamics of a bacteriophage and its satellite reveal strategies for parasitism and viral restriction. Nucl Acids Res 48:249–263

Berngruber TW, Lion S, Gandon S (2013) Evolution of suicide as a defence strategy against pathogens in a spatially structured environment. Ecol Lett 16:446–453

Chan BK, Abedon ST (2012) Bacteriophage adaptation, with particular attention to issues of phage host range. In: Quiberoni A, Reinheimer J (eds) Bacteriophages in dairy processing. Nova Science Publishers, Hauppauge, pp 25–52

Chopin MC, Chopin A, Bidnenko E (2005) Phage abortive infection in lactococci: variations on a theme. Curr Opin Mirobiol 8:473–479

Danis-Wlodarczyk KM, Cai A, Chen A, Gittrich M, Sullivan MB, Wozniak DJ, Abedon ST (2021) Friends or foes? Rapid determination of dissimilar colistin and ciprofloxacin antagonism of *Pseudomonas aeruginosa* phages. Pharmaceuticals (Basel) 14:1162

Deveau H, Barrangou R, Garneau JE, Labonté J, Fremaux C, Boyaval P, Romero DA, Horvath P, Moineau S (2008) Phage response to CRISPR-encoded resistance in *Streptococcus thermophilus*. J Bacteriol 190:1390–1400

Eriksen RS, Svenningsen SL, Sneppen K, Mitarai N (2018) A growing microcolony can survive and support persistent propagation of virulent phages. Proc Natl Acad Sci U S A 115:337–342

Frigols B, Quiles-Puchalt N, Mir-Sanchis I, Donderis J, Elena SF, Buckling A, Novick RP, Marina A, Penades JR (2015) Virus satellites drive viral evolution and ecology. PLoS Genet 11:e1005609

Fukuyo M, Sasaki A, Kobayashi I (2012) Success of a suicidal defense strategy against infection in a structured habitat. Sci Rep 2:238

Goldfarb T, Sberro H, Weinstock E, Cohen O, Doron S, Charpak-Amikam Y, Afik S, Ofir G, Sorek R (2015) BREX is a novel phage resistance system widespread in microbial genomes. EMBO J 34:169–183

Hadas H, Einav M, Fishov I, Zaritsky A (1997) Bacteriophage T4 development depends on the physiology of its host *Escherichia coli*. Microbiology 143:179–185

Hampton HG, Watson BNJ, Fineran PC (2020) The arms race between bacteria and their phage foes. Nature (London) 577:327–336

Hendrix RW, Casjens SR (2008) The role of bacteriophages in the generation and spread of bacterial pathogens. In: Hensel M, Schmidt H (eds) Horizontal gene transfer in the evolution of pathogenesis. Cambridge University Press, Cambridge, pp 79–112 van Houte S, Buckling A, Westra ER (2016) Evolutionary ecology of prokaryotic immune mechanisms. Microbiol Mol Biol Rev. 80:745–763

Hyman P, Abedon ST (2009) Practical methods for determining phage growth parameters. Meth Mol Biol 501:175–202

Iranzo J, Lobkovsky AE, Wolf YI, Koonin EV (2015) Immunity, suicide or both? Ecological determinants for the combined evolution of anti-pathogen defense systems. BMC Evol Biol 15: 43

Kronheim S, Daniel-Ivad M, Duan Z, Hwang S, Wong AI, Mantel I, Nodwell JR, Maxwell KL (2018) A chemical defence against phage infection. Nature (London) 564:283–286

Kropinski AM (2018) Practical advice on the one-step growth curve. Meth Mol Biol 1681:41–47

Labrie SJ, Samson JE, Moineau S (2010) Bacteriophage resistance mechanisms. Nat Rev Microbiol 8:317–327

Lindsay JA, Ruzin A, Ross HF, Kurepina N, Novick RP (1998) The gene for toxic shock syndrome is carried by a family of mobile pathogenicity islands in *Staphylococcus aureus*. Mol Microbiol 29:527–543

McCallin S, Oechslin F (2019) Bacterial resistance to phage and its impact on clinical therapy. In: Górski A, Miedzybrodzki R, Borysowski J (eds) Phage therapy: a practical approach. Springer, Cham, pp 59–88

Penadés JR, Christie GE (2015) The phage-inducible chromosomal islands: a family of highly evolved molecular parasites. Annu Rev. Virol 2:181–201

Ram G, Chen J, Ross HF, Novick RP (2014) Precisely modulated pathogenicity island interference with late phage gene transcription. Proc Natl Acad Sci U S A 111:14536–14541

Rostøl JT, Marraffini L (2019) (Ph)ighting phages: how bacteria resist their parasites. Cell Host Microbe 25:184–194

Tormo-Más MA, Mir I, Shrestha A, Tallent SM, Campoy S, Lasa I, Barbe J, Novick RP, Christie GE, Penades JR (2010) Moonlighting bacteriophage proteins derepress staphylococcal pathogenicity islands. Nature (London) 465:779–782

Chapter 21
Bacterial Mutation to Phage Resistance

Contents

Abstract Bacterial acquisition of resistance, resistance to phages as well as other antagonistic agents, can occur by two distinct mechanisms, mutation and horizontal gene transfer. Acquisition of resistance via mutation can be a much more immediate response, however, as it does not require that bacteria come into contact with DNA encoding resistance systems as sourced from other bacteria. Resistance mutations also tend to act by modifying the targets of antagonistic agents rather than targeting the antagonistic agent themselves. As targets of antagonistic agents often are essential or at least consistently useful to the bacterium, their genes tend to be encoded on the bacterial chromosome, hence chromosomal resistance. Given the relative simplicity by which resistance is acquired by mutation, we can easily measure the rates at which mutation to resistance can occur in the laboratory, particularly if the resistance should result in a prevention of bacterial killing by the antagonistic agent. It turns out, though, that a majority of routes toward such avoidance of bacterial killing seem to be achieved by preventing phage adsorption. This chapter considers both why it is that this avoidance of phage infections by bacteria should dominate as well as rates of bacterial mutation to phage resistance generally.

> When a pure bacterial culture is attacked by a bacterial virus, the culture will clear after a few hours due to destruction of the sensitive cells by the virus. However, after further incubation for a few hours, or sometimes days, the culture will often become turbid again, due to the growth of a bacterial variant which is resistant to the action of the virus. This variant can be isolated and freed from the virus and will in many cases retain its resistance to the action of

S. T. Abedon, *Bacteriophages as Drivers of Evolution*,
https://doi.org/10.1007/978-3-030-94309-7_21

> the virus even if subcultured through many generations in the absence of the virus. While the sensitive strain adsorbed the virus readily, the resistant variant will generally not show any affinity to it.—Salvador E. Luria and Max Delbrück (1943, p. 491)

In this chapter, we focus on chromosomally encoded mutations in bacteria that give rise to phage resistance. Most mutations that have been studied interfere in some manner with phage infection activity by preventing phage adsorption (Korona and Levin 1993), though interference with phage functions that operate after the adsorption step are also possible. These mutations generally are associated with: (1) Loss of bacterial gene-product function in combination with a loss of phage ability to interact with those gene products. For example, changes in gene-product shape, modification of gene-product chemistry, truncation of gene-product length, or outright gene deletion. Less common (2) can be loss of bacteria ability to specifically interact with a phage but in combination with retention of gene-product function, e.g., modification of a key amino acid that a phage interacts without completely interfering with gene-product activity (Zheng et al. 2004; Seed et al. 2014). Alternatively, (3) is regulatory modification of gene expression such as down-regulation of expression of phage receptors (Seed et al. 2014). In addition, (4) is up-regulation of the production by bacteria of extracellular polymers (van Houte et al. 2016). Such mutational changes are rather than substantial gains in gene function. Gains in gene function—as typically associated with various phage-resistance systems such as restriction-modification systems, CRISPR-Cas systems, and abortive infection systems—instead tend to be acquired via horizontal gene transfer (Koonin et al. 2020).

Considered in particular in this chapter are targets and rates of bacterial chromosomal mutation to phage resistance. In the following chapter (Chap. 22) pleiotropic costs of bacterial mutation to phage resistance are considered.

21.1 A Plethora of Possible Targets of Mutation?

> . . .mutations that sabotage phage production can occur at any point in the growth cycle where the phage depends on interactions with host cell components, and many such examples are known.—Roger W. Hendrix and Sherwood R. Casjens (2008, p. 88)

We can distinguish bacterial mutations that can impact phage resistance in terms of when during the phage life cycle that interference occurs. This can be, for example, during the extracellular search (thereby affecting free phages) versus intracellularly (thereby affecting the functioning of the virocell; Chap. 4). Thus, a phage diffuses through the extracellular environment, encounters the bacterial surface, attaches to that surface, translocates its genome into the bacterial cytoplasm, expresses its genes, goes through an eclipse period, during the post-eclipse period accumulates progeny virions intracellular (at least for lytic phages during productive infections), and then virions are released. Literally at any of these steps a bacterium could interfere with the progression of a phage's life cycle. But are they all equivalently likely?

21.1.1 Why Everything but Mutation to Avoidance Is Rare

One way of distinguishing among mutations leading to phage resistance is in terms of their ability to lead to avoidance, negation, self-sacrifice, or delay (Chap. 18). As discussed in Chap. 19, especially avoidance and negation may be subject to positive selection acting on bacteria, but that is not the case for either self-sacrifice or delay (Chap. 20). As most assays for bacterial resistance to phages rely on positive selection, i.e., increasing resistance-allele prevalence while bacteria are growing in the presence of an excess of phages, this means that bacterial mutations to phage resistance that will tend to be identified will be those giving rise to one of these two categories of mechanisms, avoidance or negation.

The key to achieving positive selection for phage resistance is to block the progression of a phage infection prior to their production of bacterium-inhibiting phage gene products. These phage gene products likely are common during lytic cycles. From Wagemans and Lavigne (2012), p. 120, "Bacterial viruses… have evolved novel proteins that either inhibit or co-opt critical bacterial metabolic processes to ensure proper viral development." I submit that successfully interfering with phage expression of *all* of its genes that have antibacterial activities should be easier to achieve the fewer phage genes that are expressed. As phage avoidance by bacteria operates extracellularly, before any phage gene expression has occurred, the result is an absolute minimum of phage genes that are expressed. Phage-infection negation, by contrast, acts intracellularly. Therefore, for negation to be achieved mutationally, i.e., in terms of loss or modification of a bacterial gene product that a phage requires for infection, then that bacterial gene product presumably would need to have been required for phage gene expression early on during infections, e.g., such as being involved in early phage gene transcription or subsequent phage mRNA translation. As such functions should be basic to bacterial functioning as well, we can speculate that it is difficult (meaning, probabilistically less likely) for bacteria to mutationally acquire changes in genes that will tend to interfere with earlier phage gene expression without also substantially interfering with bacterial functioning as well.

Indeed, presumably phages have been selected for not requiring bacterial gene products that are easily mutated to phage resistance without those mutations also substantially impacting bacterial functioning. Of course, we can suppose that phages should be selected to avoid using bacterial gene functions that are easily mutated to phage resistance generally. Logically, however, bacterial macromolecules that function on the outside of bacteria likely can be more easily dispensed with, at least in the laboratory (Chaps. 22 and 25), than core macromolecules involved in bacterial gene expression. It therefore should be no surprise that the vast majority of bacterial mutations to phage resistance that have been identified following positive selection have been found to effect phage avoidance, and indeed have been found especially to block phage attachment to the surface of bacteria, e.g., see the quotation at the start of this chapter. This would be so particularly to the extent that losing surface molecules is easier to achieve by bacteria, perhaps particularly under laboratory conditions, than up-regulating extracellular polymer production or interfering with phage-genome translocation into the bacterial cytoplasm.

21.1.2 Blocking Lysin Action toward Self-Sacrifice?

As noted previously (Chaps. 19 and 20), we might still expect that mutation to self-sacrifice (or delay) should be easier to achieve than mutation to phage negation—though not necessarily easier than mutation to phage avoidance—because it should be easier to interfere with the progression of a phage infection while not also saving the infected bacterium, that is, while avoiding bacterial self-sacrifice. Nonetheless, and also as noted, positive selection for self-sacrifice phenotypes is not readily achieved such as in the laboratory. Thus, even if mutations to avoidance could be separated from bacterial mutations affecting phage functioning after phage-genome uptake, we still would not expect to see many mutations to bacterial self-sacrifice in the laboratory following standard selection experiments for phage resistance simply because those bacteria by definition would not survive phage infection. An important example of the latter could be bacterial mutation to modification of the targets of phage endolysins. Specifically, from the bacterium's perspective, why should they bother modifying these targets? In this section I attempt to answer that question.

Perhaps the most discussed possible mutation to result in phage resistance, other than those altering phage receptor molecules, would be ones that interfere with the ability of phages to digest bacterial cell walls (peptidoglycan). This is of interest because of the use or proposed use of phage-encoded endolysins as so-called "Enzybiotics", a.k.a., lysins, e.g., (Shen et al. 2012; Heselpoth et al. 2017; Danis-Wlodarczyk et al. 2021). These are enzymes that in a concentrated form can be added to bacteria to lyse those cells "from without", though without an associated phage infection (Abedon 2011). A common argument is that bacteria in fact have difficulty evolving resistance to these enzymes, an absence which is easily demonstrated (LeMieux 2020). In fact, however, under the normal circumstances of phage infection there may be little incentive for bacteria to evolve such abortive infection mechanisms, as posited in the previous paragraph. This is because by the time a phage has reached a point that it is digesting the bacterial cell wall, not only should the bacterium already be quite dead, including due to the opening of phage holin proteins in the bacterial plasma membrane (Young 1992; Young and Wang 2006; Dennehy and Abedon 2021), but virions should already have been produced even if they are still trapped, at least temporarily, within the unlysed bacterium. The alternative but not mutually exclusive explanation is that the target of lysins, i.e., cell walls, simply are difficult to mutationally modify to resistance (LeMieux 2020). In light of the likely difficulties and perhaps also costs of altering the peptidoglycan structure of a bacterium's cell wall, along with the potential to instead interfere with phages at so many other stages of infection, this may just be one area—modifying cell wall structure toward achieving phage resistance—where evolvability is simply not highly useful to bacteria.

Still, the fact that lysins tend to be highly specific to different bacterial types implies that some degree of divergent evolution in terms of cell wall composition must have occurred between bacterial types. It is also possible that this diversity is at least useful with regard to resistance to phages as it could serve to further limit the productive host range of these phages. That is, especially should a phage population give rise to a host-range mutation which allows it to adsorb a different bacterial species, then those infections still may fail to be productive, or at least see their lysis substantially delayed, due to chemical differences in bacterial peptidoglycan in combination with a presumed emphasis in the evolution of most lytic phages on encoding of endolysins that effect efficient lysis rather than endolysins that have a potential to digest the cell walls of a wide-range of bacterial species. Though not involving the cell wall, Wais and Goldberg (1973) found that the normally *E. coli* phage T4 when forced to infect spheroplasts of *Aerobacter aerogenes* in fact was not able to readily end infections, suggesting at the very least that releasing virions in at least this case is a host-specific activity. Note that it is also possible that the diversity seen in cell wall structures between bacterial types could be useful toward avoidance of phages that normally target different bacterial species since phages not only need to span bacterial cell walls toward virion release but also need to span bacterial cell walls toward uptake of their genomes into the bacterial cytoplasm (Danis-Wlodarczyk et al. 2021 and see also Chap. 19, Sect. 19.1.2).

21.2 Rates of Mutation to Resistance

In addition to its pleiotropic costs (Chap. 22), a key consideration in the acquisition of phage resistance by bacteria is how readily that resistance can occur. Generally, and as noted, we can assume that mutations to loss of function (e.g., loss of phage receptors) will occur more readily than mutations to gain of function (e.g., gain of greater capsular material), which in turn may occur with higher likelihood over time than acquisition of new phage resistance systems via horizontal gene transfer. In this section we take a look especially at rates of mutation to loss of function.

21.2.1 Rates Versus Frequencies

It is important to distinguish rates of bacterial mutation to phage resistance from the frequency of bacterial mutants that are phage resistant as found within a given culture. This is because the latter (frequency) is a function of a combination of mutation rates, timing of mutant origin, mutant growth rate, and the size of a culture, with smaller sizes resulting in greater stochastic effects, that is, in greater variance. Nevertheless, mutant frequencies are much easier to measure, even if resulting numbers are expected to be subject to a great deal of "Fluctuation" (Luria and Delbrück 1943). Mutation frequencies as found within cultures certainly, as noted, are a function at least in part of mutation rates. It is just that it is important when doing experiments such as these that one begins with new cultures as initiated as close as possible to from individual bacteria for each round of analysis, since for a given culture the frequency of bacterial mutants could be more or less constant from sample to sample. Also from Luria and Delbrück, p. 504:

> It will be seen that in every experiment the fluctuation of the numbers of resistant bacteria is tremendously higher than could be accounted for by the sampling errors, in striking contrast to the results of plating from the same culture. . .

In addition, a single mutant genotype may dominate a culture, should purification be attempted, resulting in repeated isolation of bacteria possessing the same mutation if one doesn't start with new cultures. Both of these problems—excessive consistency in mutant frequency and excessive isolation of the same mutant genotypes—are avoided when starting with not just new but especially newly bottlenecked cultures, i.e., as initiated from a single colony. That is, overnight cultures should not be reused for biological repeats (versus technical repeats) when studying mutations.

The importance of rates of bacterial mutation to phage resistance is that this number describes the potential for a bacterial population to *evolutionarily* escape decimation upon exposure to a phage to which it is susceptible, contrasting ecologically escaping. If the mutation rate is high, then a relatively large fraction of the initial bacterial population as present upon phage exposure likely will survive. If the mutation rate is low, then fewer bacteria will be likely to survive. Indeed, if the rate of bacterial mutation to phage resistance is zero, then no physically phage-reachable bacteria will survive. Perhaps of even greater relevance, the rate of bacterial mutation to phage resistance will determine what bacterial population size will be necessary to assure with reasonable likelihood that the population will contain at least one bacterium that is phage resistant. This number is roughly the reciprocal of the mutation rate. Thus, if the rate of mutation to phage resistance is 10^{-7} per cell division, then a population of 10^7 would be expected to contain roughly one phage-resistant bacterium. If the bacterial population size instead were 10^6, then on average that population would be expected to contain only 0.1 phage-resistant mutants, which therefore would be relatively unlikely to exceed zero.

21.2.2 *Measurements*

In practice, what constitutes high versus low bacterial mutation rates to phage resistance, or at least high versus low mutation frequencies? Demerec and Fano (1945) calculated mutation *rates* for *E. coli* B to resistance to phage T3 that ranged from about 2×10^{-8} up to about 3×10^{-7} mutations per bacterium per bacterial generation. This worked out to an average of 7.6×10^{-8} mutations per bacterium per bacterial generation ($n = 7$; $\pm 7.3 \times 10^{-8}$ standard deviation), which has a reciprocal of roughly 10^7, i.e., about ten million. Phage T3 thus would be expected to wipe out a population of 10^6 *E. coli* B bacteria with no resistant survivors, at least on average, since this is less than 10^7 bacteria and the reciprocal numbers from their mutation rate data ranged from 3×10^6 to 6×10^7. Interestingly, their data seem to suggest that prior mutations to especially phage T1 but not to phage T6 seem to have the effect of somewhat *increasing* rates of mutation to T3 resistance, i.e., to $1.9 \times 10^{-7} \pm 5.9 \times 10^{-8}$ ($n = 3$) from the above-indicated 7.6×10^{-8}, which is about a two-fold increase.

Demerec and Fano (1945) then extended these results to phages T1, T4, T5, T6, and T7. Note that phage T2 is missing from that list because *E. coli* B possesses somewhat odder mutation properties to phage T2 resistance than to the other T phages, as involving two steps rather than one (Lenski 1984). For those phages except phage T2, mutation rates were measured as $1.7 \times 10^{-8} \pm 1.5 \times 10^{-8}$ for phage T1 ($n = 10$; reciprocal of 5.9×10^7), $7.6 \times 10^{-8} \pm 7.3 \times 10^{-8}$ for phage T3 ($n = 7$; reciprocal of 1.3×10^7, i.e., as summarized as well in the previous paragraph), $1.6 \times 10^{-7} \pm 1.9 \times 10^{-7}$ for phage T4 ($n = 4$; reciprocal of 6.3×10^6), $2.3 \times 10^{-8} \pm 4.2 \times 10^{-8}$ for phage T5 ($n = 6$; reciprocal of 4.3×10^7), $2.8 \times 10^{-7} \pm 3.5 \times 10^{-7}$ for phage T6 ($n = 5$; reciprocal of 3.6×10^6), and $1.3 \times 10^{-7} \pm 8.7 \times 10^{-8}$ for phage T7 ($n = 6$; reciprocal of 7.7×10^6). This corresponds to a minimum average value of 1.7×10^{-8} (as measured for phage T1; keeping in mind that for phage T2 this would be even lower) and a maximum average value of 2.8×10^{-7} (for phage T6). Individual experiments, though, saw as high as 10^{-6} mutations per bacterium per bacterial generation. All of the mutation rates were based on two different approaches to calculation by Demerec and Fano, though calculations were not always duplicated; thus, the numbers summarized above should be viewed as guides only.

Luria and Delbrück (1943), both predating and as well as motivating the Demerec and Fano (1945) study, measured mutation rates ranging from 10^{-8} to 4×10^{-8} also working with *E. coli* B but with phage α, which appears to be phage T1 (Abedon 2000). In other systems, Avrani et al. (2011) measured mutation frequencies to phage resistance in the range of 5×10^{-6} in *Prochlorococcus*, which is nearly the same frequency seen for *E. coli* BW25113 and resistance to phage U136B (Burmeister et al. 2020). Using phage ϕ6 and *Pseudomonas syringae*, Lythgoe and Chao (2003) estimated mutation rates to resistance of between 5.7×10^{-7} and 1.7×10^{-6}. Alternatively, the frequency of mutation to resistance to phage DMS3*vir* by *P. aeruginosa* strain P14 was found to be close to 10^{-4} (Chevallereau et al. 2019). Also in *P. aeruginosa* strain PAO1, Wright et al. (2018) measured mutation frequencies in the range of 10^{-3} against phage PA5P2, though of the phages they tested this was exceptionally high. Mutation rates thus can vary between different phage-host systems by many orders of magnitude, though mostly do not seem to be nearly as high as that observed with phages DMS3*vir* or PA5P2 of *P. aeruginosa*.

21.3 A Wee Bit of Advice

Note that the previous discussion points directly to the need for sufficient numbers of bacteria if phage-resistant mutations are going to occur with reasonable likeliness within a bacterial population. Thus, it is important to always make sure when assaying for phage-resistant mutants that one starts with sufficiently large populations of bacteria. This can be particularly an issue when doing assays in 96-well microtiter plates with their typical 100 or 200 μl usable volumes. If you start even with a bacterial concentration of 10^8 colony-forming units (CFUs)/ml, then an, e.g., 100 μl volume will contain only 10^7 CFUs. This can be inadequate if mutation rates to phage resistance are equal to 10^{-8} mutations per bacterium per bacterial generation. Researchers, reviewers, and readers thus beware.

Another issue is the definition of colony-forming unit, which is not necessarily equivalent to a single bacterium. Thus, if CFUs consist, e.g., of ten individual bacteria, then mutation frequencies or rates measured in CFU units could be tenfold higher than would be the case if the measurement instead were truly per bacterium. That is, mutations per bacterium are not identical units to mutations per CFU, unless CFUs are individual bacteria.

There actually is a broader issue here, and that is that it is important to think through your protocols before just doing them, or at least to think them through during debugging, as what works for one system is not necessarily fully translatable to another. In addition, it is possible that whoever you may be copying also had not thought through their protocols along with their impact on whatever results they have reported, even given peer review. A second issue is that negative results are inherently problematic as 'proof' of something, and using too few bacteria when assessing mutation rates is a sure way to achieve false-negative results. That is, always beware the possibility that (p. 303, Ruse 1999):

> A negative result comes about because you have not done the experiment properly, or because you have chosen the wrong organism to which your model cannot apply, or because your equipment is dirty or your statistics inappropriate, or some such thing.

Remember, rigorous validation is how science is properly done, and that doesn't mean just rigorously validating the work done by others. To paraphrase Allchin (1999), we must be confident in our own conclusions, particularly that they are not "ambiguous or inconclusive" (p. 304). I've published a number of chapters and articles on this subject of the (obvious?) importance of scientific rigor, particularly as relevant to the phage use as antibacterial agents (Abedon 2012, 2016, 2017, 2018; Abedon et al. 2021).

References

Abedon ST (2000) The murky origin of Snow White and her T-even dwarfs. Genetics 155:481–486

Abedon ST (2011) Lysis from without. Bacteriophage 1:46–49

Abedon ST (2012) Phage therapy best practices. In: Hyman P, Abedon ST (eds) Bacteriophages in health and disease. CABI Press, Wallingford, pp 256–272
Abedon ST (2016) Phage therapy dosing: the problem(s) with multiplicity of infection (MOI). Bacteriophage 6:e1220348
Abedon ST (2017) Information phage therapy research should report. Pharmaceuticals (Basel) 10: 43
Abedon ST (2018) Phage therapy: various perspectives on how to improve the art. Meth Mol Biol 1734:113–127
Abedon ST, Danis-Wlodarczyk KM, Wozniak DJ, Sullivan MB (2021) Improving phage-biofilm in vitro experimentation. Viruses 13:1175
Allchin D (1999) Negative results as positive knowledge, and zeroing in on significant problems. Mar Ecol Prog Ser 191:303–305
Avrani S, Wurtzel O, Sharon I, Sorek R, Lindell D (2011) Genomic island variability facilitates Prochlorococcus-virus coexistence. Nature (London) 474:604–608
Burmeister AR, Fortier A, Roush C, Lessing AJ, Bender RG, Barahman R, Grant R, Chan BK, Turner PE (2020) Pleiotropy complicates a trade-off between phage resistance and antibiotic resistance. Proc Natl Acad Sci USA 117:11207–11216
Chevallereau A, Meaden S, van Houte S, Westra ER, Rollie C (2019) The effect of bacterial mutation rate on the evolution of CRISPR-Cas adaptive immunity. Philos Trans R Soc Lond Ser B Biol Sci 374:20180094
Danis-Wlodarczyk KM, Wozniak DJ, Abedon ST (2021) Treating bacterial infections with bacteriophage-based enzybiotics: in vitro, in vivo and clinical application. Antibiotics 10:1497
Demerec M, Fano U (1945) Bacteriophage-resistant mutants in Escherichia coli. Genetics 30:119–136
Dennehy JJ, Abedon ST (2021) Phage infection and lysis. In: Harper D, Abedon ST, Burrowes BH, McConville M (eds) Bacteriophages: biology, technology, therapy. Springer, New York, pp 341–383
Hendrix RW, Casjens SR (2008) The role of bacteriophages in the generation and spread of bacterial pathogens. In: Hensel M, Schmidt H (eds) Horizontal gene transfer in the evolution of pathogenesis. Cambridge University Press, Cambridge, pp 79–112
Heselpoth RD, Swift SM, Linden SB, Mitchell MS, Nelson DC (2017) Enzybiotics: endolysins and bacteriocins. In: Harper DR, Abedon ST, Burrowes BH, Mc Conville M (eds) Bacteriophages: biology, technology, therapy. Springer International Publishing AG, Cham, Switzerland, pp 989–1030
van Houte S, Buckling A, Westra ER (2016) Evolutionary ecology of prokaryotic immune mechanisms. Microbiol Mol Biol Rev 80:745–763
Koonin EV, Makarova KS, Wolf YI, Krupovic M (2020) Evolutionary entanglement of mobile genetic elements and host defence systems: guns for hire. Nat Rev Genet 21:119–131
Korona R, Levin BR (1993) Phage-mediated selection and the evolution and maintenance of restriction-modification. Evolution 47:556–575
LeMieux J (2020) Lysins unlimited Phages' secret weapon. Gen Eng Biotechnol News 40:54, 57–59
Lenski RE (1984) Two-step resistance by *Escherichia coli* B to bacteriophage T2. Genetics 107:1–7
Luria SE, Delbrück M (1943) Mutations of bacteria from virus sensitivity to virus resistance. Genetics 28:491–511
Lythgoe K, Chao L (2003) Mechanisms of coexistence of a bacteria and a bacteriophage in a spatially homogeneous environment. Ecol Lett 6:326–334
Ruse M (1999) When is a negative result anomalous? Mar Ecol Prog Ser 191:302–303
Seed KD, Yen M, Shapiro BJ, Hilaire IJ, Charles RC, Teng JE, Ivers LC, Boncy J, Harris JB, Camilli A (2014) Evolutionary consequences of intra-patient phage predation on microbial populations. elife 3:e03497
Shen Y, Mitchell MS, Donovan DM, Nelson DC (2012) Phage-based enzybiotics. In: Hyman P, Abedon ST (eds) Bacteriophages in health and disease. CABI Press, Wallingford, pp 217–239

Wagemans J, Lavigne R (2012) Phages and their hosts, a web of interactions – applications to drug design. In: Hyman P, Abedon ST (eds) Bacteriophages in health and disease. CABI Press, Wallingford, pp 119–133

Wais AC, Goldberg EB (1973) An extended growth cycle in T4-infected *A. aerogenes*. Virology 55:397–399

Wright RCT, Friman VP, Smith MCM, Brockhurst MA (2018) Cross-resistance is modular in bacteria-phage interactions. PLoS Biol 16:e2006057

Young R (1992) Bacteriophage lysis: mechanisms and regulation. Microbiol Rev 56:430–481

Young R, Wang I-N (2006) Phage lysis. In: Calendar R, Abedon ST (eds) The bacteriophages. Oxford University Press, Oxford, pp 104–125

Zheng Y, Roberts RJ, Kasif S (2004) Identification of genes with fast-evolving regions in microbial genomes. Nucl Acids Res 32:6347–6357

Chapter 22
Pleiotropic Costs of Phage Resistance

Contents

Abstract With pleiotropies, a single gene is responsible for more than one phenotype. As a consequence, different alleles of a single gene can differ phenotypically in more than one way. Different alleles also can give rise to different phenotypes as measured under different conditions or at different times. As a result of these pleiotropies, alleles of a single gene can be associated with improvements in one associated phenotype but also a corresponding worsening of another phenotype. The two phenotypes controlled by this one gene in a sense thereby are antagonistic to each other, hence this situation being described as an antagonistic pleiotropy. In this chapter, we again consider bacterial mutation to resistance to bacteriophages, but instead of emphasizing the primary phenotype, i.e., the phage resistance phenotype, we instead consider secondary consequences of that resistance, which we measure in general terms as bacterial evolutionary fitness in the absence of selecting phages. Thus, improvements in bacterial fitness that are seen when phages are present (condition 1), that are a consequence of bacterial mutation to phage resistance, can be associated with corresponding declines in bacterial fitness as measured especially in the absence of these phages (condition 2).

> A bacterial mutation can bring about changes in a number of different characters (pleiotropic mutations). Changes in growth characteristics will affect the ability of the mutant to grow in competition with the normal type.—Salvador E. Luria (1947, p. 27)

In addition to what stage of a phage infection might be affected by bacterial mutation to phage resistance (Chaps. 18, 19, and 20) there also can be differences in the cost of those mutations to harboring bacteria, costs which also can be described as tradeoffs

S. T. Abedon, *Bacteriophages as Drivers of Evolution*,
https://doi.org/10.1007/978-3-030-94309-7_22

(Mangalea and Duerkop 2020). Particularly, achievement of phage resistance can be moot if as a consequence a bacterium becomes too 'sick' to be ecologically competitive, or otherwise, in the case of phage therapy, e.g., (Abedon et al. 2011, Kutter et al. 2010, Abedon 2017, Abedon et al. 2021, Alves et al. 2023), too sick to continue to contribute to the pathology of a bacterial infection. Indeed, due to this latter potential for phage resistance to be associated with lower bacterial virulence, it is often assumed that phage-resistant bacteria may be more easily removed by immune systems than phage-susceptible wild-type bacteria (Roach et al. 2017; Altamirano et al. 2021). Recently, this has also been proposed as a useful, deliberate end point to phage therapies, where phages are chosen that intentionally result in phage-resistance mutations that display lower virulence or which reduce a bacterium's potential to display antibiotic resistance, what has been called "Phage steering" (Gurney et al. 2020). See as well, however, the discussion of Smith and Huggins (1982), as considered also below, along with the long history of this same general concept as discussed by Gurney et al. (2020). Alseth et al. (2019) suggests furthermore that these costs may be especially high when bacteria are found in complex environments in which they are forced to compete with other, especially not mutated bacteria, while Buckling et al. (2006) suggest that costs can also be higher for mutations to phage resistance given the existence of previously existing bacterial mutations; see also Scanlan et al. (2015). Note, though, that Bohannan et al. (1999) saw essentially the opposite epistatic effect, with less cost than expected with multiple bacterial mutations to especially phage resistance, though with the caveat that the specific example they studied might not be typical. In any case, what is discussed in this chapter primarily are the pleiotropic costs of bacterial mutations to phage resistance.

22.1 Antagonistic Pleiotropies

In general terms, the combination of two phenotypes associated with a single allele is known as a pleiotropy. In this case, this could be a phenotype of phage resistance in combination with (possibly) phenotypes of reduced bacterial growth rates, reduced bacterial competitive ability, or reduced bacterial virulence. For relationships such as these, in which one consequence of a mutation is beneficial to the carrier (phage resistance) but another consequence of the same mutation is detrimental to the carrier (reduced competitiveness), the idea of an antagonistic pleiotropy is often invoked, a.k.a., tradeoffs (Williams 1957; Elena and Sanjuán 2003; Presloid et al. 2008; Koskella and Meaden 2013; Goldhill and Turner 2014; Rodriguez-Verdugo et al. 2014). Thus, while phage resistance could be considered to be a 'good' thing, lower bacterial virulence might instead be considered to be a 'bad' thing, at least from the bacterium's perspective. A related issue is that in some cases bacterial evolution of phage resistance can result in resensitization to antibiotics, i.e., an increased bacterial susceptibility to those chemical agents (Chan et al. 2016, 2018; Burmeister et al. 2020; Altamirano et al. 2021). Though not always the case, nevertheless this too can be viewed as an antagonistic pleiotropy, with a bacterium's escape from one antibacterial agent resulting in an increased sensitivity (or 'resensitization') to another antibacterial agent.

Importantly, though they are often both invoked in association with bacterial mutation to phage resistance, the concepts of 'antagonistic pleiotropy' and 'antagonistic coevolution' are unrelated. Indeed, antagonistic pleiotropies conceivably can interfere with antagonistic coevolution to the extent that the costs to bacteria of evolving phage resistance blocks that evolution, thereby preventing the occurrence of multi-round phage-bacterial arms races (i.e., antagonistic coevolution). We return to the concept of antagonistic coevolution in Chap. 25.

22.2 Some History

Awareness of bacterial resistance dates from early on in the study of phages, particularly since as a practical matter, especially in phage therapy, it was important to recognize which bacterial host a given phage could treat and which bacterial hosts it could not. For example, from d'Herelle (1922), p. 181, "The bacteriophage was active at the onset, the bacterium did not acquire a resistance, and its growth was quickly suppressed." More quantitative study of the evolution of phage resistance in bacterial populations, however, seems to have begun with the work of Luria and Delbrück (1943). As noted in the previous chapter (Chap. 21, Sect. 21.2.2), that work seems to have led directly to the efforts a short time later of Demerec and Fano (1945). Demerec and Fano not only studied rates of bacterial mutation to phage resistance, but the growth-rate costs of those mutations as well, as is the emphasis of this chapter.

22.2.1 Demerec and Fano 1945

The phage literature contains quite a bit of comparison of the properties of phage-resistant versus phage-sensitive bacteria, with an early example being that of Demerec and Fano (1945). They looked at both the appearance of phage-resistant colonies and the competitive ability of these strains against their wild-type, *E. coli* B parent—of historical importance, the phages used were the original T coliphages, i.e., T1, T2... T6, and T7 (Abedon 2000), and Demerec and Fano (1945) represents the original collection of those phages into that grouping. Especially, in terms of antagonistic pleiotropies, some colonies of phage resistant strains were small, i.e., half the diameter of the wild-type strain, or instead consisted of "Tiny colonies". Furthermore, reductions in colony size tended to be associated with resistance to some phages but not to others.

For competition experiments, as performed in broth, they took advantage of an ability to use phages as a selective agent during plating, i.e., with strains that are resistant to the selecting phage being able to form colonies in the presence of those phages (implying acquisition by those strains of avoidance or negation, but probably mostly avoidance; Chaps. 18 and 19) but fully sensitive bacteria unable to form colonies under these same conditions. Assays were complicated, however, by the possible passage of bacteria through three different physiological stages during competitions, corresponding to the familiar lag phase, log phase, and stationary phase, so we can't tell exactly why a strain may have been less competitive, only that it was less competitive, and indeed the authors point this out.

To fans of the idea of phage resistance displayed by bacteria resulting in antagonistic pleiotropies, Demerec and Fano's (1945) results, as they summarized them, nevertheless are perhaps disappointing:

> On the whole, growth rates for [phage-resistant] mutants are lower than for the [*E. coli* B] strain; but no large competitive disadvantage seems to be systematically associated with the phenomenon of phage resistance. This seems to hold even when a multiple mutant strain is examined, and even when the strain's colonies on agar are much smaller than the colonies of the original strain.

These conclusions do not mean, though, that antagonistic pleiotropies were not observed, but instead that their effects often were relatively small in the laboratory using this assay, at least as according to Demerec and Fano. A perhaps better conclusion from their results instead might be that in some cases these antagonistic pleiotropies are quite significant, but not in all cases. On the other hand, in some cases, mutants actually seem to even out-compete wild type! I'll get back, however, to why especially the latter observation may not be as relevant as one might think (Chap. 25, Sect. 25.5.3 and also below in this chapter), but first will look more closely at the actual competition results.

22.2.2 *Some Historical Re-calculations*

Fortunately, it is possible to be more quantitative in our assessment of the Demerec and Fano (1945) data. Specifically, what we would like to calculate is the relative fitness (w) of individual bacterial mutants in comparison to wild type. This can be accomplished by taking ratios of initial and final numbers of bacteria of the two types, taking the natural logarithm of each ratio individually, and then in turn taking the ratio of resulting transformed ratios (Wiser and Lenski 2015). That is, $w = \ln(M_f/M_i)/\ln(W_f/W_i)$, with the subscripts referring to final and initial, respectively, and M and W meaning mutant and wild type, also respectively. Unfortunately, while Demerec and Fano almost provide all of the necessary information for us to do this analysis, what they don't tell us is how much of an increase in cell counts was seen between initial and final. That is, they don't tell us what dilutions they used in arriving at colony counts. They do, however, provide a few clues, and those are that initial concentrations in growth tubes were "a few thousand per cc" while ending concentrations were "approximately 10^8 per cc" (which, though, admittedly seems low). If you take the ratio of these two numbers, then you arrive at 10^5, which is the number that I multiplied final colony counts by, but not the initial counts. Whether one chooses 10^4 or 10^6 as the multiplier instead, however, that does not make all that much of a difference in terms of calculated results.

The result of the above efforts are shown Fig. 22.1, which graphs the individual Demerec and Fano (1945) competition results re-calculated as relative fitness values. Any value below 1 is indicative of wild type outcompeting a phage-resistant mutant under phage-free conditions. Overall, in 37 of 50 cases this value was less than 1. This result is suggestive that under at least the test conditions employed, phage resistance can often impart a cost, an antagonistic pleiotropy, on the carrying bacterium—a cost again that is measured when phages are not present. On average, phage resistance seems to result under these test conditions in an approximately 10% decline in bacterial relative fitness.

Contrary to conventional expectations, in 13 of 50 instances the phage-resistant mutant displayed a higher relative fitness than wild type, including higher by 5% or more in 9 cases. Again, though, these higher fitnesses observed probably are not as interesting from an evolutionary ecological perspective as at first they might seem. In particular, most of and the greater fitness gains were associated with resistance evolution to phage T1 and T5, both of which are known to use the same surface receptor, the loss of which is known as well for not always giving rise to fitness losses under laboratory conditions despite its presumptive utility in the environment as a ferrichrome transporter (Hyman 2017).

What these experiments imply overall is that mutation to phage resistance can be costly some of the time explicitly under the relatively rich (nutrient broth) conditions used for these competition experiments. What we don't know, however, is how these costs may manifest under more realistic (Meaden et al. 2015) or simply other (Bohannan et al. 1999) environmental conditions, including when considering factors other than just rates of bacterial population growth. Thus, it is entirely possible, and perhaps even likely, that a greater fraction of these mutants would display reductions in Darwinian fitness under more realistic conditions. I am hesitant to claim that all of them will display reduced fitness, however, as the possibility always remains that at least some of the underlying mutations represent selectively neutral changes in gene sequence, e.g., as giving rise to evasion of phage recognition without loss of gene product function (Burmeister et al. 2020). Nevertheless, it is clear from these early experiments that mutation to phage resistance by bacteria is not inherently cost free.

22.3 A Sampling of More Modern Studies

The highest declines in fitness observed by Demerec and Fano (1945) were, for six phage-resistant strains, in excess of 30%, with a high of 39% (Fig. 22.1). These reductions were similar to those seen by Gómez and Buckling (2011) during soil rather than broth-based phage resistance-evolution experiments (36%). In both of these cases, though, fitness declines were not necessarily a consequence of only

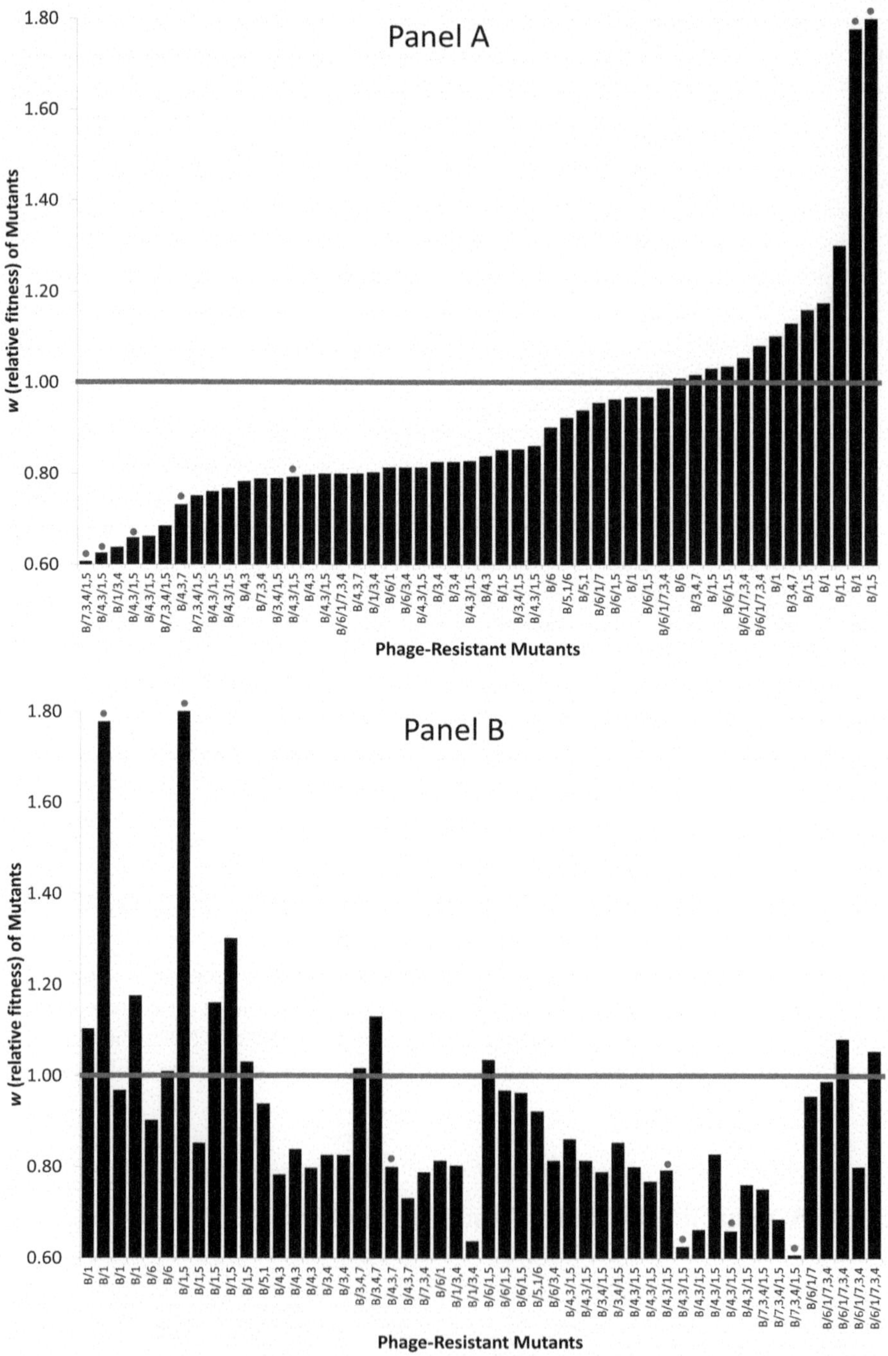

Fig. 22.1 Relative fitness (w) of phage-resistant mutants compared to *E. coli* B (wild type) as re-calculated from data provided by Demerec and Fano (1945). In terms of the names of phage-resistant bacterial mutants (x axis), the "/" symbol (pronounced as 'bar') indicates both 'selection

single mutations. See also the broth results of Wright et al. (2018). In addition, superficially similar mutants, i.e., same phage-resistance patterns but representing different isolates, sometimes produced somewhat different measured relative fitnesses, e.g., see B/4,3/1,5 in Fig. 22.1. This suggests a possible diversity of mutation spectra giving rise to similar phage-resistance phenotypes. This has indeed been confirmed by Lenski (1988) with mutations resulting in *E. coli* B resistance to phage T4, though as phage T4 infecting *E. coli* B absorbs to lipopolysaccharide (LPS), which can mutate to different lengths with different properties, it is difficult to tell how generally confirmatory this is especially in comparison to protein-receptor loss-of-function mutations to phage resistance. Beyond Gómez and Buckling (2011) and Wright et al. (2018), the post-Demerec and Fano literature contains numerous examples of phage-associated bacterial antagonistic pleiotropies, as I briefly review below.

Smith and Huggins (1982) used phages that were "anti-K1" antigen, an important virulence determinant of the *E. coli* bacteria strain they were therapeutically targeting. Importantly, they found that bacterial mutants that were resistant to these anti-K1 phages were K1-minus, for which they expected a lower bacterial virulence as based on their previous studies. Levin and Bull (1996) subsequently took an in-depth reexamination the Smith and Huggins study. See also Altamirano et al. (2021). Similarly, in an *in vivo* model, Oechslin et al. (2017) noted a loss of infectivity in phage-resistant *P. aeruginosa*. The results of Sumrall et al. (2019) and Trudelle et al. (2019) show declines in virulence associated with *Listeria monocytogenes* phage-resistant mutants. See Azam and Tanji (2019) for a recent review of this subject of reduced bacterial virulence associated with phage-resistant bacteria. Markwitz et al. (2021) looked at the costs of resistance in terms of growth rates for a variety of phages, a variety of phage receptors, and a variety of *P. aeruginosa* bacterial strains; interestingly and exceptional, for one of the

Fig. 22.1 (continued) against' and 'resistance to', with numbers indicating phage types (e.g., a "3" for phage "T3"). Multiple bars indicate multiple selection events and thus presumably multiple mutations conferring phage resistance. (Panel A) For these head-to-head, otherwise non-selective (no phages present) broth competitions, a value of less than one (1.000; highlighted in gray) indicates that wild-type won (37 instances) while a value of greater than one indicates that the mutant won (13 instances). Values of 1/100 of the corresponding value were used instead of zeros in certain instances to include results, as is seen (going from left to right) with columns 1, 2, 4, 7, 14, 49, and 50, and as indicated with gray dots (these thus are less certain calculations). As the information was not otherwise explicitly provided, an assumption was made that a 10^5-fold increase in bacterial densities took place between initial and final determinations of colony-forming units. The average of the calculated and shown values is 0.92. If 10^4-fold or 10^6-fold increases in cell densities are assumed instead, then the values are 0.91 or 0.93, respectively. If the values indicated with dots are ignored, then with an assumption of a 10^5-fold increase, the average value instead is 0.90. Thus, there would appear to be a tendency for phage-resistant mutants to on average display about a 10% decrease in fitness relative to wild-type over the course of competitive growth in only a single tube, going from lag through stationary phases, though this tendency toward lower-than-wild-type relative fitnesses of phage-resistant mutants in the absence of phages is not consistently the case. (Panel B) Same as panel A but with bacteria sorted by type rather than by relative fitness

P. aeruginosa strains tested, *P. aeruginosa* A5803, there were few costs to such mutations, whereas mutations conferring resistance to multiple phages often resulted in greater reductions in fitness including for this strain; see too (Hall et al. 2012; Koskella et al. 2012). Bohannan and Lenski (2000) reviewed costs of resistances, also in T phages, under different growth conditions, considering especially the impact of these costs on phage-bacterium community dynamics.

In an interesting twist on the antagonistic pleiotropy theme, Chan et al. (2016, 2018) used a phage that targeted an antibiotic reflux pump, thereby resulting in an association of phage resistance with antibiotic sensitivity that was in combination with antibiotic resistance being associated with phage sensitivity. Wright et al. (2018), in the above-noted study, determined broth growth fitness of 263 phage-resistant mutations against 27 different phages as relative to wild type, for which all showed a relative fitness of somewhat less than 1. Wang et al. (2019) showed that phage-resistant strains of a bacterial plant pathogen grew more slowly in competition experiments and to lower densities alone than their wild-type parents. Greater reductions were seen given exposure of bacteria to more than one phage type, though not all resistant strains showed these reductions. Mangalea and Duerkop (2020) have reviewed this subject with regard to the usefulness of phages in phage therapy, differentiating among bacterial fitness, colonization, and virulence costs. In an almost paradoxical example of antagonistic pleiotropy involving phage resistance, Avrani et al. (2011) found not only reduced growth rates in phage-resistant mutants of *Prochlorococcus*, but mutations that conferred resistance to some phages resulted in increased susceptibility, in terms of adsorption rates, to other phages.

This is just a sampling of studies that have considered antagonism between fitness gains due to phage resistance and fitness declines relative to wild-type bacteria as determined absent phages. A notable exception, indeed as noted in the previous section (Sect. 22.2), are mutations to phage T5 resistance (Lenski and Levin 1985; Bohannan and Lenski 2000; Hyman 2017), which has allowed these mutations to serve as a selectively neutral marker in evolution experiments (Dykhuizen 1990). Bacterial mutations that result in phage resistance due to loss of function should be detrimental in at least one environment that these bacteria evolved in, however (Chap. 25, Sect. 25.5.2), as is also the case with phage T5 resistance. Overall, it seems that it can be at least supposed that bacterial mutations to phage resistance are rarely cost free and at the very least should be limiting in terms of what environments phage-resistant bacterial mutants will remain competitive in.

References

Abedon ST (2000) The murky origin of Snow White and her T-even dwarfs. Genetics 155:481–486

Abedon ST (2017) Bacteriophage clinical use as antibactertial "drugs": utility, precedent. Microbiol Spectr 5:BAD-0003-2016

Abedon ST, Danis-Wlodarczyk K, Alves DR (2021) Phage therapy in the 21st century: evidence of phage-mediated clinical efficacy. Pharmaceuticals (Basel) 14:1157

Abedon ST, Kuhl SJ, Blasdel BG, Kutter EM (2011) Phage treatment of human infections. Bacteriophage 1:66–85

Alseth EO, Pursey E, Lujan AM, McLeod I, Rollie C, Westra ER (2019) Bacterial biodiversity drives the evolution of CRISPR-based phage resistance. Nature (London) 574:549–552

Altamirano FG, Forsyth JH, Patwa R, Kostoulias X, Trim M, Subedi D, Archer SK, Morris FC, Oliveira C, Kielty L, Korneev D, O'Bryan MK, Lithgow TJ, Peleg AY, Barr JJ (2021) Bacteriophage-resistant *Acinetobacter baumannii* are resensitized to antimicrobials. Nat Microbiol 6:157–161

Alves D, Danis-Wlodarczyk K, Abedon ST (2023) Phage therapy: a clinician's guide. In: Liu D (ed) Handbook of Molecular Biology. CRC Press, Boca Raton

Avrani S, Wurtzel O, Sharon I, Sorek R, Lindell D (2011) Genomic island variability facilitates *Prochlorococcus*-virus coexistence. Nature (London) 474:604–608

Azam AH, Tanji Y (2019) Bacteriophage-host arm race: an update on the mechanism of phage resistance in bacteria and revenge of the phage with the perspective for phage therapy. Appl Microbiol Biotechnol 103:2121–2131

Bohannan BJM, Lenski RE (2000) Linking genetic change to community evolution: insights from studies of bacteria and bacteriophage. Ecol Lett 3:362–377

Bohannan BJM, Travisano M, Lenski RE (1999) Epistatic interactions can lower the cost of resistance to multiple consumers. Evolution 53:292–295

Buckling A, Wei Y, Massey RC, Brockhurst MA, Hochberg ME (2006) Antagonistic coevolution with parasites increases the cost of host deleterious mutations. Proc R Soc Lond B Biol Sci 273: 45–49

Burmeister AR, Fortier A, Roush C, Lessing AJ, Bender RG, Barahman R, Grant R, Chan BK, Turner PE (2020) Pleiotropy complicates a trade-off between phage resistance and antibiotic resistance. Proc Natl Acad Sci U S A 117:11207–11216

Chan BK, Sistrom M, Wertz JE, Kortright KE, Narayan D, Turner PE (2016) Phage selection restores antibiotic sensitivity in MDR *Pseudomonas aeruginosa*. Sci Rep 6:26717

Chan BK, Turner PE, Kim S, Mojibian HR, Elefteriades JA, Narayan D (2018) Phage treatment of an aortic graft infected with *Pseudomonas aeruginosa*. Evol Med Pub Health 1:60–66

Demerec M, Fano U (1945) Bacteriophage-resistant mutants in *Escherichia coli*. Genetics 30:119–136

d'Herelle F (1922) The bacteriophage: its role in immunity. Williams and Wilkins Co./Waverly Press, Baltimore

Dykhuizen DE (1990) Experimental studies of natural selection in bacteria. Ann Rev Ecol Syst 21: 373–398

Elena SF, Sanjuán R (2003) Climb every mountain? Science (New York, N Y) 302:2074–2075

Goldhill DH, Turner PE (2014) The evolution of life history trade-offs in viruses. Curr Opin Virol 8: 79–84

Gómez P, Buckling A (2011) Bacteria-phage antagonistic coevolution in soil. Science (New York, N Y) 332:106–109

Gurney J, Brown SP, Kaltz O, Hochberg ME (2020) Steering phages to combat bacterial pathogens. Trends Microbiol 28:85–94

Hall AR, De VD, Friman VP, Pirnay JP, Buckling A (2012) Effects of sequential and simultaneous application of bacteriophages on populations of *Pseudomonas aeruginosa in vitro* and in waxmoth larvae. Appl Environ Microbiol 78:5646–5652

Hyman P (2017) Phage receptor. In: Reference module in life sciences. Elsevier, Amsterdam

Koskella B, Lin DM, Buckling A, Thompson JN (2012) The costs of evolving resistance in heterogeneous parasite environments. Proc Biol Sci 279:1896–1903

Koskella B, Meaden S (2013) Understanding bacteriophage specificity in natural microbial communities. Viruses 5:806–823

Kutter E, De Vos D, Gvasalia G, Alavidze Z, Gogokhia L, Kuhl S, Abedon ST (2010) Phage therapy in clinical practice: treatment of human infections. Curr Pharm Biotechnol 11:69–86

Lenski RE (1988) Experimental studies of pleiotropy and epistasis in *Escherichia coli*. I. Variation in competitive fitness among mutants resistant to virus T4. Evolution 42:425–432

Lenski RE, Levin BR (1985) Constraints on the coevolution of bacteria and virulent phage: a model, some experiments, and predictions for natural communities. Am Nat 125:585–602

Levin BR, Bull JJ (1996) Phage therapy revisited: the population biology of a bacterial infection and its treatment with bacteriophage and antibiotics. Am Nat 147:881–898

Luria SE (1947) Recent advances in bacterial genetics. Bacteriol Rev 11:1–40

Luria SE, Delbrück M (1943) Mutations of bacteria from virus sensitivity to virus resistance. Genetics 28:491–511

Mangalea MR, Duerkop BA (2020) Fitness trade-offs resulting from bacteriophage resistance potentiate synergistic antibacterial strategies. Infect Immun 88:e00926-19

Markwitz P, Lood C, Olszak T, van Noort V, Lavigne R, Drulis-Kawa Z (2021) Genome-driven elucidation of phage-host interplay and impact of phage resistance evolution on bacterial fitness. ISME J 16:533–542

Meaden S, Paszkiewicz K, Koskella B (2015) The cost of phage resistance in a plant pathogenic bacterium is context-dependent. Evolution 69:1321–1328

Oechslin F, Piccardi P, Mancini S, Gabard J, Moreillon P, Entenza JM, Resch G, Que YA (2017) Synergistic interaction between phage therapy and antibiotics clears *Pseudomonas aeruginosa* infection in endocarditis and reduces virulence. J Infect Dis 215:703–712

Presloid JB, Ebendick-Corp BE, Zarate S, Novella IS (2008) Antagonistic pleiotropy involving promoter sequences in a virus. J Mol Biol 382:342–352

Roach DR, Leung CY, Henry M, Morello E, Singh D, Di Santo JP, Weitz JS, Debarbieux L (2017) Synergy between the host immune system and bacteriophage is essential for successful phage therapy against an acute respiratory pathogen. Cell Host Microbe 22:38–47

Rodriguez-Verdugo A, Carrillo-Cisneros D, Gonzalez-Gonzalez A, Gaut BS, Bennett AF (2014) Different tradeoffs result from alternate genetic adaptations to a common environment. Proc Natl Acad Sci U S A 111:12121–12126

Scanlan PD, Hall AR, Blackshields G, Friman VP, Davis MR Jr, Goldberg JB, Buckling A (2015) Coevolution with bacteriophages drives genome-wide host evolution and constrains the acquisition of abiotic-beneficial mutations. Mol Biol Evol 32:1425–1435

Smith HW, Huggins MB (1982) Successful treatment of experimental *Escherichia coli* infections in mice using phage: its general superiority over antibiotics. J Gen Microbiol 128:307–318

Sumrall ET, Shen Y, Keller AP, Rismondo J, Pavlou M, Eugster MR, Boulos S, Disson O, Thouvenot P, Kilcher S, Wollscheid B, Cabanes D, Lecuit M, Grundling A, Loessner MJ (2019) Phage resistance at the cost of virulence: *Listeria monocytogenes* serovar 4b requires galactosylated teichoic acids for InlB-mediated invasion. PLoS Pathog 15:e1008032

Trudelle DM, Bryan DW, Hudson LK, Denes TG (2019) Cross-resistance to phage infection in *Listeria monocytogenes* serotype 1/2a mutants. Food Microbiol 84:103239

Wang X, Wei Z, Yang K, Wang J, Jousset A, Xu Y, Shen Q, Friman VP (2019) Phage combination therapies for bacterial wilt disease in tomato. Nat Biotech 37:1513–1520

Williams GC (1957) Pleiotropy, natural selection, and the evolution of senescence. Evolution 11: 398–411

Wiser MJ, Lenski RE (2015) A comparison of methods to measure fitness in *Escherichia coli*. PLoS One 10:e0126210

Wright RCT, Friman VP, Smith MCM, Brockhurst MA (2018) Cross-resistance is modular in bacteria-phage interactions. PLoS Biol 16:e2006057

Part VI
Natural Selection

Chapter 23
Concepts of Natural Selection in Light of Phage Exposure

Contents

Abstract This chapter provides a transition from more specific aspects of natural selection—selection for phage resistance in bacteria along with secondary consequences of that resistance—to a more general look at natural selection acting on bacteria as caused by phages. In a sense, this is looking especially at ways of thinking about the natural selection that can lead to the evolution of phage resistance in bacterial populations rather than thinking just about the resulting resistance. Discussed are how we consider the impact of natural selection in terms of organism fitness, the biased reproductive success of some alleles over others, and the resulting adaptation of an organism to its environment. We then consider how more complex adaptations, ones requiring more than one mutation to acquire, can be slowed or prevented from coming to fruition due to intermediate forms displaying lower fitnesses than parental forms, which are limitations on evolution described as historical contingencies. We then consider the differences between natural selection acting in such a way that population survival is threatened (hard selection) versus only competitive ability between conspecifics being of concern (soft selection).

> Each bacterial strain possesses an individual resistance, particularly when freshly isolated, which renders it more or less resistant to a bacteriophage accustomed to an *in vitro* existence. Later we will see that this resistance increases by a phenomenon of natural selection.— F. d'Herelle (1922, p. 73)

Though natural selection inescapably has been addressed throughout this monograph, and a brief introduction was provided in Chap. 3 (Sect. 3.2.4), in this chapter I further elaborate on this subject as a phage-mediated driver of evolution. This is in

S. T. Abedon, *Bacteriophages as Drivers of Evolution*,
https://doi.org/10.1007/978-3-030-94309-7_23

anticipation of two final chapters, one focusing on frequency-dependent selection (Chap. 24) and the other on the back and forth 'arms races' that are associated with the natural selection of antagonistic coevolution (Chap. 25). In all cases, note that natural selection is simply an evolutionary force, typically as supplied by a population's environment, that can modify a population's existing allele frequencies, doing this modifying in a biased/deterministic manner rather than randomly/stochastically. Though the forces of natural selection act on individual organisms, it is populations that are modified in terms of their allele frequencies.

Natural selection acts especially to reduce the frequency within populations of less evolutionarily fit alleles, particularly alleles that are detrimental (Fig. 23.1). One product of natural selection is the spread of adaptive alleles through populations and, at least potentially, an increase in the absolute fitness associated with the collective genotypes making up a population. Related is the concept of spread of adaptive genotypes rather than just of individual alleles particularly within clonal populations. In this chapter, we consider a few key concepts that are relevant especially to a better understanding of natural selection for phage-resistance in bacterial populations exposed to phages, a topic which is again considered in both Chaps. 24 and 25, with evolution of phage resistance by bacteria also covered in the five previous chapters (Chaps. 18 through 22).

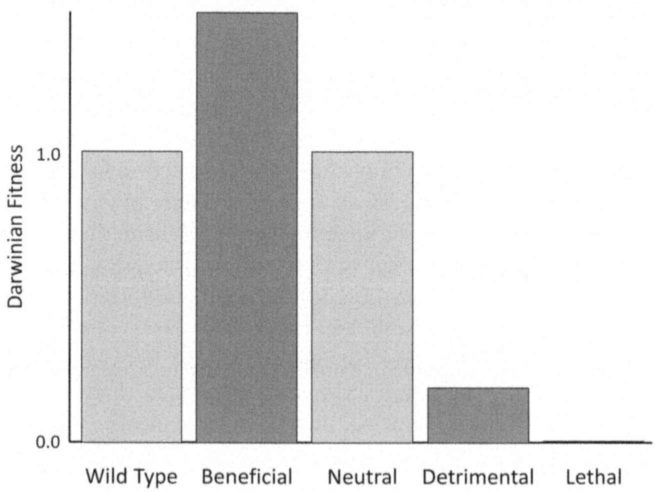

Fig. 23.1 Relative fitness of the various alleles as distinguished by their descriptors. Most of the heights are arbitrary except that the fitness of neutral alleles are at least similar to those of wild-type alleles. The fitness associated with beneficial alleles is higher than that of wild type while the fitness of detrimental alleles is lower than that of wild type, though how high or how low will vary. Included among detrimental alleles are alleles that are lethal to the carrying organism, for which, if consistently lethal, then Darwinian fitness is reduced to zero

23.1 Higher Fitness

The outcome of natural selection should be one of an increased frequency of alleles, or of whole genotypes, that bestow higher fitness on their associated organisms, i.e., alleles or genotypes which are beneficial in some manner (Fig. 23.1). This higher fitness should be context specific, at a minimum being observed only with the specific environmental conditions under which natural selection had been acting. Higher fitness in response to environmental factors, however, is not always easily achieved in terms of modification of individual genotypes (see Historical contingencies, Sect. 23.1, and also tradeoffs and antagonistic pleiotropies as discussed in Chap. 22). Nevertheless, there generally are three components that may be associated with a higher fitness, often described as life-history characteristics. These are (1) shorter generation times, (2) a greater potential for individuals to survive until they reproduce, and (3) greater numbers of progeny organisms produced per individual (though with individual bacteria, given their reproduction via binary fission, this latter component generally will remain unchanged) (Fig. 23.2). It can be difficult to enhance these different components of higher fitness simultaneously, however, and different organisms will tend to emphasize different aspects, e.g., producing progeny sooner versus producing progeny that individually possess greater survival potentials.

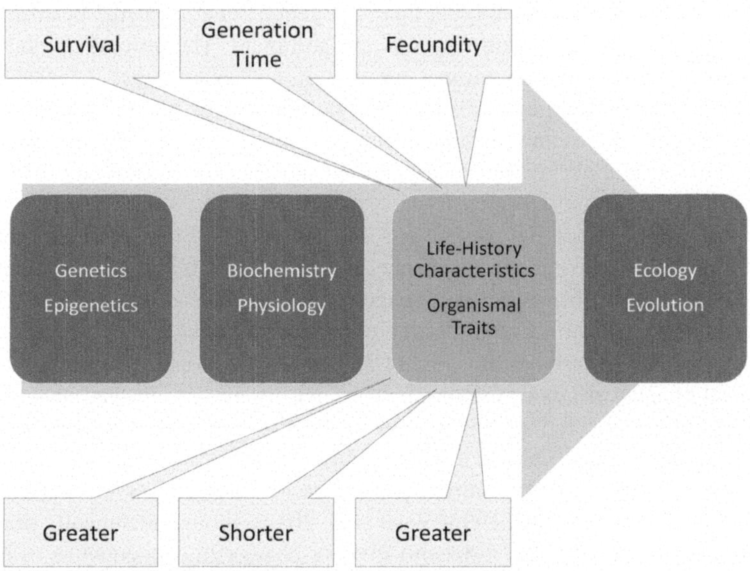

Fig. 23.2 Life-history characteristics, their general underlying bases, and their impacts. General categories of life-history characteristics are shown across the top and how these may be modified toward improving Darwinian fitness is shown across the bottom. This figure is derived from those of Abedon (2009) and Dennehy and Abedon (2021)

Given exposure of bacteria to phages, clearly greater potentials to survive in light of phage attack should be selected for, with phage resistance thereby leading to higher fitness, at least while selecting phages persist in a bacterial population's environment. As we have seen (Chap. 22), however, higher fitness in the presence of phages is not necessarily associated also with higher fitness in the absence of phages. Especially, as a tradeoff, phage-resistant bacteria often will display slower replication than the wild-type bacteria from which they descended. Thus, longer bacterial generation times tend to be measured in association with the evolution of phage resistance, as can interfere with the fitness gains of phage resistance, but so too potentially reduced bacterial durability may be seen in association with bacterial mutation to phage resistance. These issues of antagonistic pleiotropies, in a context of antagonistic coevolution, are considered as well in Chap. 25.

23.2 Beneficial Alleles and Adaptation

Beneficial mutations and beneficial alleles are ones that by definition are favored by natural selection. Beneficial, that is, implies a relative fitness of greater than one in comparison to wild type, versus less than one for detrimental and effectively one for neutral (Fig. 23.1). A beneficial allele equivalently is adaptive, and potentially will underlie adaptations in combination with other beneficial alleles. An adaptation (noun) is a variation on a phenotype, that is, a genetic character that bestows on an organism a higher fitness within a specific environment. The process of adaptation, i.e., adapting (verb), is the mutational creation of such phenotypes, or instead their acquisition by horizontal gene transfer. That is followed by natural selection acting on the underlying alleles and thereby, absent substantial genetic drift, increasing the frequency of those alleles within populations (Fig. 23.3), or at least preventing their decrease in frequency.

In terms of phage-bacterial interactions, the most obvious of adaptations are those that confer to bacteria a phage resistance, but also those that confer to phages an overcoming or escape from bacterial phage-resistance mechanisms (Chap. 25). The latter can involve reestablishment of an ability to infect what previously had been a phage-susceptible bacterial strain (including phage mutational escape from control by bacterial phage resistance systems such as CRISPR-Cas) or instead phage gain of an ability to infect a bacterial strain that previously had not been susceptibility to that phage (the latter as due to phage host range broadening or switching). There also is phage epigenetic escape from control by restriction-modification systems (Arber and Linn 1969, Korona and Levin 1993, Vasu et al. 2012), though technically that is not an evolutionary adaptation as it doesn't involve change in the sequence of nucleotides. Similarly, phenotypic changes in bacteria that allow them to resist phage attack, if lacking in underlying genetic changes, also are not evolutionary adaptations, though these changes could themselves be aspects of previously acquired adaptations (Bull et al. 2014).

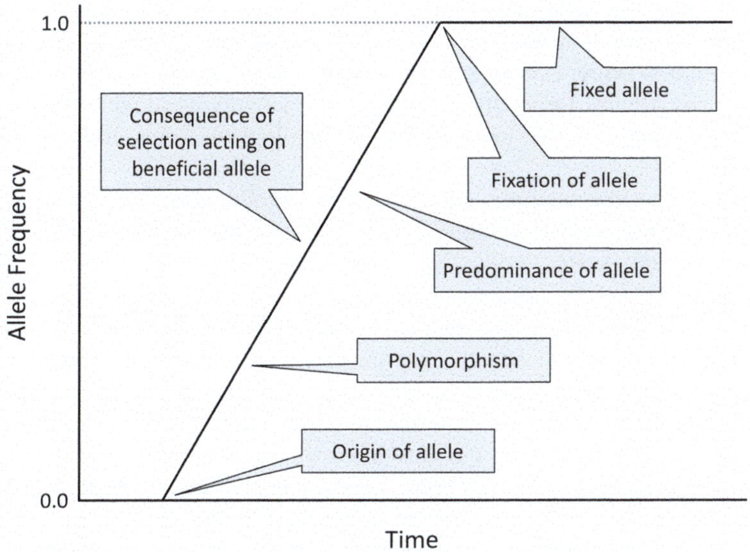

Fig. 23.3 Impact of natural selection on the frequency of a beneficial allele over time. The frequency of alleles within populations general start at very low levels, e.g., especially just a single allele. As frequencies increase, specific alleles can be said to be contributing to a polymorphism. At higher levels, we might describe the allele as being predominant within the population though still contributing to a polymorphism, i.e., with other alleles present at other than trivial frequencies. Lastly, with fixation, other alleles found at the same locus indeed have become trivial in their frequencies across a population

23.3 Historical Contingencies

> . . .modification of existing genes can explore functions that are available within small numbers of mutational steps from the existing sequences. This is because, in population genetic terms, the genes do not provide robust functions during transition from one adaptive peak to another, making exploration of the adaptive landscape problematic.—Jeffrey G. Lawrence and Heather Hendrickson (2008, p. 6)

While acquisition of some adaptations requires only a single underlying beneficial mutation, in other cases the development of a single adaptation requires multiple underlying mutations, whether in a single gene or instead across multiple genes. In principle, these multiple mutations may be sequentially acquired by an organism, or much more rarely they may be more or less simultaneously acquired, but sequential acquisition generally will efficiently occur only if each mutation individually also is adaptive, that is, supplying a higher fitness to the organism. It is when these necessary individual mutations are not beneficial, and potentially even detrimental, that we see the concept of historical contingencies (Blount et al. 2008). Basically, going from functional genotype A to functional as well as adaptive genotype B can be constrained given the existence of historical contingencies and this means that not all of the required mutational steps between the genotypes A and B are necessarily adaptive. The organism, in other words, can be adaptively 'stuck' because natural selection does not favor the intermediate genotypes that must be passed through to achieve an adaptive response to some aspect of the environment. At an extreme, whales for example have difficulty evolving an ability to walk on land, i.e., as many of the intermediate morphological forms likely would be maladaptive.

As a more on-topic example, not all bacteria are able to evolve resistance to all phages, e.g., as has been seen with *S. aureus* strain Newman and a phage dubbed PYO (Dickey and Perrot 2019). This failure should be seen as due to historical contingencies. That is, for whatever reason, mutations knocking out the ability of a phage to infect these bacteria either are effectively unobtainable (e.g., such as because a phage is able to adsorb to more than one receptor molecule found on that bacterium) or instead because the same mutations that bestow phage resistance also knock out the ability of the bacterium to survive even in the absence of phages (Chap. 25, Sect. 25.5.3). Such circumstances of bacteria being unable to evolve some form of resistance to a given phage type are relatively rare, however. Also, in principle bacteria can overcome limitations on their ability to evolve resistance to specific phages by acquiring phage-resistance systems via horizontal gene transfer, including as consisting of multi-gene adaptations. In at least certain instances, this is especially via conjugation (Daly et al. 1996; Moineau 1999; Coffey and Ross 2002). Alternatively, some phage-resistance systems are also prophage encoded (Chap. 16).

Phages, too, display historical contingencies with regard to their ability to impact bacteria. Specifically, the potential for phages to acquire mutations that change or broaden their host ranges, so that they become able to infect otherwise phage-resistant bacteria, is not necessarily infinitely high (Chap. 25, Sect. 25.5.2). Again, this means that there exist barriers to the acquisition of necessary adaptive mutations, in this case ones that are required to allow a phage to infect a novel host. An important means by which phages can overcome such historical contingencies limiting their host-range evolution is by recombining their genomes with those of other phages (Håggard-Ljungquist et al. 1992; Bouchard and Moineau 2000; Burrowes et al. 2019) or by additional means other than via point mutations (Tétart et al. 1996; Liu et al. 2002).

23.4 Hard Selection and Soft Selection

Bacterial adaptations of phage resistance, and phage adaptations overcoming especially bacterial resistance mechanisms that operate after irreversible virion attachment (see Chap. 18 for an overview of such bacterial adaptations), are both responses to hard selection, e.g., (Chao et al. 2000). Hard selection results from imposition of environmental factors that have the effect of reducing the fitness displayed by a population. Most dramatically, this involves actual killing of substantial numbers of a population, such as mediated by phages on phage-susceptible bacteria (Fig. 23.4) or as mediated on phages by post-adsorption phage-resistance mechanisms. In the

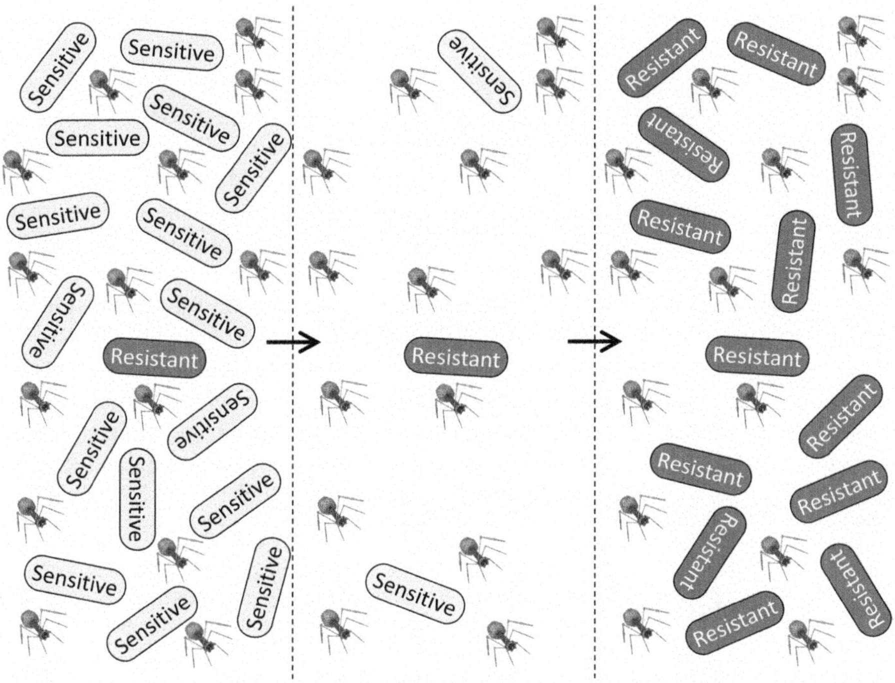

Fig. 23.4 Example of hard selection as mediated by a population of bacteriophages. Note that the fewer bacteria to the right than on the left is intentional, reflecting that phage-resistant bacteria may not be as evolutionarily fit as phage-sensitive bacteria given an absence of phages

absence of countering adaptations, hard selection especially can result in extinction of a population. Individuals displaying adaptations that overcome hard selection may or may not display fitnesses equal to those observed in populations prior to the imposition of hard selection, i.e., as due to tradeoffs/antagonistic pleiotropies (Chap. 22).

Much of the consideration of natural selection regarding the evolution of phage resistance as well as the overcoming of phage resistance (Chap. 25) involves hard selection. Competition between members of the same species, i.e., as exemplified here in terms of disruptive frequency-dependent selection (Chaps. 24, Sects. 24.3 and 24.4), does not, by contrast, involve hard selection. Instead, at least in principle, competition within populations results in gains in fitness by some members of the population at the expense of declines in fitness by other members of the same population, and this is rather than necessarily declines in the fitness of the entire population. This can be described instead as soft selection, where the fitness of the population as a whole is not reduced even though the fitness of some members of the population is reduced. An example of soft selection can be for lysogen allelopathy, or allelopathy generally, as a bacterial competitive strategy (Chap. 24, Sect. 24.4), i.e., where gains in fitness by one bacterial genotype come at the expense of the fitness one or more competitor genotypes, but not in terms of gains in the overall fitness of the bacterial population.

An interesting twist to these ideas is that in the presence of hard selection, soft selection can be less severe. Particularly, if hard selection has the effect of substantially reducing the size of a population, then there can be less competition within populations (Buckling and Rainey 2002) and thereby reduced soft selection. From p. 498, "Parasite-induced mortality may relax resource competition. This may reduce the selective advantage of newly arisen types capable of exploiting alternative resources, and hence [negatively impact] diversification." Rather than simply due to reductions in the size of bacterial populations as a route toward reduced intra-population competition, it is possible that constraints associated with phage-bacterium antagonistic coevolution also could play a role. Particularly, the hard selection of phage-mediated predation may directly interfere with the potential for populations to evolve the genetic diversity necessary for soft selection to take place, though the authors provided evidence against that possibility (Buckling and Rainey 2002).

23.5 Coevolution

Coevolution, *sensu stricto*, is adaptive evolution involving two specific, directly interacting species. Specifically, individual coevolutionary steps result in both species, at least one at a time if not both at the same time, displaying higher fitness in the presence of the other. The nature of the underlying adaptations depends on the specifics of the interactions between the two organism types, and generally those interactions can be described as falling within a spectrum (Jalasvuori and Koonin 2015) ranging from cooperative (mutualisms; Obeng et al. 2016, Shapiro and Turner 2018) to antagonistic (exploiter-victim relationships), the latter as covered in more detail in Chap. 25.

Prophages and their host bacteria potentially will display cooperative coevolutionary interactions, and these are other than the mutualistic interactions of bacteria supplying prophages with a 'home' and prophages supplying bacteria with beneficial phenotypes (for latter, see Chaps. 15 and 16) or potentially domesticable genes (Chap. 17). These coevolutionary interactions can involve the occurrence, for example, of compensatory mutations that allow the phenotypes encoded by the prophage and bacterium to better mesh physiologically, or in the case of the chronically infecting phage M13 (Shapiro and Turner 2018, p. 711), "how mutualism evolves from maladaptive parasitism." Generally, the number of genes prophages express is small, however, suggesting that the number of prophage targets for compensatory mutations also may be small. Perhaps, though, this small number of targets reflects past coevolutionary interactions with hosts, ones that selected for tight control of prophage gene expression specifically to limit interference with host metabolic processes. Bacterial genes, too, can evolve to better accommodate prophage presence.

Note that prophage inactivation in terms of ability to produce virions prevents subsequent coevolutionary interactions from occurring between species since the prophage thereby is no longer a separate organism from its host. Nevertheless, natural selection will still tend to be ongoing, favoring modifications to the now defective prophage genome that serve to enhance host fitness. Indeed, coevolution can occur not just between loci found within different species but also can occur between loci found within the same species.

Antagonistic coevolution, rather than serving to increase levels of cooperation between two organisms, instead involves selection for adaptations that can be described as giving rise to arms races (Chap. 25). These arms races that are seen with phage-bacterial antagonistic coevolution, however, are asymmetrical. Specifically, for bacteria as victims, natural selection will favor better defenses against phages, while for phages as exploiters, natural selection will favor their overcoming bacterial defenses. Thus, when bacterial and phage populations meet, there will be selection for phage adaptations to more effectively infect (thereby increasing phage fitness as well as phage antibacterial 'virulence') while at the same time there will be selection on bacteria for adaptations that prevent phages from effectively infecting (thereby increasing bacterial fitness, at least while still in the presence of sufficient densities of the selecting phage).

For different combinations of phages and bacteria, there seem to exist limitations on their abilities to coevolve. Mechanistically, these limitations presumably are due to historical contingencies, but coevolutionarily phages presumably are selected to take advantage of bacterial adaptations that are more useful to bacteria and thereby less easy for bacteria to mutationally modify or lose. Alternatively, phages may be selected to take advantage of bacterial adaptations that are less easily varied between bacterial strains toward phage resistance. This perhaps results in it being more difficult for bacteria to mutationally evolve mechanisms of phage negation (Chap. 19, Sect. 19.2). Conversely, bacteria likely are selected to minimize the susceptibility of their adaptations to utilization by phages, but whether that translates into phages thereby being less able to coevolutionarily adapt to exploiting alternative bacterial adaptations, i.e., such as by mutating to use different surface receptors, is an open question.

References

Abedon ST (2009) Phage evolution and ecology. Adv Appl Microbiol 67:1–45

Arber W, Linn S (1969) DNA modification and restriction. Annu Rev. Biochem 38:467–500

Blount ZD, Borland CZ, Lenski RE (2008) Historical contingency and the evolution of a key innovation in an experimental population of *Escherichia coli*. Proc Natl Acad Sci USA 105: 7899–7906

Bouchard JD, Moineau S (2000) Homologous recombination between a lactococcal bacteriophage and the chromosome of its host strain. Virology 270:65–75

Buckling A, Rainey PB (2002) The role of parasites in sympatric and allopatric host diversification. Nature (London) 420:496–499

Bull JJ, Vegge CS, Schmerer M, Chaudhry WN, Levin BR (2014) Phenotypic resistance and the dynamics of bacterial escape from phage control. PLoS One 9:e94690

Burrowes BH, Molineux IJ, Fralick JA (2019) Directed *in vitro* evolution of therapeutic bacteriophages: the Appelmans protocol. Viruses 11:241

Chao L, Hanley KA, Burch CL, Dahlberg C, Turner PE (2000) Kin selection and parasite evolution: Higher and lower virulence with hard and soft selection. Q Rev. Biol 75:261–275

Coffey A, Ross RP (2002) Bacteriophage-resistance systems in dairy starter strains: molecular analysis to application. Antonie van Leeuwenhoek J Microbiol 82:303–321

d'Herelle F (1922) The bacteriophage: its role in immunity. Williams and Wilkins Co./Waverly Press, Baltimore

Daly C, Fitzgerald GF, Davis R (1996) Biotechnology of lactic acid bacteria with special reference to bacteriophage resistance. Antonie van Leeuwenhoek J Microbiol 70:99–110

Dennehy JJ, Abedon ST (2021) Phage infection and lysis. In: Harper D, Abedon ST, Burrowes BH, McConville M (eds) Bacteriophages: biology, technology, therapy. Springer, New York, pp 341–383

Dickey J, Perrot V (2019) Adjunct phage treatment enhances the effectiveness of low antibiotic concentration against *Staphylococcus aureus* biofilms *in vitro*. PLoS One 14:e0209390

Håggard-Ljungquist E, Halling C, Calendar R (1992) DNA sequences of the tail fiber genes of bacteriophage P2 evidence for horizontal transfer of tail fiber genes among unrelated bacteriophages. J Bacteriol 174:1462–1477

Jalasvuori M, Koonin EV (2015) Classification of prokaryotic genetic replicators: between selfishness and altruism. Ann N Y Acad Sci 1341:96–105

Korona R, Levin BR (1993) Phage-mediated selection and the evolution and maintenance of restriction-modification. Evolution 47:556–575

Lawrence JG, Hendrickson H (2008) Genomes in motion: gene transfer as a catalyst for genome change. In: Schmidt H, Hensel M (eds) Horizontal gene transfer in the evolution of pathogenesis. Cambridge University Press, Cambridge, pp 3–22

Liu M, Deora R, Doulatov SR, Gingery M, Eiserling FA, Preston A, Maskell DJ, Simons RW, Cotter PA, Parkhill J, Miller JF (2002) Reverse transcriptase-mediated tropism switching in *Bordetella* bacteriophage. Science 295:2091–2094

Moineau S (1999) Applications of phage resistance in lactic acid bacteria. Antonie van Leeuwenhoek J Microbiol 76:377–382

Obeng N, Pratama AA, Elsas JD (2016) The significance of mutualistic phages for bacterial ecology and evolution. Trends Microbiol 24:440–449

Shapiro JW, Turner PE (2018) Evolution of mutualism from parasitism in experimental virus populations. Evolution 72:707–712

Tétart F, Repoila F, Monod C, Krisch HM (1996) Bacteriophage T4 host range is expanded by duplications of a small domain of the tail fiber adhesin. J Mol Biol 258:726–731

Vasu K, Nagamalleswari E, Nagaraja V (2012) Promiscuous restriction is a cellular defense strategy that confers fitness advantage to bacteria. Proc Natl Acad Sci USA 109:E1287–E1293

Chapter 24
Frequency-Dependent Selection in Light of Phage Exposure

Contents

Abstract A special form of natural selection involves the evolutionary fitness of different alleles or genotypes varying with their frequencies within populations. Thus, fitness may be higher when an allele is rarer or instead fitness may be lower when an allele rarer. The resulting frequency-dependent selection can be stabilizing versus disruptive, respectively, in either case of polymorphisms: Either rarer alleles are protected by natural selection from being driven to extinction or instead rarer alleles are driven to extinction by natural selection (not just genetic drift) particularly as they become rarer. With frequency-dependent selection as driven by phages, I suggest that it often is not so much the frequency of bacterial alleles or genotypes that are most relevant but instead actual densities, that is, absolute bacterial prevalence rather than relative bacterial allele or genotype prevalence. Thus, either high concentrations of one genotype drives the extinction of a rarer genotype (disruptive) or instead a genotype by existing at a low concentration is protected from extinction (stabilizing). In this chapter, I also consider how thinking about these sorts of things evolutionarily versus ecologically can differ. I then describe what can be a purely ecological process that is similar to stabilizing frequency-dependent selection, i.e.,

S. T. Abedon, *Bacteriophages as Drivers of Evolution*,
https://doi.org/10.1007/978-3-030-94309-7_24

killing the winner, and also how stabilizing frequency-dependent selection could result in Muller's ratchet.

> . . .more general models of theoretical biology are used to deduce the form of possible solutions, rather than to predict future states of the system being modelled.—Peter J. Wangersky (1978).

In this chapter, we consider frequency-dependent selection, which strictly speaking is an evolutionary phenomenon, i.e., as a form of natural selection. Logically, however, it also can be quite similar to kill the winner (or killing the winner), which metaphorically can be either ecological or evolutionary depending on how many bacterial species are involved. In either case—frequency-dependent selection or killing the winner—what is going on is that change in relative numbers of organism types is driven by the relative frequencies of those types. In terms of frequency-dependent selection, either rarer types tend to increase in frequency because they are rare (resulting in a *stabilizing* frequency-dependent selection) or instead common types tend to increase in frequency because they are common (resulting in a *disruptive* frequency-dependent selection). Before separately considering these three phenomena (disruptive frequency-dependent selection, stabilizing frequency-dependent selection, and killing the winner, in that order) we first will delve a bit further into the issue of distinguishing ecological from evolutionary phenomena.

24.1 Ecological Versus Evolutionary: Three Distinctions

Ecological phenomena tend to substantially manifest over shorter time scales than evolutionary phenomena, with *substantially* a key word in this sentence. Thus, if substantial change is occurring within a single generation, then it often is ecological in its nature, e.g., in terms of where organisms live or when they mate, and this can be so even if it results in changes in allele frequencies. Alternatively, even ongoing and otherwise constant ecological phenomena can lead over multiple generations to substantial evolutionary change, e.g., as populations adapt to new environments (directional selection). Evolution also can be occurring over multiple generations even if the population is not changing and under constant ecological circumstances, as due to stabilizing selection. Importantly, ecology and evolution are both distinguishable and not identical, though in terms of natural selection, the ecological generally is driving the evolutionary.

An alternative perspective is seen when distinguishing between phenomena that occur between species, as these give rise to ecological outcomes including changes in the relative numbers of individuals making up different species, and phenomena that occur within species, as giving rise to changes in the relative numbers of alleles (evolution). Frequency-dependent selection by definition is a process of natural selection that occurs within species and therefore is inherently evolutionary in its nature, though this is with a bit of a twist (next paragraph). On the other hand, there is killing the winner, which as noted is quite similar in its underlying logic to certain aspects of frequency-dependent selection. To the extent that killing the winner is something that is occurring between rather than within species, however, then kill the winner is inherently ecological rather than evolutionary. Thus, as we consider both frequency-dependent selection and killing the winner in this chapter, it is important to keep in mind that they are distinctly different phenomena in terms of their outcomes, evolutionary and ecological, respectively.

As noted, natural selection almost always occurs as a function of a population's ecology, with that ecology either consistently or inconsistently experienced by individual members of the affected population. So too, frequency-dependent selection is an inherently ecological phenomenon and this is so even though it involves allele frequencies, that is, with the relevant ecological phenomena being driven by allele frequencies rather than simply allele frequencies being modified by ecology. In this chapter, however, I will suggest that there is even more to frequency-dependent selection than just its inherent ecological aspects and these are that, just as with killing the winner, it is not just frequencies of different organism types that are important, but their absolute prevalences as well. In other words, selection for or against either rare or highly prevalent alleles or species or strains can be dependent not just on frequencies of types but also on concentrations of types, i.e., absolute prevalences, with concentration an inherently ecological rather than evolutionary measure. This, I believe, is highly relevant to phage-driven frequency-dependent selection and certainly is highly relevant to phage-driven killing the winner.

24.2 Stabilizing, Disruptive, Polymorphic, Monomorphic

Levin (1988), in describing selection acting on bacterial populations, distinguished frequency-dependent selection into stabilizing, also described as 'negative', versus disruptive or 'positive' forms. These I explain below with the basic distinction illustrated in Fig. 24.1. What is being stabilized or disrupted are polymorphisms.

A polymorphism is a genetic locus within a population's gene pool for which more than one allele exists. For example, for humans this could be the gene locus encoding eye color. A population consisting of only blue-eyed individuals is monomorphic at this eye-color locus. A population consisting of both blue- and brown-eyed individuals instead would be described as polymorphic. In that specific example, the blue allele is recessive to the brown allele, allowing us to easily recognize that a solely blue-eyed population is monomorphic, but otherwise dominant-recessive allele relationships have nothing to do with the concepts of monomorphic and polymorphic. All that matters is whether more than one allele is present across the population at the locus in question (polymorphic) or instead effectively no more than one allele is present (monomorphic).

Stabilizing frequency-dependent selection is a form of natural selection that leads to retention of polymorphisms within populations, that is, the retention of genetic

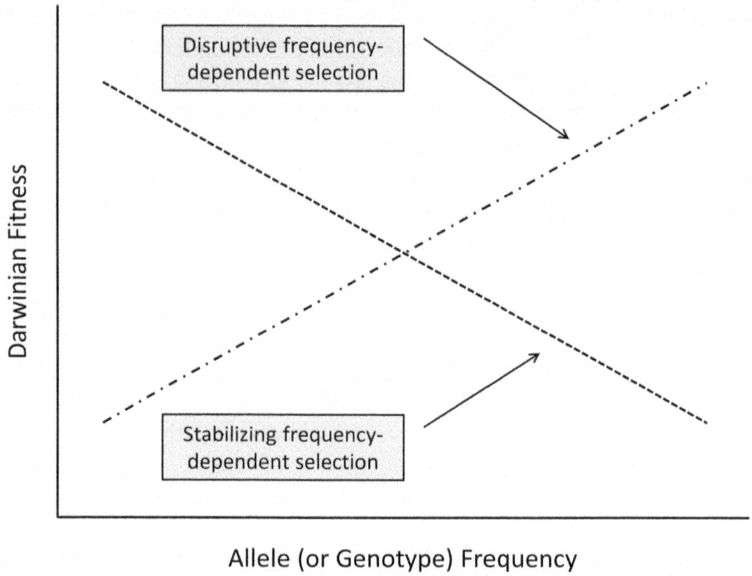

Fig. 24.1 Frequency-dependent selection impacts the retention of polymorphisms as a consequence of allele or genotype fitnesses varying as functions of allele or genotype frequencies. Disruptive frequency-dependent selection 'disrupts' polymorphisms due to positive correlations between fitness and frequency. Stabilizing frequency-dependent selection 'stabilizes' polymorphisms due to negative correlations between fitness and frequency

variation at a given locus. Disruptive frequency-dependent selection, by contrast, is a form of natural selection that leads to loss of polymorphisms within populations, i.e., a loss of genetic variation. What's left in the polymorphism's place following disruptive frequency-dependent selection when that has continued to its logical conclusion is a monomorphism, which is another way of saying that an allele has become fixed within the population at its locus. It is important to keep in mind that other mechanisms exist that can either stabilize or disrupt polymorphisms, but with frequency-dependent selection, key is that the strength of this stabilizing or disrupting natural selection varies as a function of allele frequencies, with either rare alleles selected for (stabilizing) or instead rare alleles selected against (disruptive). Note, however, that there are strong distinctions between rare alleles being actively selected against (disruptive frequency-dependent selection) and rare alleles being more likely to be lost due to genetic drift (Chap. 7), where the latter explicitly is not directly driven by natural selection and therefore is not an issue considered further in this chapter. In any case, the key difference between disruptive and stabilizing frequency-dependent selection as mediated by phages is whether or not the phage-producing bacterial strain is resistant to the phages it is producing.

24.3 Disruptive Frequency-Dependent Selection—Extracellular Toxins

With disruptive frequency-dependent selection, the tendency as noted is for polymorphic populations to become monomorphic. That is, for alleles to be lost, i.e., to go extinct, at least locally. Furthermore, the tendency is for these losses to be frequency dependent in a disruptive manner, in this case meaning that natural selection will favor more prevalent alleles over less prevalent ones. Antagonistic mechanisms where offensive 'weapons' are genetically linked to protection from the same weapons will tend to give rise to disruptive frequency-dependent selection. Thus, for example, from Levin (1988), if a bacterium encodes an extracellularly acting antibacterial toxin or bacteriocin, i.e., as affects the same species of bacterium, then toxin-producing genotypes will tend to eliminate toxin non-producing genotypes from the same local area. Again, for this to take place, the toxin production must be genetically linked to toxin protection, i.e., the producer must also be resistant. (Also, keep in mind that 'antagonistic' here has little to do with either antagonistic pleiotropies, Chap. 22, or antagonistic coevolution, Chap. 25.)

The bacterial genotype encoding toxin production and protection will tend toward fixation in this scenario, but why is that tendency frequency dependent? The answer would have to do with limitations on the ability of toxin-producing bacteria to individually affect toxin non-producing bacteria. Thus, if we hold the number of bacteria of a given species that are found within a given volume constant, then the more toxin-producing bacteria that are present, the more that toxin-susceptible bacteria will be negatively affected, i.e., killed. This is simply because more toxin will be produced and therefore toxin concentrations ultimately will be higher the more toxin-producers that are present. The toxin-producing bacteria thus will increase in prevalence (frequency) within the bacterial population at the expense of the toxin-susceptible bacteria. Note in this scenario that we are not considering other bacterial species nor indeed any other aspects of the focus species other than in terms of toxin production and susceptibility. Nevertheless, within at least this simple, one-species, two-allele system—and particularly given spatial structure (Chao and Levin 1981; Levin 1988; Hawlena et al. 2010)—we have an expectation that the production by bacteria of toxins that can act on conspecific bacteria will tend to give rise to disruptive frequency-dependent selection.

24.4 Disruptive Frequency-Dependent Selection—Induced Prophages

What does any of this have to do with phages? We can consider the prophages themselves as representing a *potentially* extracellularly acting antibacterial 'toxin', i.e., as following prophage induction. These same phages also encode protection from the virions they produce, that is as superinfection immunity (Chap. 16, Sect.

16.1). This is a process which Levin has also considered (Stewart and Levin 1984), describing it as "Allelopathic effects", and what I like to describe as 'lysogen allelopathy' (Abedon and LeJeune 2005; Breitbart et al. 2005); see Harrison and Brockhurst (2017) for a relatively recent short review. Allelopathic effects are associated especially with the production of substances by one organism that negatively affect the growth of another organism, and also not substantially negatively impact the producer. Here, though, this allelopathic suppression of the growth of organisms would be associated not with a 'chemical' but instead with the production of anti-bacterial virions.

Specifically, consider two alleles, a 'minus' allele corresponding to the prophageless state (a bacterium not infected by the prophage in question) and a 'plus' allele corresponding to harboring of the prophage. In this case, prophage induction, virion release, and subsequent lytic infections of minus bacteria will result not only in some degree of elimination of the minus bacteria, but also some degree of conversion of the minus bacteria to plus bacteria (lysogenic cycles in the newly infected bacteria). Thus, bacterial lysogens have a potential to increase their frequency within a bacterial population as their prophages display productive cycles. This is also because the lysogens display superinfection immunity and thus are resistant to the virions they release (Chap. 16, Sect. 16.1), and indeed in terms of spatial structure, it also may be relevant that the likelihood of display of superinfection immunity by a bacterial population tends to increase as bacteria are exposed to more homoimmune phages, rather than decrease, something I describe as high-multiplicity lysogeny decisions (Abedon 2017b; Igler and Abedon 2019; Abedon 2019).

This increase in lysogen frequency can be viewed as a form of disruptive frequency-dependent selection because the greater the frequency of the plus bacteria within environments, then the greater the numbers of these phages that are produced, which should lead to greater rates of phage acquisition and infection of minus bacteria. Once all of the phage-susceptible bacteria within the local environment have become infected by that one phage type, whether killed lytically or now hosting a lysogenic cycle, then the bacterial population can be considered to be monomorphic with regard to this prophage 'allele'. Lysogeny within that bacterial population, as effected by this prophage, would now be genetically fixed.

24.5 Frequency Dependence or Instead Density Dependence?

What if not all bacteria within an environment are susceptible to an extracellular toxin, or to an extracellular phage, and these toxin-resistant bacteria are not even conspecific to the bacterium producing the toxin? In that case, what we would expect is a reduction in the numbers of one species, i.e., as due to the impact of the toxin (or phage) on susceptible conspecifics, but no directly consequent change in numbers of the not-impacted species. Reduction in the prevalence of a given species within a community of species is an ecological change rather than evolutionary one, though within the affected species we would see an evolutionary change as well, i.e., as toxin producers increase in frequency. With higher frequencies of toxin producers within the affected bacterial population, then ecological as well as evolutionary effects should be smaller since there will be fewer susceptible bacteria to eliminate, whereas with lower frequencies of toxin producers, ecological change can be greater, since then there would be more susceptible bacteria to lose, i.e., as discussed immediately above as disruptive frequency-dependent selection.

24.5.1 Density Dependence

The disruptive effects of extracellular antibacterial agents are dependent not just on the relative prevalence of producing bacteria, that is, the frequency of 'plus' genotypes, but also on the absolute *prevalence* of toxin-producing bacteria, that is, their density within environments. It is only if we assume that a bacterial population remains constant in density that frequency would directly describe the absolute prevalence of antibacterial-producing bacteria. Thus, if toxin producers cannot ecologically substantially increase their numbers having reduced toxin-susceptible bacteria in number, such as given competition with other species (Blazanin and Turner 2021), then evolutionary effects will not strictly be frequency dependent, as the strength of selection would not thereby change with allele frequency.

Just as with Lotka-Volterra predator-prey dynamics, that is, what is important is densities of predators and this is rather than, e.g., ratios of predators to prey (Bohannan and Lenski 1997). Indeed, toxin-producing bacteria can very well be viewed as all-but-predators of toxin-susceptible bacteria, though with the predation possibly more for 'sport' rather than for 'food'. More rigorously stated, both predators and toxin-producing bacteria are exploiters of victim individuals, and the impact of exploiters on victims will be dependent on the density of exploiters rather than strictly on the frequency of exploiters among species generally. Disruptive *frequency*-dependent selection in this instance, in other words, might be viewed as something of a misnomer.

24.5.2 The Lysogens Are the Exploiters in Disruptive Frequency-Dependent Selection

It is important to keep in mind with these musings that explicitly it is toxin-producing bacteria, or prophage-carrying lysogens, that we should be viewing as the exploiters in the above example. The strength of the resulting disruptive frequency-dependent selection is a function of lysogen density, therefore, and this is because the more lysogens that are present within an environment, then the greater the number of 'toxic' phages that can be produced, and phages, or toxins, impact bacteria based on their densities, or based on their titers as is the equivalent for phages.

The latter is a point that I often try to emphasize: It is not the ratio of phages to susceptible bacteria that is the primary determinant of the impact of phages on bacteria but instead it is the titers of phages that determines the magnitude of their impact (Abedon 2016). You can have 1000 phages for every bacterium, but if that adds up to only, e.g., 10^5 phages/ml, then you are not going to end up having much of an impact of phages on bacteria (Abedon 2011; Danis-Wlodarczyk et al. 2020). So too, if you don't have sufficient numbers of bacteria present to produce sufficient numbers of phages, then the phage impact will be small even if ratios of phage-producing to phage-susceptible bacteria is high. Again, ecologically, what is important here is the absolute prevalence of alleles and this is rather than their frequency.

I should point out that it is not as though Levin was unaware of the importance of density in frequency-dependent selection (Levin 1988). Indeed, he makes this point better or at least pithier than I do (p. 470): "Ecological interactions between populations are almost invariably density-dependent, and frequency-dependence is a common hand-maiden of density-dependence."

24.5.3 Or Neither Lysogen Frequency Nor Lysogen Density Dependence?

> . . .for bacteriocins and clinically used antibiotics . . .the evidence that in natural populations the inhibition of competitors is the evolutionary *raison d'être* for these compounds is, at best, underwhelming.—J. P. W. Young and B. R. Levin (1993, p. 174)

There is another ecological issue with temperate-phage mediated disruptive frequency-dependent selection, also with bacterial absolute prevalence being relevant. This is that the generation of new phages by lysogens is expected to occur at some more or less constant background rate. This means that when lysogens are rare in terms of absolute numbers, then it may be unlikely that any induction event is observed, while as these lysogens become more common, then the likelihood of an induction event will increase. So far, that is consistent with the idea of disruptive frequency-dependent selection. The problem, though, is that in either case it is uncertain whether the number of new virions produced will reach sufficient titers to substantially impact the phage-sensitive bacterial population, that is, unless the phage-sensitive population supports the further propagation of those phages. Related to this issue, burst sizes can be smaller given spontaneous inductions versus inductions associated with exposure of lysogens to DNA damaging agents, e.g., up to an extrapolated 50-fold greater maximum burst size as measured with *A. calcoaceticus* (Berenstein 1986). That is, absent amplification by sensitive bacteria, phage titers produced will be a function of both the number of induction events, which may be small if only spontaneously occurring, and the resulting burst sizes, which also may be small if only spontaneously occurring.

The ability of a bacterial population to support phage propagation to substantially disruptive titers is a function of that population's absolute prevalence, that is, its density (Chap. 7). As phage-sensitive bacteria become less prevalent, due either to lysis or their conversion to lysogens, then the potential for those bacteria to support substantial phage propagation should decline as well. Thus, the higher the starting ratios of phage-producing to phage-sensitive bacteria, then potentially the lower the prospect for lysogen-associated allelopathic effects to be disruptive, that is, to have substantial negative impacts on the phage-sensitive bacteria (Brown et al. 2006; Li et al. 2017). In a perhaps more realistic ecological scenario, however, phage-producing lysogen exploiters and phage-sensitive bacterial victims will encounter each other at some respective population densities. At the point of such an encounter, there may or may not be enough of the former (exploiters) present for production of the antagonistic phages to occur. Should prophage induction occur, however, then there also still may or may not be sufficient densities of phage-sensitive bacteria (victims) to result in phage amplification to levels needed to be substantially disruptive, i.e., to "trigger an epidemic among susceptible competitors" (Brown et al. 2006), p. 2048.

That lysogen allelopathic effects truly will increase with lysogen frequency, in other words, is not necessarily true. I suspect, however, that prophage induction should be disruptive particularly if released phages encounter victim bacteria to infect over very short, sub-millimeter distances, i.e., before virions are diluted into the larger environment (Abedon 2017a), i.e., from Young and Levin (1993), p. 174, "The diffusing allelopathic substance kills sensitive bacteria in the vicinity of the colony producing it". In this case, the more lysogens present, then the more induction events and virions released, thereby the greater the overall impact on immediately adjacent phage-susceptible bacteria. Indeed, bacteria production of antibacterial substances generally, including bacteriocins as well as antibiotics, could very well have as a primary utility the inhibition of susceptible, invading bacteria over very short distances. Antibacterial activities could be concentrated, for example, at the surfaces of especially bacterial microcolonies toward inhibition of adjacent colonization by other bacteria, such as (p. 550) "by crowding up against target bacteria within biofilms" (Abedon 2015). See Fig. 24.2 for illustration.

24.6 Stabilizing Frequency-Dependent Selection Involving Phages

Contrasting disruptive frequency-dependent selection, where polymorphisms will tend toward monomorphisms, with stabilizing frequency-dependent selection, polymorphisms instead will tend to be selectively maintained (Fig. 24.3). With disruptive frequency-dependent selection, it is extracellular toxins and their ilk as produced by

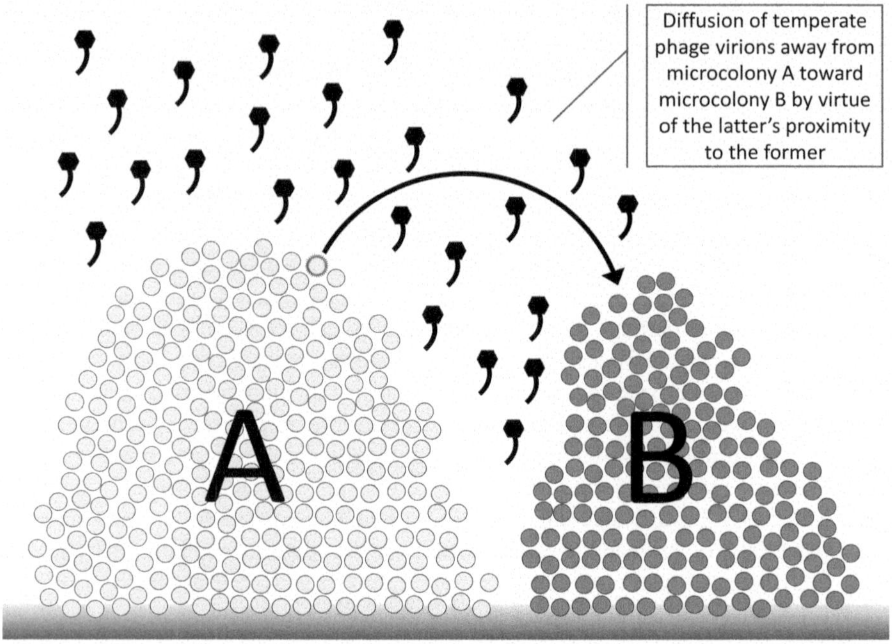

Fig. 24.2 Lysogen allelopathy as envisioned between bacterial microcolonies of the same species but as formed by clonally distinct bacterial strains. Microcolony A consists of clones of a bacterial lysogen (cocci) with induction having occurred in its upper, right (i.e., at the base of the arrow). Microcolony B consists of bacteria (also cocci) which are not immune to the released phages. Bacteria and phages are not drawn to scale as virions are indicated as being one or two orders of magnitude larger than would otherwise be the case

bacteria and acting upon conspecifics that drives polymorphisms toward monomorphisms. With stabilizing frequency-dependent selection, it also is extracellularly located antibacterial antagonists that represent the selective driving force. The difference is that when frequency-dependent selection is stabilizing, then antibacterial antagonists are not associated with a given, exploiter bacterial genotype. Instead it is now lytic phages as exploiters of bacteria that should give rise to the stable maintenance of polymorphisms among bacteria. Thus, frequency-dependent selection is disruptive (at least arguably, above) if bacterial lysogens are doing the attacking as well as resisting but stabilizing if instead it is the phages alone that are doing the attacking.

In humans, loci encoding major histocompatibility complex (MHC) proteins can be highly diverse. The common explanation for this diversity has to do with the role of MHCs in combatting infections, such as viral infections, e.g., (Bodmer and Bodmer 1978; Petersen et al. 2003; Migalska et al. 2019). Were a population monomorphic for MHC, then a given viral pathogen need only overcome the impact of that MHC type to overcome the cell-mediated immunity effected by the entire

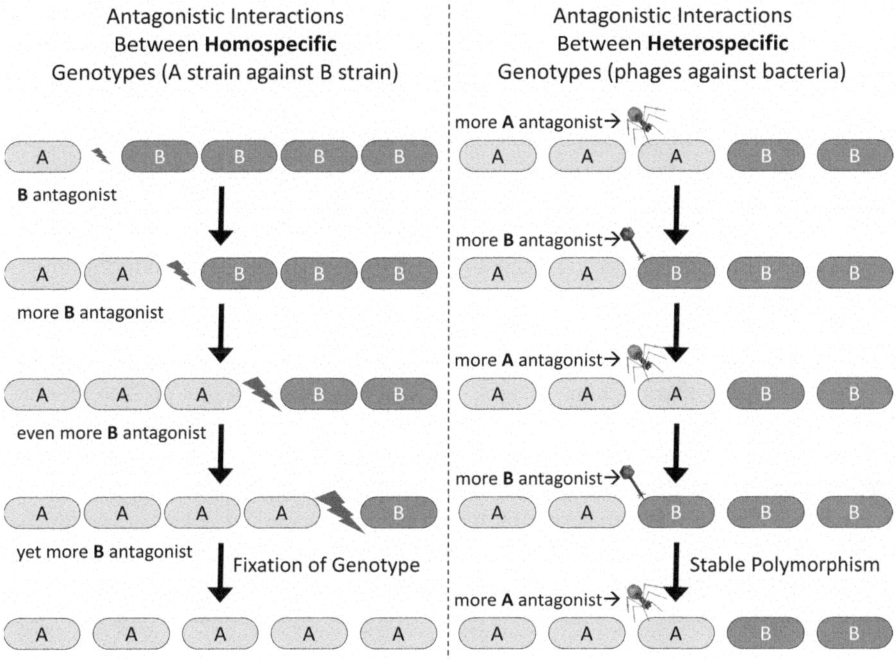

Fig. 24.3 Disruptive (left) versus stabilizing (right) frequency-dependent selection. Antagonism within species can lead to elimination of an antagonized genotype (bottom-left), e.g., as in the case of bacteriocins (indicated as 'lightning bolts'), here as produced by strain A. Antagonism affecting multiple species or strains can lead to selection against the most prominent strains and thereby selection for the less prominent genotype. This occurs particularly if predators prefer more-prominent to less-prominent strains. In the case of phages as predators, it would be different phage types with preferences for different bacterial strains, and greater bacterial-strain numbers giving rise to more of the phage antagonists against those strains, i.e., a within-species form of killing the winner that would give rise to a stabilizing frequency-dependent selection

host population. When MHC is diverse, indeed not just polymorphic but associated with large numbers of distinct alleles, then the evolution of a virus to overcome one person's MHC will not necessarily translate into an ability to evade another person's MHC. Thus, protection from pathogens tends to select for immune system diversity, which in turn is reflected in MHC diversity: stabilizing frequency-dependent selection.

Bacteria have immune system equivalents, both innate and acquired (Abedon 2012) (Chap. 18). Were bacterial strains identical in their anti-phage defenses, then a phage that has evolved to overcome those defenses in one bacterial strain will have equivalently evolved to overcome those defenses in other bacterial strains. Within a given bacterial species, however, not all anti-phage defenses, nor vulnerabilities, will necessarily be identical. The result is selection mediated by specific phage types *against* those bacterial strains that are susceptible to those specific phage types, and selection as well *for* those bacterial strains that are not susceptible to those specific phage types. Levin (1988) in his review of "Frequency-dependent selection in bacterial populations" used restriction-modification systems as his primary example of a bacterial anti-phage defense, arguing that exposure to a diversity of phage types should select for a diversity bacterial phage-defense types, in this case displaying different restriction modification systems that are able to resist different phages, including epigenetically modified phages (Arber and Linn 1969; Korona and Levin 1993; Vasu et al. 2012). Overall, we have an expectation that selection would not favor different bacterial types sharing especially strictly lytic phages among themselves within the same environment.

As a bacterial type becomes more absolutely prevalent within an environment, they will be able to support the propagation of those phages to which they are sensitive to higher titers (Chap. 7, Sect. 7.3.2). Those bacteria that are sensitive to these amplified phages as a consequence will tend to become reduced in numbers, including even bacteria that are not contributing substantially to that phage amplification, thereby in effect selecting for rarer bacterial phage-resistance types, which is another way of saying that stabilizing frequency-dependent selection would be occurring. Note though again that this is as much an absolute prevalence issue (density) as it is a relative prevalence issue (frequency), and very likely more density than frequency driven.

24.6.1 Again, Frequency Dependence or Density Dependence?

As with disruptive frequency-dependent selection, with stabilizing frequency-dependent selection it especially is absolute prevalence which is the driver rather than necessarily frequency—absolute prevalence of specific bacterial strains but also resulting absolute prevalence of specific phage types. That is, with stabilizing frequency-dependent selection, the higher the density of a specific bacterial strain within an environment, then the higher the resulting titers of phages that may be propagated by that bacterial strain (Fig. 24.4), though this statement is ignoring that at sufficiently high densities bacteria can enter into a phage infection-refractory stationary phase. Rarity by contrast represents a 'numerical refuge' from predation (Chap. 7, Sect. 7.3.2) so is selected for, with 'numerical' defined explicitly in terms of absolute bacterial densities within an environment rather than frequencies (Chao et al. 1977); thus, from p. 375, "This density-dependent response, in effect, protects the bacterial population from over-predation, a situation we shall refer to as a 'numerical refuge.'" Indeed, changing the frequency of phage sensitivity within a bacterial population should have no impact at all on its numerical refuge status so long as bacterial prevalence and physiology are not altered.

Still, if more-prevalent phage-sensitive bacteria are reduced in number, then any bacteria that are resistant to the particular attacking phage should automatically

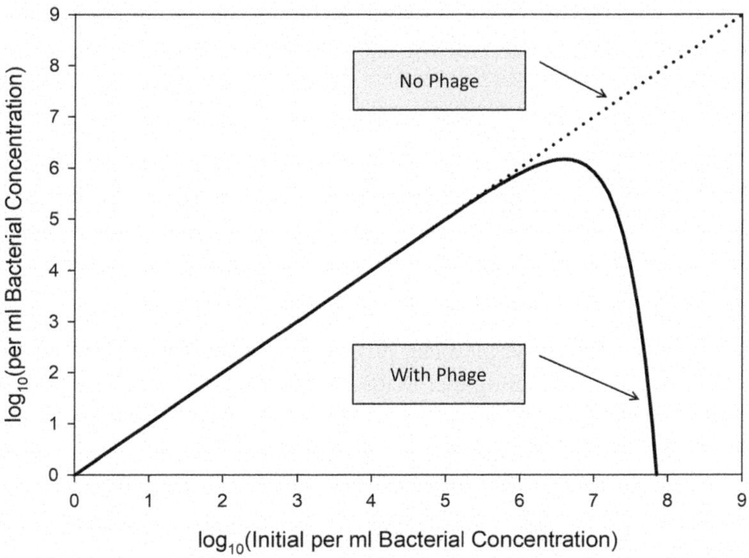

Fig. 24.4 Potential consequence of the concentration of a bacterial strain on the phage impact on that strain. Assumed is that the phage concentration is found at a titer that is equal to the initial bacterial concentration, with the latter the *x* axis. Shown is that initial bacterial concentration (dotted line) as well as the concentration of not yet phage adsorbed bacteria after 10 min of phage exposure assuming an ongoing generation of phages and therefore no phage depletion over time stemming from adsorption (solid line). An adsorption rate constant of 2.5×10^{-9} ml^{-1} min^{-1} is assumed. Note that the phage impact, given the dependence of its titer on bacterial concentrations, is effectively unnoticeable in this case until initial bacterial densities and therefore phage densities have reached nearly 10^7 per ml

increase in frequency. Nevertheless, only if that increase in allele or genotype frequency is accompanied also by increases in bacterial strain absolute prevalence should those bacteria too come to be able to support extensive propagation of new phage types, so long as phages to which those bacteria are sensitive are also present, resulting in those previously resistant bacteria thereby also being substantially negatively affected. Ultimately, so long as there are enough phage types present that are specific to individual bacterial types, then no one bacterial genotype should be able to rise in frequency to fixation. Thus, with decreasing bacterial genotype rarity, particularly with individual bacterial genotypes thereby escaping numerical refuges, then there should be increased vulnerability to specific attacking phages (Weitz et al. 2005). Notwithstanding these various arguments against the relevance of frequency—and just as with disruptive frequency-dependent selection—if we were to hold the size of a bacterial population constant, then frequency would indeed directly determine density and therefore any genotypes which were present at higher frequencies should be more affected by phages, all else held constant. It is just that the extent to which specific bacterial genotypes can support the propagation of phages ultimately will be a function of the absolute (i.e., actual) densities of those bacteria rather than their relative densities.

24.7 Killing the Winner

We can consider more than just frequency-dependent selection within bacterial populations and instead concentrate on how phages can impact the relative prevalence of different bacterial species. At this point we have dropped any pretense that we are talking solely about evolutionary processes and instead would be considering an ecological process, something that in the literature has come to be well known in phage biology as "Killing the winner" or "Kill the winner" (Rodriguez-Brito et al. 2010; Winter et al. 2010; Diaz-Munoz and Koskella 2014; Jacquet et al. 2018), or, as described for more closely related bacterial strains, "Constant-diversity dynamics" (Rodriguez-Valera et al. 2009). For an early expression of the concept, or at least that is my interpretation, see Wiggins and Alexander (1985, p. 22):

> It is not certain how many natural ecosystems contain populations of individual species or strains greater than this threshold. . . .bacteriophages probably do not account for the frequent disappearance of species introduced into such [low bacterial density] environments. Conversely, in environments in which high densities of hosts exist, bacteriophages would flourish and could affect the number and activity of the bacteria present.

Killing the winner is just like stabilizing frequency-dependent selection except that rather than competition between alleles occurring within a single bacterial population, instead competition is occurring between essentially clonal bacterial lineages within a bacterial community. These bacteria can be and often are considered to be different bacterial species, thus, again, making this an ecological phenomenon rather than an evolutionary one. In actuality, however, a concept of bacterial phage-susceptibility types (van Berkum et al. 1993) can be used instead, where I define 'phage-susceptibility type' based on Thingstad et al. (2008), p. 265, as, "the group of hosts susceptible to a specific type of phage". Thus, within a single bacterial population, bacteria displaying different phage-resistance mechanisms could be viewed as having different phage susceptibilities, whereas within a single bacterial community different bacterial species also often can be viewed as having different phage susceptibilities. All of these distinctions are predicated on the idea that individual phages often display relatively narrow host ranges (Chap. 10) and thus phages exist—perhaps most of them—that are able to infect and kill just a limited number of bacterial phage-susceptibility types.

Also as with stabilizing frequency-dependent selection, the net consequence of killing the winner is that more prevalent phage-susceptibility types, i.e., bacteria that are found within environments at greater densities, will tend to be more susceptible to catastrophic phage impacts than less prevalent phage-susceptibility types. These more prevalent bacterial types are the so-called 'winners' and, all else held constant, phages can propagate on bacterial winners to higher densities than they can infecting non-winner hosts, resulting in greater phage-induced declines of winners than of non-winners. The suggestion therefore is that when kill the winner operates efficiently, then the diversity of bacterial communities, at least in terms of phage-susceptibility types, will be higher, though the same caveats apply as with stabilizing frequency-dependent selection. That is, this proposed efficiency requires that bacteria be able to replicate to sufficiently high densities that they can support the production of sufficiently high phage titers to have a substantial impact on the bacterial populations supporting their growth.

Overall, then, the more bacteria that are present, especially in terms of the prevalence of individual phage-susceptibility types, then the more vulnerable those bacteria should be to phage attack, all other variables held constant. This is so if considering just different bacterial strains within a single bacterial population, and therefore what legitimately can be described as giving rise to stabilizing frequency-dependent selection, but it also is true if instead of a single bacterial species we consider multiple bacterial species. Unfortunately, for all of this musing, distinguishing among bacteria in terms of phage-susceptibility types in practice is time consuming, i.e., see the concept of phage typing (Williams and LeJeune 2012), and is particularly so to the extent that phage susceptibility can differ within species by single nucleotides. This makes explicit testing of these various ideas, except under well-controlled conditions, somewhat difficult. Still, all else held constant and as a default assumption, we expect those bacteria that are most prevalent within environments to also be those bacteria that are most vulnerable to phage attack. If that is not the case, then it is of interest to ask why that is so, e.g., such due to a lack of appropriate phages, though alternatively it could be due to substantial micro-diversity within a single bacterial strain that corresponds to a diversity of phage-susceptibility types, which can be challenging to become experimentally aware of.

24.8 Stabilizing Frequency-Dependent Selection Mimicking Muller's Ratchet?

One interpretation of killing the winner is that the most fit bacterial genotypes, absent phage presence, can be the most susceptible given phage presence. This is explicitly the case to the extent that 'winning' is equated with higher rates of bacterial population growth (Jacquet et al. 2018). To me, as applied to individual bacterial populations, this is recognizably similar to Muller's ratchet, except that rather than the most-fit genotypes being removed by genetic drift, instead the most-fit genotypes are being removed—or at least dramatically reduced in number—by natural selection. Thus, if the bacterial type with the least number of detrimental mutations is the fastest growing, thinking especially in terms of individual bacterial populations, then it will be genotypes that are at least arguably closer to wild type that will tend to be reduced in frequency by phage-mediated predation according to the kill-the-winner/stabilizing frequency-dependent selection model. Furthermore, if those detrimental mutations are associated with phage-resistance pleiotropies (Chap. 22), then winner populations may be even more directly selected by phages toward possessing otherwise deleterious mutations. Thus, as a consequence of this frequency-dependent selection within individual bacterial populations, both the otherwise most-fit and the most-phage-susceptible genotypes can be lost from bacterial populations.

This analogy came to me while reading Avrani et al. (2011) with their suggestion (p. 607) of an "'arms race' between bacteria and their viruses leads to the emergence of resistant bacteria in both a sequential and accumulative process, resulting in a continuum of cyanobacteria with different but overlapping ranges of viral susceptibility." In other words, different numbers of potentially pleiotropically detrimental phage-resistance mutations can exist within the individuals making up a population, with more or less the most mutation free bacteria the more effectively replicating and therefore most likely to be lost to subsequent phage predation. Just as with Muller's ratchet, however, wild-type or, at least, least-mutated genotypes may still be reconstituted via horizontal gene transfer between these bacteria, and it therefore is perhaps telling that Avarni et al. also map a large fraction of mutations to phage resistance to presumably highly mobile genomic islands (Chap. 13) within *Prochlorococcus*. Alternatively, excessive linkage between these mutations, i.e., as within individual genomic islands, might hinder rather than help toward restoring wild-type genotypes, unless the actual genotype-restoring homologous recombination is seen within islands rather than in terms of island acquisition as a whole. A complication on this Muller's ratchet-like scenario, however, is that in the real world, competition between species also should exist, presumably interfering with the potential for somewhat mutated bacterial strains to reach winner densities, thereby returning to the actual killing the winner scenario. Still, and again, to the extent that wild-type genotypes may be fully or partially restored through horizontal gene transfer, then so too a potential to re-attain winner capabilities within bacterial populations might be restored as well through horizontal gene transfer.

References

Abedon ST (2011) Envisaging bacteria as phage targets. Bacteriophage 1:228–230
Abedon ST (2012) Bacterial 'immunity' against bacteriophages. Bacteriophage 2:50–54
Abedon ST (2015) Ecology of anti-biofilm agents I. antibiotics versus bacteriophages. Pharmaceuticals (Basel) 8:525–558
Abedon ST (2016) Phage therapy dosing: the problem(s) with multiplicity of infection (MOI). Bacteriophage 6:e1220348
Abedon ST (2017a) Active bacteriophage biocontrol and therapy on sub-millimeter scales towards removal of unwanted bacteria from foods and microbiomes. AIMS Microbiol 3:649–688
Abedon ST (2017b) Commentary: Communication between viruses guides lysis-lysogeny decisions. Front Microbiol 8:983
Abedon ST (2019) Look who's talking: T-even phage lysis inhibition, the granddaddy of virus-virus intercellular communication research. Viruses 11:951
Abedon ST, LeJeune JT (2005) Why bacteriophage encode exotoxins and other virulence factors. Evol Bioinform Online 1:97–110

Arber W, Linn S (1969) DNA modification and restriction. Annu Rev Biochem 38:467–500
Avrani S, Wurtzel O, Sharon I, Sorek R, Lindell D (2011) Genomic island variability facilitates Prochlorococcus-virus coexistence. Nature (London) 474:604–608
Berenstein D (1986) Prophage induction by ultraviolet light in *Acinetobacter calcoaceticus*. J Gen Microbiol 132:2633–2636
Blazanin M, Turner PE (2021) Community context matters for bacteria-phage ecology and evolution. ISME J 15:3119–3128
Bodmer WF, Bodmer JG (1978) Evolution and function of the HLA system. Br Med Bull 34:309–316
Bohannan BJM, Lenski RE (1997) Effect of resource enrichment on a chemostat community of bacteria and bacteriophage. Ecology 78:2303–2315
Breitbart M, Rohwer F, Abedon ST (2005) Phage ecology and bacterial pathogenesis. In: Waldor MK, Friedman DI, Adhya SL (eds) Phages: Their Role in Bacterial Pathogenesis and Biotechnology. ASM Press, Washington DC, pp 66–91
Brown SP, Le Chat L, De Paepe M, Taddei F (2006) Ecology of microbial invasions: amplification allows virus carriers to invade more rapidly when rare. Curr Biol 16:2048–2052
Chao L, Levin BR (1981) Structured habitats and the evolution of anticompetitor toxins in bacteria. Proc Natl Acad Sci USA 78:6324–6328
Chao L, Levin BR, Stewart FM (1977) A complex community in a simple habitat: an experimental study with bacteria and phage. Ecology 58:369–378
Danis-Wlodarczyk K, Dąbrowska K, Abedon ST (2020) Phage therapy: the pharmacology of antibacterial viruses. Curr Issues Mol Biol 40:81–16
Diaz-Munoz SL, Koskella B (2014) Bacteria-phage interactions in natural environments. Adv Appl Microbiol 89:135–183
Harrison E, Brockhurst MA (2017) Ecological and evolutionary benefits of temperate phage: what does or doesn't kill you makes you stronger. Bioessays 39:170012
Hawlena H, Bashey F, Lively CM (2010) The evolution of spite: population structure and bacteriocin-mediated antagonism in two natural populations of *Xenorhabdus* bacteria. Evolution 64:3198–3204
Igler C, Abedon ST (2019) Commentary: A host-produced quorum-sensing autoinducer controls a phage lysis-lysogeny decision. Front Microbiol 10:1171
Jacquet S, Zhong X, Peduzzi P, Thingstad TF, Parikka KJ, Weinbauer MG (2018) Virus interactions in the aquatic world. In: Hyman P, Abedon ST (eds) Viruses of Microorganisms. Caister Academic Press, Norwich, pp 115–141
Korona R, Levin BR (1993) Phage-mediated selection and the evolution and maintenance of restriction-modification. Evolution 47:556–575
Levin BR (1988) Frequency-dependent selection in bacterial populations. Philos Trans R Soc Lond Ser B Biol Sci 319:459–472
Li XY, Lachnit T, Fraune S, Bosch TCG, Traulsen A, Sieber M (2017) Temperate phages as self-replicating weapons in bacterial competition. J R Soc Interface 14:20170563
Migalska M, Sebastian A, Radwan J (2019) Major histocompatibility complex class I diversity limits the repertoire of T cell receptors. Proc Natl Acad Sci USA 116:5021–5026
Petersen JL, Morris CR, Solheim JC (2003) Virus evasion of MHC class I molecule presentation. J Immunol 171:4473–4478
Rodriguez-Brito B, Li L, Wegley L, Furlan M, Angly F, Breitbart M, Buchanan J, Desnues C, Dinsdale E, Edwards R, Felts B, Haynes M, Liu H, Lipson D, Mahaffy J, Martin-Cuadrado AB, Mira A, Nulton J, Pasic L, Rayhawk S, Rodriguez-Mueller J, Rodriguez-Valera F, Salamon P, Srinagesh S, Thingstad TF, Tran T, Thurber RV, Willner D, Youle M, Rohwer F (2010) Viral and microbial community dynamics in four aquatic environments. ISME J 4:739–751
Rodriguez-Valera F, Martin-Cuadrado AB, Rodriguez-Brito B, Pasiç L, Thingstad TF, Rohwer F, Mira A (2009) Explaining microbial population genomics through phage predation. Nat Rev Microbiol 7:828–836

Stewart FM, Levin BR (1984) The population biology of bacterial viruses: Why be temperate. Theor Pop Biol 26:93–117

Thingstad TF, Bratbak G, Heldal M (2008) Aquatic phage ecology. In: Abedon ST (ed) Bacteriophage Ecology. Cambridge University Press, Cambridge, pp 251–280

van Berkum P, Kotob SI, Basit HA, Salem S, Gewaily EM, Angle JS (1993) Genotypic diversity among strains of *Bradyrhizobium japonicum* belonging to serogroup 110. Appl Environ Microbiol 59:3130–3133

Vasu K, Nagamalleswari E, Nagaraja V (2012) Promiscuous restriction is a cellular defense strategy that confers fitness advantage to bacteria. Proc Natl Acad Sci USA 109:E1287–E1293

Wangersky PJ (1978) Lotka-Volterra population models. Ann Rev Ecol Syst 9:189–218

Weitz JS, Hartman H, Levin SA (2005) Coevolutionary arms races between bacteria and bacteriophage. Proc Natl Acad Sci USA 102:9535–9540

Wiggins BA, Alexander M (1985) Minimum bacterial density for bacteriophage replication: implications for significance of bacteriophages in natural ecosystems. Appl Environ Microbiol 49:19–23

Williams ML, LeJeune JT (2012) Phages and bacterial epidemiology. In: Hyman P, Abedon ST (eds) Bacteriophages in Health and Disease. CABI Press, Wallingford, pp 76–85

Winter C, Bouvier T, Weinbauer MG, Thingstad TF (2010) Trade-offs between competition and defense specialists among unicellular planktonic organisms: the "killing the winner" hypothesis revisited. Microbiol Mol Biol Rev 74:42–57

Young JPW, Levin BR (1993) Adaptation in bacteria: unanswered ecological and evolutionary questions about well-studied molecules. In: Berry RL, Crawford TJ, Hewitt GM (eds) Genes in Ecology. Blackwell Scientific Publications, London, pp 169–192

Chapter 25
A Primer on Phage-Bacterium Antagonistic Coevolution

Contents

Abstract The primary emphasis of this monograph has been on how one population, particularly of a given phage type, can drive the evolution of a second population, particularly of a given bacterium type. Less simplisitic, instead two populations can drive each other's evolution and thereby co-evolve. When two populations are coevolving in a conflicting manner, we can describe this as antagonistic coevolution. It can be more straightforward, however, to envision two cooperating populations coevolving, e.g., an uninduced prophage and its host, and for them to coevolve with substantial exclusivity. With antagonistic coevolution, by contrast, there often can be little guarantee that the antagonistic interactions will be limited to between only two populations. Still, this chapter considers some basics regarding antagonistic coevolution as in principle this could occur between only one particular, especially strictly lytic phage, and one particular at least initially susceptible bacterium. The phage will tend to evolve toward having a greater negative impact on the bacterial population while the bacterium will tend to evolve toward interfering with that phage-mediated negative impact. More generally, bacterial variants will tend to be selected toward being less negatively affected by phages while phages will tend to be selected to overcome those bacterial tendencies. There can, however, be a number real-world complications on such processes and their interpretation, as this chapter considers.

S. T. Abedon, *Bacteriophages as Drivers of Evolution*,
https://doi.org/10.1007/978-3-030-94309-7_25

> . . .the phage can mutate to a form able to attach to the resistant bacterium. The process can continue by the occurrence of a new bacterial mutant resistant to the new phage, and another phage mutant capable of growth on the latest bacterial strain. One can imagine an indefinite series of such fresh invasions of previously resistant hosts, followed by the emergence of correspondingly new bacterial types.—Allan Campbell (1961).

25.1 Preamble

Throughout this monograph, I have striven to avoid delving too deeply into either ecology or bacteriophage evolutionary biology, or especially molecular evolution. This is because the intention has been to keep the focus on general principles of what evolution phages are *causing*. This also has been rather than what evolution phages have been undergoing, unless, e.g., phage evolution could contribute to bacterial evolution, such as in the case of phage morons (Chap. 14). Only occasionally as well does evolution driven by phages appear to be most explainable based on somewhat specific ecological scenarios (e.g., Chap. 24). Mostly, also, I have done my best to avoid molecular considerations altogether, both for reasons of constraints on space and constraints on my personal interests, though of course it is impossible to ignore molecules completely when considering microorganisms (Young and Levin 1993).

In this chapter, I will attempt to continue this approach of emphasizing basic principles of what evolution phages are driving. Thus, even though the topic of antagonistic coevolution—also described as phage-bacterium arms races—is both inherently ecological and involves a large amount of phage evolution, I have done my best to keep much of the focus on bacteria and their evolution. The caveat, though, is that avoiding talking about either phage ecology or phage evolution in the context of antagonistic coevolution isn't really possible, e.g., see De Sordi et al. (2019) for instead substantial consideration of both phage ecology and phage evolution, and indeed also phage molecular evolution, all in a context of phage-bacterial antagonistic coevolution. An additional issue is that much of the bacteria side of coevolution is phage resistance, as has been covered in multiple previous chapters (Chaps. 18 through 24). Therefore, in this chapter I consider the concept of antagonistic coevolution without delving too deeply into the specifics of phage resistance, or the specifics of phage evolution, or much of the ecology of antagonistic coevolution, while at the same not really ignoring any of these issues. Though that statement may sound a bit bizarre, and substantially limiting, in fact a fair amount of thinking on what phage-bacteria antagonistic coevolution is all about can be and still is presented.

25.2 Introduction to Coevolution and Antagonistic Coevolution

Coevolution refers to changes in allele frequencies over time, which is evolution, but coevolution is evolution as it can occur in association with two different but interacting entities. The idea, basically, is that genetic-based phenotypic changes that occur in one of these two entities can motivate—via natural selection acting on existing genetic variation—the occurrence of one or more genetic-based phenotypic changes in the other entity. These different entities can be, e.g., associated with different loci found within the same species or instead different loci located within different species. The defining characteristic of coevolution thus is that frequencies of alleles change in response to changes in the frequencies of alleles found at one or more different loci, and vice versa. Thus, the structures of two different proteins within the same bodies can coevolve so that they better interact with each other, or the anatomical structures of two different bodies can coevolve also to better interact, the latter, e.g., such as in terms of the physical processes involved in mating.

Dennehy (2012) reviewed the theory underlying these processes with some emphasis on phage-bacterial coevolution; for a shorter version, see Box 2 of de Jonge et al. (2019). See also Brockhurst et al. (2021), and for a more general discussion see Brockhurst and Koskella (2013). Below, though, the emphasis is less on integrating this theory than simply attempting to understand the dynamics of what might be going on in laboratory experiments, and by extension in nature, in terms of phage-bacterial coevolution. First, though, I consider just what is meant by coevolution and antagonistic coevolution.

25.2.1 Evolution of Interactions

Coevolution is typically seen with mutualistic interactions, that is, where relations between two species can improve toward some sort of reciprocal benefit, or interactions instead can be antagonistic. Fitness gains are, as always, favored within populations, but for a given mutation, these fitness gains can either be shared between the two interacting species—as a mutualism, with both species benefiting as one allele increases in frequency within one population—or instead fitness gains by one species can occur at the expense of the other species (antagonism). For the latter case, one species benefits as one allele increases in frequency within that species' population while the other species, at the same time, is harmed.

In either case, for this to be coevolution between two different species, then generally evolution should not be occurring in response to multiple aspects of the environment (not 'diffuse'; Sect. 25.5.1). Instead, coevolution explicitly should be a product of interactions going on between only two species. The result can be chains of genetic modification; for example a change at locus *A* in species 1 motivating a change at locus *B* in species 2, which in turn motivates a change at locus *C* in species

1, and so on. This I will describe as an 'iterative coevolution'; see equivalently, "Iterative dynamics" (Obeng et al. 2016). With coevolution more generally, the changes that are selectively beneficial to the carrier of the coevolving alleles are constrained by either being responsible for the same phenotype, e.g., as may be seen with changes to different subunits of the same protein, or instead by being responsible for changes in separate phenotypes that in some manner interact.

25.2.2 *Antagonistic Coevolution*

Antagonistic coevolution occurs between two specific species where individual members of course antagonize or at least oppose each other in some manner. By 'antagonism', this generally means that they can cause harm to each other, though that harm typically is manifest in terms of the individuals of one species in some manner attacking individuals of the other species and individuals of the other species in some manner protecting themselves. An obvious example of antagonism is seen with predation or herbivory. Often, however, antagonistic interactions between species involve more than two species, e.g., with predators harming more than one species of prey and prey being preyed upon by more than one species predator (here we can again use the word, 'diffuse'). In practice, this means that a species that is being antagonized, i.e., the victim, may evolve defenses that are less specific to the species that is doing the antagonizing, i.e., the exploiter, while the exploiter evolves means to overcome defenses that are not necessarily specific to a given victim species. For example, the evolution of armor in aquatic environments versus the countering evolution of armor-piercing adaptations in predators such as the latter's possessions of large teeth in strong jaws (Morris 1885), to cite a familiar evolutionary example. That, over longer time scales, may at best have been products of diffuse antagonistic coevolution rather than antagonistic coevolution occurring between just two different species. For bacteria and phages, though, we might equivalently cite the evolution of the bacterial cell wall as a form of general anti-phage armor (Prangishvili 2013) (Chap. 19).

By contrast, in simper ecosystems and over shorter time scales, it indeed can be the case that there is only one predator species and only one prey species. Since viruses, or parasites more generally, often are both harmful and specific to their hosts, the occurrence of antagonistic coevolution between only two species, the virus and the host, is more easily envisaged. Even in this example, however, it is true that host species often have to contend with multiple viruses, and indeed contend with multiple parasite species, while to some degree viruses—along with parasites more generally—can have host ranges that span more than one population or even more one species; e.g., for phages possessing host ranges that span more than one bacterial species, see (Hyman and Abedon 2010, Koskella and Meaden 2013, de Jonge et al. 2019, Ross et al. 2016, Hyman 2019). Nonetheless, in considering antagonistic coevolution in this chapter, we will keep things simple by assuming that we are working with one bacterial virus *population* and one bacterial *population*, i.e., so that we can identify basic principles of phage-bacterium antagonistic coevolution.

25.3 Short Historical Overview of Phage-Bacterium Antagonistic Coevolution

Note the quotation from Campbell (1961) found at the start of this chapter, which clearly is describing phage-bacterium antagonistic coevolution, but neither of those words appear to be used in that publication. Consequently, I started my look at the history of phage-bacterium antagonistic coevolution, as a phrase, with a Google Scholar search on "Antagonistic coevolution", in quotation marks, in combination with various forms of "Phage". Thus, "antagonistic coevolution" AND (Phage OR Phages OR Bacteriophage OR Bacteriophages), was my search. I then narrowed down the search using the "Custom range..." function. The results, in terms of the early history of the use of this term, were interesting indeed, indicating an early appreciation that then failed to catch on for nearly two decades. Keep in mind, though, that *study* of phage-bacterium antagonistic coevolution does not require use of the term, antagonistic coevolution; again, see the quotation at the start of this chapter. Nonetheless, given the now popular use of 'antagonistic coevolution' in the phage literature, it is interesting to explore the origin of its use to describe phage-bacteria interactions.

From the above Google Scholar search, I was able to see that the oldest hit is from 1983, in a chapter by Bruce R. Levin and his then postdoc, Richard E. Lenski (Levin and Lenski 1983). This is titled, "Coevolution in bacteria and their viruses and plasmids". In that chapter, there are three mentions of "Antagonistic coevolution". The following quotation is from p. 126, and is its last explicit mention in the chapter. The first emphasis is theirs and the second is mine:

> The route that is taken for over-replication will depend on the genetic and physiological constraints on the replicon and its host and on the environment of the bacterial population. With respect to the latter, we believe that *population density of the host is the primary factor in determining the form of over-replication.* In high density populations of bacteria, replicons with effective mechanisms of horizontal transmission would be favored. *The extreme of this would be virulent phage and the resulting antagonistic coevolution.* As the density of the host population declined, so would the intensity of selection for infectious transmission. Niceness would become increasingly important for the persistence of the autonomous replicon, and mutualistic coevolution would result.

See also Lenski and Levin (1985) for an extension of their 1983 chapter, focusing exclusively on strictly lytic phage-bacterial coevolution, though which doesn't explicitly use the phrase, 'antagonistic coevolution'. After that, over the succeeding years, Google Scholar searches become less straightforward as hits do not consistently consider antagonistic coevolution and phages in the same context. So far as I was able to tell, however, there was an almost 20-year gap between the first use of the term antagonistic coevolution by Levin and Lenski within a phage context and the next publication using the term, also in a phage context. This next publication is that of Buckling and Rainey (2002), "Antagonistic coevolution between a bacterium and a bacteriophage", which I consider in detail at the end of this chapter.

It is mostly or perhaps even entirely from this later publication that 'antagonistic coevolution' began to enter into the phage biology lexicon, and this perhaps is especially the case since not even Buckling and Rainey cite the 1983 Levin and Lenski chapter. They certainly acknowledged, though, that (p. 931), "The most frequently used microbial system for the study of antagonistic coevolution is bacteria and virulent phage." They also cited Lenski (1984), who noted (p. 322) that, "Rodin & Ratner also assume that coevolution of bacteria and phage is necessarily antagonistic." Rodin and Ratner (1983a) in turn stated that (p. 189), "This ecosystem consists of two antagonistic populations: bacteria (preys) and virulent phages (predators)." Rodin and Ratner (1983b) also state (p. 206) that, "Antagonistic interactions between phage and bacteria strains coexisting (and may be coevolving) in chemostat conditions…" Not surprisingly, given that they were published in the same year, neither of those Rodin and Ratner publications cite Levin and Lenski (1983). In any case, clearly we can trace application of at least the term, 'antagonistic', to coevolving phage and bacterial populations to the early 1980s, though this concept, by the name, antagonistic coevolution, does not seem to have really caught on until the early 2000s.

Notwithstanding this history of the use of 'antagonistic coevolution' in the phage literature, it is worth keeping in mind that the first use of 'antagonistic' in this literature is found in what historically was the first phage publication to be recognized as a phage publication (d'Herelle 1917, Summers 1999, d'Herelle 2007, d'Herelle 2011): "Sur un microbe invisible antagoniste des bacilles dysentériques" (On an invisible microbe antagonistic to [or toward] dysentery bacilli). The word 'coevolution' in a phage context, by contrast, seems to date especially to Horne (1970): "Coevolution of *Escherichia coli* and bacteriophages in chemostat culture", which most certainly is an article studying antagonistic coevolution even if not explicitly so described. Again as noted in the quotation at the start of this chapter, the idea of antagonistic coevolution in the phage literature, if not the actual terminology, dates at least as far back as Campbell (1961). See also Stent (1963), as quoted further below.

25.4 Brief Primer on Phage-Bacterium Antagonistic Coevolution

Before proceeding further, see Fig. 25.1 for my overview of phage-bacterial antagonistic coevolution, particularly as viewed from the phage perspective. Note that we can differentiate the consequences of this coevolution—particularly in terms of improvements in phage performance—from two different perspectives. These are (1) how phage evolution can result in greater numbers of phages (i.e., titers),

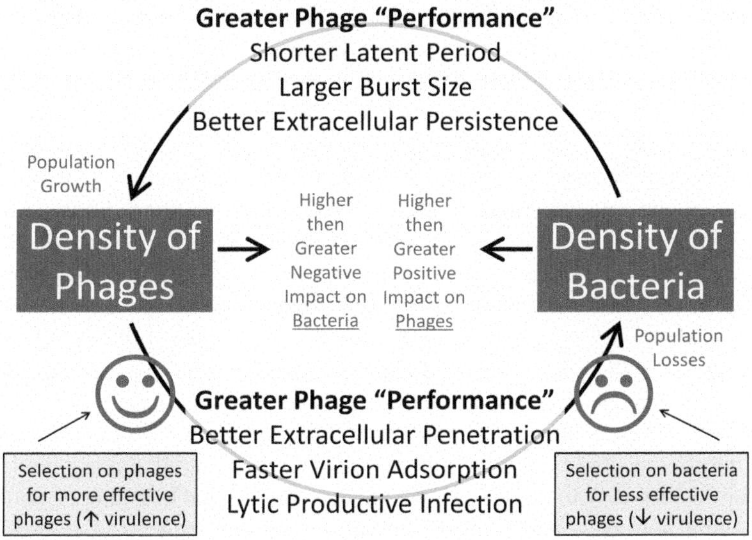

Fig. 25.1 Ecological bases of phage-bacterium antagonistic coevolution, as presented especially from a phage perspective. Anything that increases phage densities (top, center) or the ability of individual phages to reduce the fitness of individual bacteria (bottom, center), will result in greater antagonistic selective pressures acting on bacteria. Though not indicated other than as a gray frowny face in the figure, anything that bacteria can do to resist these phage tendencies will tend to serve as selective pressures acting on phages to counter those bacterial mechanisms. 'Extracellular penetration' refers to the effectiveness of virion movement through environments toward bacteria such as within biofilms (Dennehy and Abedon 2021) and many of the presented ideas otherwise can be derived from discussions by Levin and Lenski (1983)

measured especially in terms of the production of new phage virions (top in the figure) in combination with how bacteria can come to resist this, and (2) how phage evolution can result in a greater *per capita* impact of phages on bacteria (bottom in the figure), the latter being what can be described as one aspect of phage antibacterial virulence (Horne 1970, Levin and Lenski 1983), and again how bacteria can come to resist this. The extent of the phage impact on a bacterial population thus should be dependent on these two things, i.e., on how many phages are produced and the ability of each phage that is produced to individually harm bacteria (Levin and Lenski 1983), while bacterial resistance evolution (Chaps. 18 through 20) should result in some degree of reduction of either or both. Overall, the total numbers of phages produced, however, is also a function, ecologically, of bacterial densities (see the quotation from Levin and Lenski in Sect. 25.3 for the latter, as well as Fig. 24.4 in Chap. 24, and also Chap. 7).

As indicated, bacteria in turn evolve to resist this phage onslaught, with the strength of selection for the resulting phage resistance presumably a function of the potential for the phage population to negatively impact the bacterial population (Abedon 2011, 2012) (Chap. 24). Evolution of phage resistance by the bacterial population should then motivate phage populations to overcome that resistance. This results, if the phages are successful in that evolution, in increases in phage numbers and/or increases in the ability of phages to reach and then infect bacteria, and so on. These are points that we will return to especially in Sect. 25.5.

The result of antagonistic coevolution can be increased specialization by both phages and bacteria. For phages, the changes that result in phages becoming better infectors generally can be specific to a targeted bacterial population. This can also give rise to phage antagonistic pleiotropies to the extent that phage improvements in infecting one host population, such as one host strain, results in phages that are less effective at infecting a different host population, such as a different bacterial strain (Dennehy 2012). It is important to note at this point that these abilities are not black and white, which is to say that less effective abilities do not necessarily mean no ability at all but rather, e.g., perhaps resulting in smaller burst sizes when infecting the host strain that they have not evolved on. In addition, this is not to say that phage improvements in infecting a bacterial strain it is coevolving with cannot result as well in improvements in abilities to infect a different bacterial strains—and indeed phages also can evolve to be less specialized in terms of what hosts they are infecting though again not necessarily without pleiotropic costs (Dennehy 2012)—it is just that phage phenotypic improvements in the course of phage-bacterial coevolution are expected to be more bacterial strain specific than not.

Bacteria, by contrast, can become in a sense specialists in evading specific phages. This is explicitly true in the case of acquisition of new CRISPR spacer sequences (Chap. 5, Fig. 5.4), since those sequences exactly match unique or at least somewhat infrequent proto-spacer sequences found on the target phage. This is true as well with temperate phages and resulting lysogenic cycles, since those bacteria with their superinfection immunity (Chap. 16, Sect. 16.1) are more or less specialists in evading phages that have the same immunity types as their prophages. Less obviously, though, so too phage resistance can be an example of an antagonistic pleiotropy in terms of bacterial fitness as measured in the absence of phages (Chap. 22), but less commonly might also mean that decreased susceptibility to one phage is associated with increased susceptibility to another (Avrani et al. 2011). In terms of the former consideration (again, Chap. 22), bacteria in a sense become phage-evasion specialists rather than, e.g., population growth-rate specialists. Thingstad et al. (2008) describe this latter tradeoff as being between a bacterium being (p. 262) a "defense specialist (e.g., as against viruses)" versus being a "competition specialist (e.g., as for nutrients)".

Note that Janzen (1980), whom we shall return to below (Sect. 25.5.1), provides a number of cautions when interpreting something as coevolutionary. Restated from a perspective of phages and bacteria, basically these can be summed up as 'don't just assume that either bacterial resistance to a phage or a phage's ability to overcome a bacterial resistance mechanism are products of coevolution'. In Dąbrowska and Abedon (2019), within a context of phage therapy, we make a similar suggestion in distinguishing between "Community resistance" by bacteria to phages (not necessarily coevolutionary) and "Treatment resistance" (which instead is, at least to a degree, coevolutionary). A related issue is just how diverse especially hosts must be before we consider coevolution to be diffuse. The contrasting question is whether individual phage coevolution with just a single clonal bacterial population ever occurs in nature. Furthermore, a single phage population evolving to become 'generalists' in their infection of a single but mutationally diversifying clonal bacterial population, e.g., Sect. 25.6, can at the same time can be evolving to become specialists at infecting that same clonal bacterial population. Thus, even when coevolution is both occurring and less diffuse (initially only one phage type and only one bacterium type), a phage population still can evolve to be both less effective at infecting other bacterial strains and more effective at infecting the single diversifying bacterial strain with which it is coevolving. Antagonistic coevolution, in other words, is not necessarily as straightforward as the simple 'arms race' metaphor might suggest. In the following section (Sect. 25.5), I take a further look at just what is and what is not phage-bacterium antagonistic coevolution.

25.5 Different Faces of Phage-Bacterium Antagonistic Coevolution

Though constrained by what it is that evolves and how, the underlying basis of antagonistic coevolution nevertheless still is evolution. Thus, bacteria are changing in some manner mutationally, these changes are selected especially by the negative consequences of phage infections, and phages are motivated to evolve in response to those bacterial changes, particularly coming to be better infectors of specific, targeted bacteria. Indeed, Buckling and Rainey (2002) define antagonistic coevolution within this context simply as "Reciprocal evolution of host resistance and parasite infectivity." Beyond those basic ideas, in my mind there are two highly relevant additional issues regarding antagonistic coevolution and these are (1) the nature of the evolutionary changes in combination with the associated ecology and (2) the extent to which phages in nature tend to be sufficiently limited in terms of what hosts they are infecting for true coevolution to actually occur. That is, what is actually going on and, at least in nature, can we actually call it coevolution? I start with the latter question, basically one of how long chains of reciprocal evolutionary change must be before we should call it coevolution. I then turn to the considerations of the former, especially addressing first whether phages might be more limited than bacteria in terms of their ability to undergo coevolutionary change (short answer: yes, but not necessarily always) and then the impact of different types of bacterial mutations. The latter is particularly in terms of how they affect bacterial fitness in the absence of phages, i.e., as also covered though from a slightly different perspective in Chap. 22.

25.5.1 How Long Must Chains Be?

Ideally, coevolution consists not just of evolution that occurs in response to a second species but instead evolution that occurs cumulatively, i.e., with new coevolutionary mutations occurring in individuals which are already carrying old coevolutionary mutations. This thus would be 'cumulative' evolution, or what typically is described in the antagonistic coevolution literature as 'arms race dynamics'. Strictly speaking, though, coevolution is not necessarily always such an iterative process. Indeed, in the earliest reference to coevolution between phages and bacteria, which as noted is an example of antagonistic coevolution, Horne (1970) commits only to the suggestion (p. 993, emphasis mine) that, "Coevolution of bacterial resistance and phage virulence *may* involve changes in both bacteria and phage." Janzen (1980), by contrast, explicitly defines coevolution (p. 611, emphasis also mine) as "an evolutionary change in a trait of the individuals in one population in response to a trait of the individuals of a second population, *followed by an evolutionary response by the second population to the change in the first*." Horne thus suggests that phage-bacterium coevolution could be cumulative and Janzen seems to suggest that it *ought to be* before we call it coevolution.

Alternatively, if a population of phages is evolving to overcome phage-resistance associated with one bacterial population, and then evolving to overcome resistance associated with a different bacterial population, then that is not iterative coevolution, even if those mutations are building upon each other. That is, in this case evolutionary change indeed may be cumulative, but it is not exactly coevolutionary in the Janzen sense of the term. Similarly, if bacteria are evolving to be resistant to one phage population and then evolving to be resistance to a different, unrelated phage population, that too is not iterative coevolution, even if both mutations again are found in the same genome (again, even if the evolution is cumulative, it is not necessarily iterative *co*evolution). As I've hinted at above, Janzen (1980) suggested a concept of "Diffuse coevolution", which he defined as occurring when coevolving entities are each defined by more than one population, e.g., which could be multiple bacterial strains interacting with multiple phage types. Such diffuse coevolution between bacteria and phages could result, I suppose, in the evolution of more general mechanisms of phage resistance among multiple bacteria such as perhaps thicker capsules or, over longer time scales, more general means by which phages overcome bacterial resistance systems (Hampton et al. 2020), such as the restriction-modification or CRISPR-Cas systems that are found in different bacterial strains.

In any case, we still can consider single-round rather than iterative coevolution. This would involve a single phage population overcoming bacterial resistance displayed by a single bacterial population, or a single bacterial population evolving phage resistance in response to exposure to a single phage population (the latter as equivalent to 'treatment resistance'; Dąbrowska and Abedon 2019). The traits that are being overcome, however, in either case would not have evolved in the course of association between the same two phage and bacterial populations. This scenario would not violate Janzen's (1980) caution, as actual evolution by one population in response to a trait associated with a second population would be occurring, though Dennehy (2012) does suggest (p. 4) that, "If only the hosts are responding, it is merely evolution, not coevolution." This rather would be a bacterium being resistant to a given phage population upon the first contact between those populations and, at least arguably, the phage then *coevolutionarily* responding, or the bacterium being sensitive to a phage upon first contact and then, also at least arguably, the bacterium *coevolutionarily* responding. Left at just that, regardless of whether one agrees whether coevolution was occurring, these scenarios certainly would not be iterative.

We also can consider scenarios consisting of two rounds of coevolution, where evolution of phage-resistance in a bacterial population is countered by the evolution in a phage population of mechanisms overcoming that newly acquired phage-resistance. Or, alternatively, where the phage population first evolves to better impact the bacterial population, with that greater phage virulence then countered by the evolution of phage-resistance in the bacterial population. This certainly gets us closer to Janzen's (1980) definition of coevolution as an iterative process. Still, it would seem to lack any assumptions that the two populations would be coevolving in a cumulative manner since what is being described is basically only two mutational events, one in the bacterial target of a phage and another in the phage population.

What is truly interesting—in terms of how evolutionarily it occurs and the extent to which ecologically it can occur—therefore is when coevolution takes place over more than two rounds, with increased bacterial susceptibility to co-located phages followed by decreased susceptibility, which is then followed again by increased susceptibility, and with bacterial and phage populations evolving reciprocally. This is certainly iterative, and quite close to Janzen's (1980) definition, and see too the definition provided by Scanlan et al. (2015, p. 1): "Antagonistic coevolution entails reciprocal selection for host resistance and parasite infectivity, such that both host and parasite phenotypes change over evolutionary time." This too, however, does not necessarily represent an accumulation of mutations in either bacterial or phage individuals. That is, there is not necessarily in this scenario an assumption that phage-resistance-conferring mutations come to fixation within a bacterial population or at least that it will be those previously selected mutants that will come to encode also the next round (or rounds) of phage-resistance-conferring mutations; same too for the phages in overcoming those mutations. I would still agree, however, that two populations are iteratively coevolving even if coevolutionary mutations are not accumulating within the same individuals. This is a contrast between the noted arms race dynamics and what instead could be a 'fluctuating state dynamics'. That is, with fluctuating state dynamics, abilities do not accumulate in individuals even as two populations are coevolving iteratively. See Fig. 25.2 for comparison.

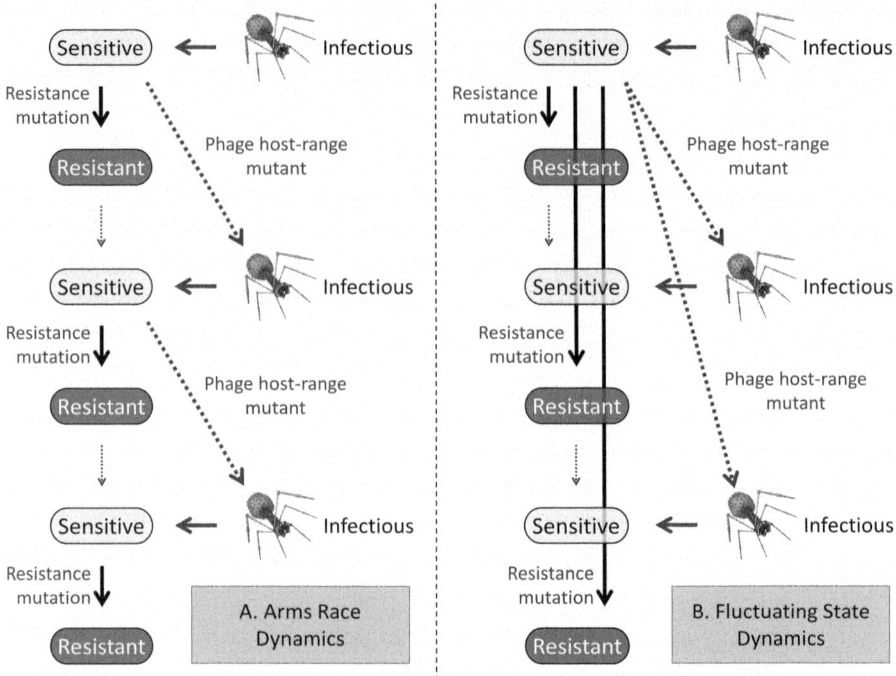

Fig. 25.2 Arms race dynamics versus fluctuating state dynamics. Solid, vertical, downward-pointing black arrows indicate bacterial mutational evolution (keeping in mind that the results of this evolution in terms of the frequency of alleles is a property of populations rather than of individuals). Solid, horizontal, leftward-pointing gray arrows indicate phage infection. Dotted, angled, larger gray arrows indicate phage mutational evolution. Dotted, smaller, vertical, downward-pointing gray arrows indicate that phage evolution is overcoming the resistance phenotype displayed by the previous bacterium, i.e., the bacterium in this case is not genotypically changing though its phage sensitivity is. (**a**) With arms race dynamics, mutations accumulate in bacteria and phages, with phage-sensitive bacteria mutationally giving rise to phage-resistant bacteria. The newly arisen phage-resistant bacteria then become phage-sensitive again due to phage host-range mutation, and those now again phage-sensitive bacteria mutationally give rise to phage-resistant bacteria, and so on. (**b**) With fluctuating state dynamics, instead mutations—to phage resistance by bacteria or phage-resistance overcoming phage host-range mutations—could be occurring, for example, within otherwise wild-type genetic backgrounds. Therefore, mutations would not accumulate within individual genomes, either bacterium or phage

25.5.2 Greater Constraints on Phage Evolution?

As Stent (1963) suggested (p. 181), "The coexistence in nature of bacteria and bacterial viruses is thus sustained by a delicate mutational equilibrium that saves both protagonists from total extinction." This is antagonistic coevolution, if not necessarily given that exact moniker. Lenski (1984), however, explicitly argued against this scenario; see also (Lenski and Levin 1985). In this section, I therefore consider Lenski's argument (p. 319, and who similarly quoted Stent) that, "In general, structural constraints on the highly site-specific phage adsorption process appear more severe than physiological constraints on resource assimilation by bacteria." That is, particularly as based on experience up to that point with coevolution between *E. coli* and various coliphages, Lenski argued against the idea of iterative antagonistic coevolution occurring between phages and bacteria in the Janzen (1980) sense, or at least argued against this coevolution being able to occur more or less indefinitely.

In any case, phage-bacterium antagonistic coevolution as it was initially envisioned—though it wasn't yet referred to strictly as antagonistic coevolution at least in the phage world—involved phages and their bacterial hosts trading back and forth (Campbell 1961) potentially pleiotropically detrimental point mutations (Levin and Lenski 1983). The host, with its larger genome, may however have a greater potential to evolve (Lenski 1984) and also a greater motivation to evolve (Chan and Abedon 2012). In short, more bacterial genes might change in more ways, and the reward to bacteria for successfully evolving in the face of phages was bacterial survival. Phages, by contrast, have fewer genes and would only lose their 'dinner' if they failed to evolve. I think, however, that these perspectives are a bit overly simplistic for a number of reasons. My concerns, as derived in part from Chan and Abedon (2012), include the following observations:

1. Higher phage per-gene mutation rates. Though phages tend to have fewer genes in their genomes than bacteria do, their per gene or per nucleotide mutation rates also tend to be higher. For instance, the per nucleotide mutation rate of coliphages λ, T2, and T4 are together roughly 0.00000005 (7.7×10^{-8} for phage λ and 2.4×10^{-8} for phages T2 and T4) while that for *E. coli* is roughly 0.0000000005 (5.4×10^{-10}). Thus, in these cases the phage per-nucleotide mutation rates, per round of replication, are about one-hundred-fold higher than the bacterial per-nucleotide mutation rate (Drake et al. 1998). Phage populations, that is, can be better able to more fully mutationally explore their sequence space than bacterial populations are able to explore the bacterial sequence space. Dennehy (2012), however, suggests that because these phage mutations by necessity will tend to require a gain or in many cases at least a retention of function (Burmeister et al. 2021), fewer of the occurring mutations may be coevolutionarily useful to phages than is the case for bacteria. Alternatively, though, to the extent that loss-of-function mutations are costly to bacteria (Chap. 22 and below), then perhaps this proposed qualitative distinction between bacteria and phages in terms of rates of occurrence of coevolutionary useful mutations—rates of gain of function versus rates of loss of function—is less relevant than the proposed quantitative differences in per-nucleotide mutation rates.
2. More phage mutation occurrences. Phages potentially can have larger population sizes than bacteria, at least at late stages of phage population growth while infecting a given bacterial population. The more individuals in a population, then the more per-nucleotide mutations that will occur, e.g., (Campbell 1961, Rodin and Ratner 1983a, Weitz et al. 2005). Thus, if a phage population were to grow to ten times the size of a bacterial population, an often bandied about ratio,

then the number of mutations per nucleotide would be one-thousand-fold greater for phages than for bacteria using the above-cited per-nucleotide mutation rates (which are really per nucleotide, per individual mutation rates). It's important to point out, however, that in smaller volumes of both bacteria and phages, mutation numbers too will be smaller regardless of bacterial and phage densities (Chap. 21, Sect. 21.3). Also, if their numbers are sufficiently small, then bacteria themselves may not mutate to phage resistance with reasonable likelihood (Lenski 1984), and nor do phage populations necessarily always exceed bacterial populations in number, e.g., (Lenski and Levin 1985, Bohannan and Lenski 1997, Meyer et al. 2012).

Perhaps more important with regard to the very last point, if a phage population does not grow to exceed that of a bacterial population in size, then both the impact of the phage population on the bacteria and the motivation for both bacteria and phages to evolve resistance or overcome resistance will be lower. In other words, if there are not enough bacteria to support phage population growth to overwhelming titers (Chap. 7, Sect. 7.3.2), essentially if bacteria are not present at 'winner' densities (Chap. 24, Sect. 24.7), then both wild-type bacteria and wild-type phages might continue to coexist without having high motivation to coevolve. Thus, if the phage numerical advantage over bacteria does not exist, then so too the need for coevolution might not be great (Levin and Lenski 1983). I return to this issue in Sect. 25.6. Another issue concerning numbers of individuals and numbers of mutations is a suggestion that more successful bacteria should also be able as populations to more readily mutate to phage resistance (Morgan et al. 2012).

3. Rarity of forced phage loss of function. CRISPR-Cas or restriction-modification systems may target regions of essential phage genes that cannot be mutationally modified without substantially impacting phage fitness (Gurney et al. 2019). It seems unlikely, though, that bacterial acquisition of CRISPR spacer sequences targeting such highly critical regions of phage genomes would be common, though it is worth keeping in mind that mutations need not be highly pleiotropically detrimental to still be detrimental. Similar to CRISPR-Cas recognition, restriction enzyme recognition sites may similarly be difficult for phages to remove mutationally without cost. Phage host-range mutations still might be detrimental, however, even without CRISPR-Cas or restriction enzyme recognition site involvement, as may be the case with so-called "Trigger-happy" mutants that can possess less selective adsorption abilities but also lower virion stability (Drexler et al. 1989). Phages needing to count on this latter option, however, should only be the case to the extent other host-range modifying strategies are unavailable. From the bacterium side, though, acquisition of resistance involving adaptive immunity, i.e., CRISPR-Cas, should be less costly than mutational loss or perhaps even change in bacterial gene function (Levin 2010).
4. Phage resistance by bacteria can still be costly to phages. Though phages might only lose a 'dinner' if they don't succeed in infecting a given bacterial host, phage virions don't necessarily remain un-inactivated forever (Chan and Abedon 2012) and they at best only get one 'dinner' per lifetime. The stakes for phages in terms

of their survival thus are perhaps only quantitatively different from those for bacteria rather than qualitatively different; that is, phages too can die if the encounter between them and a given bacterium does not lead to adsorption. This should be especially so to the extent that phage-resistant bacteria come to dominate an environment, where for phages it literally could be the case of adapt to being able to infect the new bacterial population or instead go locally extinct (Levin and Lenski 1983, Lenski 1988a). Furthermore, not all mechanisms of phage resistance result in phage survival, i.e., negation and bacterial self-sacrifice (Chaps. 18 through 20). Thus, at least the stakes for phages to coevolve are not necessarily fundamentally different from the stakes for bacteria, as both may be facing hard selection (Sect. 23.4) for coevolutionary change.

5. A role of recombination. Returning us to the question of just how many phage and bacteria types we should be considering when talking about antagonistic coevolution, we can ask whether changes in both phages and bacteria are always mutational or instead might involve recombination, and particularly sex (Bouchard and Moineau 2000, De Sordi et al. 2017). But if either bacteria or phages should modify their resistance characteristics or antibacterial virulence by acquiring genes that evolved in different lineages, would that still strictly be coevolution, or instead some form of diffuse coevolution? I will not attempt to answer that question, but instead will stress that this means that the ability of phages to counter bacterial resistance mechanisms likely is somewhat broader than their ability to achieve this increased virulence via mutation alone. Of course, bacteria can acquire phage-resistance mechanisms via horizontal gene transfer as well.

Ultimately, I think that the Lenski (1984) arguments may, for whatever reason, not be fully generalizable to all or maybe even most phage-bacterium systems. This especially is likely the case given the occurrence of recombination rather than just mutation toward phage evolution to overcome bacterial resistance. In addition, bacterial existence within more complex environments likely places additional costs and therefore limitations on the bacteria ability to fully explore possible mutations to phage resistance, i.e., from Blazanin and Turner (2021), by these environments "modifying the pleiotropic consequences of adaptation". The latter point in fact is indirectly suggested by Lenski, p. 323: "...for each extant phage, there be at least one host in at least one environment for which the metabolic functions associated with the phage adsorption site are essential to bacterial growth". Indeed, Gómez and Buckling (2011) provide evidence that bacterial populations are better able to evolutionarily sustain resistance over time in nutrient richer environments rather than nutrient poorer environments (similar results were seen by, e.g., Gómez et al. 2015). This suggests that it may be especially in nutrient poorer environments that the 'phage adsorption sites' are more useful to their displaying bacteria. In fact, Gómez and Buckling (2011) noted (p. 108) that "there was an approximately 36% reduction in relative fitness associated with [phage] resistance in soil", whereas in rich broth no costs were noted.

Rather than making such bacterial mutations impossible, however, I would suggest instead that these costs may make such mutations only less likely, particularly in terms of bacterial survival over medium time frames as involving either changes in environmental conditions or competition with other bacteria. Specifically, if bacteria must retain adequate functionality over time, then their mutational acquisition of phage resistance may require *change* to phage-required bacterial macromolecules rather than simply elimination of those molecules. This may thus make mutation to medium-term ecologically successful phage-resistance less likely but also might make phage mutation to overcoming successful resistance more probable, e.g., as involving slightly different interactions with same receptor molecules rather than needing to recognize entirely different receptors molecules (though see Sect. 25.6).

Thus, "Bacteria cannot completely block the infection of phage without also losing their ability to uptake resources", p. 9536, of Weitz et al. (2005), and certainly accumulating multiple deleterious phage-resistance mutations that result in a cumulative loss of functions should be increasingly costly to bacteria or at least increasingly difficult for bacteria to achieve (Koskella et al. 2012). Alternatively, it is possible for bacteria to evolve back the fitness that was lost with mutation to phage resistance, without necessarily loss of the original phage resistance, even in the absence of phages (Lenski 1988b). It is difficult to say how broadly applicable that observation is, however, given that the original mutation in that case was one impacting lipopolysaccharide rather than bacterial structural proteins. In any case, over medium time scales, under real-world conditions, iterative antagonistic coevolution might still occur over multiple rounds, but perhaps at rates that are slower as well as more iterative than may be the case based solely on laboratory experiments that take place within environments that are overly simplistic, not substantially changing over time, and/or involving only a certain subset of coliphages.

25.5.3 Ecological Scenarios

Thinking on phage-bacterium antagonistic coevolution traditionally did not involve either adaptive immunity or recombination. Consequently, we could readily divide bacterial resistance to phages into three different categories, as based on the fitness of the harboring bacterium as seen in the absence of phages. That is, when phages are around, these phage-resistance mutations must to some degree be selected for, else neither we nor longer-term evolution would ever see them. But when phages are not around, then what? My answer to this question is that these mutations, in the absence of phages, may be either (1) immediately costly to the harboring bacterium (i.e., as considered in Chap. 22), (2) eventually costly to the harboring bacterium (as alluded to with the phrasing 'medium time frames' in Sect. 25.5.2), and (3) not costly to the harboring bacterium. There is also a forth category where (4) mutations actually seem to be beneficial to the harboring bacterium in the absence of phages, but my assumption is that since any mutations will represent changes to otherwise evolved systems, then these mutations either will someday be costly or at least only have an overall neutral impact on bacterial fitness (Campbell 1961). In other words, if a phage-resistance mutation is beneficial in the absence of phages, and is easily attained, i.e., by a single point mutation, then if it were truly advantageous outside of the laboratory it should or at least might already be present in the bacterial population. A similar argument was made by Levin and Lenski (1983) with regard to the coevolutionary response of phages (p. 107): "If host range phage were as fit as or fitter than wild-type phage when competing for sensitive clones, there would be no wild-type phage." Therefore, I will ignore category (4) essentially as a laboratory artifact. As follows, I emphasize instead categories (1) through (3), though not in that order.

We can assume, to start, that cost-free bacterial mutations to phage resistance (3) will tend to be the rarest. That statement is more or less equivalent to my suggestion, above, that phage-resistance mutations that do not impact bacterial functioning will be less likely as they generally should involve subtle changes to gene products rather than partial or full elimination of those gene products. If phage-resistance mutations are cost free, even if only in the near term (2), then they can have the potential to drive phage populations to extinction, unless the phage population is able to evolve to overcome this bacterial mutation to resistance (Levin and Lenski 1983, Lenski 1988a). If the bacterial mutation eventually becomes costly within the same genetic background (again, example 2), perhaps the wild-type allele may be recovered via horizontal gene transfer. Alternatively, and as noted, CRISPR-Cas-based immunity may fit into this category of no-cost mutations (3). So too to some degree may the acquisition of prophages with their superinfection immunity, again forcing the susceptible phage population to either adapt or go extinct (for both, this is extinction at least locally, and with superinfection immunity it is only the virion population that is harmed). On the other hand, if a mutation to phage resistance is immediately detrimental (1), then there can be two general scenarios:

The first of these sub-scenarios (1 sub a) occurs if a mutation to phage resistance is too detrimental. In this case, then the bacterial population simply may be quickly driven to extinction. This presumably is the situation for bacteria that are 'unable' to mutate to resistance to specific phages. That is, we measure mutations to phage resistance in terms of bacterial survival, usually colony formation or in terms of broth culture turbidity, e.g., (Dickey and Perrot 2019, Berryhill et al. 2021, Saez et al. 2021). If bacteria that have mutated to phage resistance are unable to replicate, then they of course will not form colonies or turn broth turbid. In short, mutations that are easily achieved that allow these bacteria to resist infection by a given phage also might kill the bacterium whether or not the phage is present, rather than the bacteria being inherently unable to mutate to resistance (this, in other words, would be an example of an antagonistic pleiotropy in the extreme). As also would be the case with phages going extinct (above), we would not expect this scenario to lead to phage-bacterium antagonistic coevolution. It is possible that an approximation of such restraints on antagonistic coevolution is seen in the data of Buckling et al. (2006), where apparently higher fitness costs to mutation to phage resistance appear to be associated with lower abilities for bacterial populations to sustain phage resistance. As an aside, note that another route toward obtaining no phage-resistant mutants is simply starting with a culture that contains an insufficient number of bacteria, since the fewer the bacteria, then the higher the mutation rate to phage resistance must be in order to reliably detect mutations to phage resistance (Chap. 21, Sect. 21.3).

The second sub-scenario (1 sub b) is that the bacterial mutation to phage resistance, though immediately detrimental, does not otherwise kill the bacterium. If the bacterium is able to mutate to resistance while still retaining some viability, then an equilibrium of sorts may be established, one involving the mutated bacterium, the wild-type phage-susceptible bacterium, and the phage (Levin and Lenski 1983). This, however, is not necessarily an iterative phage-bacterium coevolution since the phage is not assumed to be mutating. Instead, wild-type phages may be sustained by the wild-type bacterial population, even if that population is diminished in size relative to the phage-resistance population due to phage attack. The phage population, though, might eventually evolve to infect the resistant bacterial population. That indeed would be an example of iterative coevolution, at least to a degree. Whether this or other coevolutionary scenarios will tend to progress beyond just one or two rounds is considered in the section that follows (Sect. 25.6).

25.6 An Indefinite Series of Coevolutionary Change?

The idea of phage-bacterium antagonistic coevolution being driven by the existence of refuges, physical or otherwise, is possibly consistent with the experiments of Buckling and Rainey (2002). First, they employed what they describe as static microcosms. Though these consist of broth rather than solid media or semi-solid media, they are not explicitly mixed, which in fact Buckling and Rainey suggest could have given rise to (p. 934), "spatial refuges for sensitive bacteria". Second, an average of only 10^7 bacteria and 10^4 phages were transferred, within 60 microliters, a total of 50 times. This means that generally bacteria were substantially outnumbering phages, suggesting that phages for whatever reason were not able to infect a large majority of the bacteria present, presumably because those bacteria either could not be physically reached by phages (i.e., as due to existence within static microcosms) or, if they could be reached, the bacteria were resistant to the specific phages reaching them. Levin and Lenski (1983, p. 103), similarly suggested that selection for bacterial resistance to phages is unlikely given, e.g., one thousand-fold excess of bacteria over phages: "Thus, at low phage density, a mutation rendering a cell resistant to the phage may be disadvantageous if that mutation engenders a significant reduction in the cell's intrinsic rate of increase. The same mutation will likely be advantageous, however, when phage are abundant."

To better appreciate the underlying community dynamics, I've changed these units to ones of concentrations and titers, which are roughly 2×10^8 colony-forming units/ml and 2×10^5 plaque-forming units/ml. Still, it is difficult to tell from these numbers exactly how phages were imposing much selective pressure on bacteria, unless this pressure was being exerted highly locally, e.g., as equivalent to plaque formation in semi-solid media. Even then, it is difficult to see how phage-resistant bacteria were being selected by phages across these populations, though it is easier to see how phages able to overcome any phage resistance could be selected for locally. How phage populations of this size could support the occurrence of host-range mutations especially highly locally in the vicinity of subpopulations of resistant bacteria, however, is less obvious. Notwithstanding my difficulty in seeing from these numbers how phages could possibly have exerted much selective pressure on the bacteria they were cohabitating with, bacteria did tend to evolve in these experiments so that they were better at resisting the phages they were evolving with than they were at resisting phages that were evolving in different microcosms. This would seem to imply that this quite low titer of phages was exerting a greater selective pressure on bacteria than mass action would suggest, and this would seem to be rather than being consistent with phage resistance arising as a pleiotropic effect of phage-unrelated selective pressures.

One possible explanation for how this selection may have occurred in the Buckling and Rainey (2002) experiments is that though it is average numbers of colony-forming units and plaque-forming units that were reported, perhaps variance between transfers was large, with phages mostly displaying very low titers but sometimes instead displaying more selectively meaningful titers. Even here, though, it is difficult for me to reconcile that scenario with the use of static microcosms, unless the microcosms were being mixed in other ways, as I consider below. Overall then, though this approach to the study of phage-bacterium antagonistic coevolution certainly is intriguing, and at this point in time has now been often repeated especially by Buckling and colleagues, my primary sense from the Buckling and Rainey (2002) experiments is more consistent with the occurrence of (p. 159) "an indefinite series of such fresh invasions of previously resistant hosts, followed by the emergence of correspondingly new bacterial types" (Campbell 1961) rather than that I have much idea as to how these dynamics may have been occurring. I will speculate, however, as follows.

(1) Though experiments took place over 50 transfers, perhaps selection for phage resistance tended to happen relatively infrequently over that time, e.g., in association with relatively rare overgrowths of phage numbers. (2) Perhaps more mixing took place than we imagine, e.g., such as due to mixing prior to sampling, mixing following transfer, and mixing due to bacterial motility. Variation in mixing indeed has been found to impact the dynamics of phage-bacterium coevolution (Gómez et al. 2015). Phage virions also ought to be able to fairly effectively diffuse across 6 ml of broth media over the 48 h between transfers even without mixing, though not necessarily to homogeneity given that individual phage bursts should be localized. (3) Due to the nature of the phage and bacterium used (phage SBW25ϕ2 and *Pseudomonas fluorescens* strain SBW25), the bacteria perhaps were limited in how they were able to achieve resistance, i.e., unable to simply delete phage receptors in order to block phage infection, thereby positively impacting in some manner the potential for ongoing phage evolution.

Thus, it is possible that we are not actually looking at anything even approaching "an indefinite series", with it only seeming to look that way but as based on limitations in the amount of information provided. It is also possible that the means of bacterial resistance, along perhaps with the means by which the phages were able to overcome that resistance, were such that more rounds of coevolution than had been expected were achieved, e.g., as can occur given CRISPR-Cas-based resistance mechanisms (Levin 2010). Host range evolution in phage SBW25ϕ2, however, does appear to be somewhat limited in terms of what genes are affected and those genes appear to encode virion-associated proteins (Paterson et al. 2010, Hall et al. 2011), strongly suggesting that the antagonistic coevolution occurring in this system is associated with resistance to phage adsorption, with phages then mutationally overcoming that resistance. Furthermore, an important factor driving phage host-range evolution in this system is the variable nature of the coevolving bacterial hosts (Paterson et al. 2010, Hall et al. 2011). Indeed, in continuing the above list, (4) the mostly static nature of these experiments could have limited the potential for any one phage mutant to come to fixation within phage populations, thereby allowing for a greater genetic diversity upon which subsequent phage evolution could be based, which in turn could have allowed for a greater potential for the phage population to evolve to better infect a coevolved diversity of bacterial hosts. This is not to say that arms race dynamics were not occurring, but instead that they may have been 'bushier'—greater phage genetic diversity present than necessarily seen—than would have been the case had experiments somehow been conducted instead in a continuously well-mixed environment.

Substantial phage host range evolution in terms of overcoming bacterial mutation to adsorption-resisting phenotypes is not unprecedented. As reviewed by Henning and Hashemolhosseini (1994, p. 293), "Host range mutants have been isolated from [T4-family coliphages] TuIa and TuIb which use other or additional receptors, e.g., TuIa can give rise to mutants able to recognize OmpF, OmpC, or LamB proteins, and TuIb can yield mutants using OmpC or LamB." In addition (p. 295), The T2-family phage "M1 could switch from OmpA to OmpC, from OmpC to another outer membrane protein, and then on to a yet unidentified receptor." Lastly, from p. 295, the T2-family phage Ox2 "could switch receptors, from OmpA to OmpC, from OmpC to OmpX (an outer membrane protein of unknown nature), from OmpX to *E. coli* B LPS [lipopolysaccharide] with two terminal glucose residues, and, finally, to an LPS with only one such residue." Phage λ also has been found to be able to evolve an ability to recognize a new surface receptor, switching from LamB to OmpF, though not directly from the wild-type phage but rather from evolved mutants (Meyer et al. 2012). Even phages T2, T4, and T7 are able do display some degree of coevolution with their phage-resistant mutants of *E. coli* B (Bohannan and Lenski 2000). Thus, it is entirely possible that at least some coliphages are as capable of effecting an iterative antagonistic coevolution with their *E. coli* host as phage SBW25ϕ2 can with its *P. fluorescens* host, even without CRISPR-Cas involvement.

25.7 Coda

Clearly there is more to study in terms of phages as drivers of evolution than phage-bacterial arms races, however those arms races may be manifest. My hope with this monograph is to have made some contribution to the study of those various processes, or at least toward documenting their diversity.

References

Abedon ST (2011) Envisaging bacteria as phage targets. Bacteriophage 1:228–230
Abedon ST (2012) Thinking about microcolonies as phage targets. Bacteriophage 2:200–204
Avrani S, Wurtzel O, Sharon I, Sorek R, Lindell D (2011) Genomic island variability facilitates *Prochlorococcus*-virus coexistence. Nature (London) 474:604–608
Berryhill BA, Huseby DL, McCall IC, Hughes D, Levin BR (2021) Evaluating the potential efficacy and limitations of a phage for joint antibiotic and phage therapy of *Staphylococcus aureus* infections. Proc Natl Acad Sci U S A 118:e2008007118
Blazanin M, Turner PE (2021) Community context matters for bacteria-phage ecology and evolution. ISME J 15:3119–3128
Bohannan BJM, Lenski RE (1997) Effect of resource enrichment on a chemostat community of bacteria and bacteriophage. Ecology 78:2303–2315
Bohannan BJM, Lenski RE (2000) Linking genetic change to community evolution: insights from studies of bacteria and bacteriophage. Ecol Lett 3:362–377
Bouchard JD, Moineau S (2000) Homologous recombination between a lactococcal bacteriophage and the chromosome of its host strain. Virology 270:65–75
Brockhurst MA, Koskella B (2013) Experimental coevolution of species interactions. Trends Ecol Evol 28:367–375
Brockhurst MA, Koskella B, Zhang QG (2021) Bacteria-phage antagonistic coevolution and the implications for phage therapy. In: Harper DR, Abedon ST, Burrowes BH, McConville M (eds) Bacteriophages: biology, technology, therapy. Springer, New York, pp 231–251
Buckling A, Rainey PB (2002) Antagonistic coevolution between a bacterium and a bacteriophage. Proc R Soc Lond B Biol Sci 269:931–936
Buckling A, Wei Y, Massey RC, Brockhurst MA, Hochberg ME (2006) Antagonistic coevolution with parasites increases the cost of host deleterious mutations. Proc R Soc Lond B Biol Sci 273: 45–49
Burmeister AR, Sullivan RM, Gallie J, Lenski RE (2021) Sustained coevolution of phage lambda and *Escherichia coli* involves inner- as well as outer-membrane defences and counter-defences. Microbiology 167:001063
Campbell A (1961) Conditions for the existence of bacteriophages. Evolution 15:153–165
Chan BK, Abedon ST (2012) Bacteriophage adaptation, with particular attention to issues of phage host range. In: Quiberoni A, Reinheimer J (eds) Bacteriophages in dairy processing. Nova Science Publishers, Hauppauge, pp 25–52
d'Herelle F (1917) Sur un microbe invisible antagoniste des bacilles dysentériques. C R Acad Sci Ser D 165:373–375
d'Herelle F (2007) On an invisible microbe antagonistic toward dysenteric bacilli: brief note by Mr. F. D'Herelle, presented by Mr Roux. Res Microbiol 158:553–554
d'Herelle F (2011) *On an invisible microbe antagonistic to dysentery bacilli.* Note by M. F. d'Herelle, presented by M. Roux. Comptes Rendus Academiedes sciences 1917; 165:373-5. Bacteriophage 1:3–5

Dąbrowska K, Abedon ST (2019) Pharmacologically aware phage therapy: pharmacodynamic and pharmacokinetic obstacles to phage antibacterial action in animal and human bodies. Microbiol Mol Biol Rev 83:e00012–e00019

de Jonge PA, Nobrega FL, Brouns SJJ, Dutilh BE (2019) Molecular and evolutionary determinants of bacteriophage host range. Trends Microbiol 27:51–63

De Sordi L, Khanna V, Debarbieux L (2017) The gut microbiota facilitates drifts in the genetic diversity and infectivity of bacterial viruses. Cell Host Microbe 22:801–808

De Sordi L, Lourenco M, Debarbieux L (2019) "I will survive": a tale of bacteriophage-bacteria coevolution in the gut. Gut Microbes 10:92–99

Dennehy JJ (2012) What can phages tell us about host-pathogen coevolution? Int J Evol Biol 2012: 396165

Dennehy JJ, Abedon ST (2021) Adsorption: phage acquisition of bacteria. In: Harper D, Abedon ST, Burrowes BH, McConville M (eds) Bacteriophages: biology, technology, therapy. Springer, New York, pp 93–117

Dickey J, Perrot V (2019) Adjunct phage treatment enhances the effectiveness of low antibiotic concentration against *Staphylococcus aureus* biofilms *in vitro*. PLoS One 14:e0209390

Drake JW, Charlesworth B, Charlesworth D, Crow JF (1998) Rates of spontaneous mutation. Genetics 148:1667–1686

Drexler K, Riede I, Montag D, Eschbach ML, Henning U (1989) Receptor specificity of the *Escherichia coli* T-even phage Ox2. Mutational alterations in host range mutants. J Mol Biol 207:797–803

Gómez P, Ashby B, Buckling A (2015) Population mixing promotes arms race host-parasite coevolution. Proc Biol Sci 282:20142297

Gómez P, Buckling A (2011) Bacteria-phage antagonistic coevolution in soil. Science (New York, NY) 332:106–109

Gurney J, Pleska M, Levin BR (2019) Why put up with immunity when there is resistance: an excursion into the population and evolutionary dynamics of restriction-modification and CRISPR-Cas. Philos Trans R Soc Lond Ser B Biol Sci 374:20180096

Hall AR, Scanlan PD, Buckling A (2011) Bacteria-phage coevolution and the emergence of generalist pathogens. Am Nat 177:44–53

Hampton HG, Watson BNJ, Fineran PC (2020) The arms race between bacteria and their phage foes. Nature (London) 577:327–336

Henning U, Hashemolhosseini S (1994) Receptor recognition by T-even type coliphages. In: Karam JD, Eiserling FA, Black LW (eds) The molecular biology of bacteriophage T4. ASM Press, Washington, DC, pp 291–298

Horne MT (1970) Coevolution of *Escherichia coli* and bacteriophages in chemostat culture. Science (New York, N Y) 168:992–993

Hyman P (2019) Phages for phage therapy: isolation, characterization, and host range breadth. Pharmaceuticals (Basel) 12:35

Hyman P, Abedon ST (2010) Bacteriophage host range and bacterial resistance. Adv Appl Microbiol 70:217–248

Janzen DH (1980) What is coevolution? Evolution 34:611–612

Koskella B, Lin DM, Buckling A, Thompson JN (2012) The costs of evolving resistance in heterogeneous parasite environments. Proc Biol Sci 279:1896–1903

Koskella B, Meaden S (2013) Understanding bacteriophage specificity in natural microbial communities. Viruses 5:806–823

Lenski RE (1984) Coevolution of bacteria and phage: are there endless cycles of bacterial defences and phage counterdefences? J Theor Biol 108:319–325

Lenski RE (1988a) Dynamics of interactions between bacteria and virulent bacteriophage. Adv Microbial Ecol 10:1–44

Lenski RE (1988b) Experimental studies of pleiotropy and epistasis in *Escherichia coli*. II. Compensation for maladaptive effects associated with resistance to virus T4. Evolution 42: 433–440

Lenski RE, Levin BR (1985) Constraints on the coevolution of bacteria and virulent phage: a model, some experiments, and predictions for natural communities. Am Nat 125:585–602
Levin BR (2010) Nasty viruses, costly plasmids, population dynamics, and the conditions for establishing and maintaining CRISPR-mediated adaptive immunity in bacteria. PLoS Genet 6: e1001171
Levin BR, Lenski RE (1983) Coevolution in bacteria and their viruses and plasmids. In: Futuyma DJ, Slatkin M (eds) Coevolution. Sinauer Associates, Inc., Sunderland, pp 99–127
Meyer JR, Dobias DT, Weitz JS, Barrick JE, Quick RT, Lenski RE (2012) Repeatability and contingency in the evolution of a key innovation in phage lambda. Science (New York, N Y) 335:428–432
Morgan AD, Quigley BJ, Brown SP, Buckling A (2012) Selection on non-social traits limits the invasion of social cheats. Ecol Lett 15:841–846
Morris C (1885) Attack and defense as agents in animal evolution. In: Proceedings of the Academy of Natural Sciences of Philadelphia, pp 385–392
Obeng N, Pratama AA, Elsas JD (2016) The significance of mutualistic phages for bacterial ecology and evolution. Trends Microbiol 24:440–449
Paterson S, Vogwill T, Buckling A, Benmayor R, Spiers AJ, Thomson NR, Quail M, Smith F, Walker D, Libberton B, Fenton A, Hall N, Brockhurst MA (2010) Antagonistic coevolution accelerates molecular evolution. Nature (London) 464:275–278
Prangishvili D (2013) The wonderful world of archaeal viruses. Ann Rev Microbiol 67:565–585
Rodin SN, Ratner VA (1983a) Some theoretical aspects of protein coevolution in the ecosystem of "phage-bacteria". I. The probelm. J Theor Biol 100:185–195
Rodin SN, Ratner VA (1983b) Some theoretical aspects of protein coevolution in the ecosystem "phage-bacteria". II. The deterministic model of microevolution. J Theor Biol 100:197–210
Ross A, Ward S, Hyman P (2016) More is better: selecting for broad host range bacteriophages. Front Microbiol 7:1352
Saez MD, Visram Z, Mutti M, Restrepo-Cordoba M, Hartmann S, Kremers AI, Tisakova L, Schertler S, Wittmann J, Kalali B, Monecke S, Ehricht R, Resch G, Corsini L (2021) Epsilon (2)-phages are naturally bred and have a vastly improved host range in *Staphylococcus aureus* over wild type phages. Pharmaceuticals (Basel) 14:325
Scanlan PD, Buckling A, Hall AR (2015) Experimental evolution and bacterial resistance: (co)-evolutionary costs and trade-offs as opportunities in phage therapy research. Bacteriophage 5: e1050153
Stent GS (1963) Molecular biology of bacterial viruses. WH Freeman and Co., San Francisco
Summers WC (1999) Felix d'Herelle and the origins of molecular biology. Yale University Press, New Haven
Thingstad TF, Bratbak G, Heldal M (2008) Aquatic phage ecology. In: Abedon ST (ed) Bacteriophage ecology. Cambridge University Press, Cambridge, pp 251–280
Weitz JS, Hartman H, Levin SA (2005) Coevolutionary arms races between bacteria and bacteriophage. Proc Natl Acad Sci U S A 102:9535–9540
Young JPW, Levin BR (1993) Adaptation in bacteria: unanswered ecological and evolutionary questions about well-studied molecules. In: Berry RL, Crawford TJ, Hewitt GM (eds) Genes in ecology. Blackwell Scientific Publications, London, pp 169–192

Glossary

Abiotic Not living, and indeed not necessarily from a living source. Thus, physical aspects of Environments are abiotic as too are water, air, and minerals. Contrast with Biotic.

Abortive infection Strictly a phage infection of a bacterium that gives rise to low phage Efficiencies of Plating, but here used more specifically to describe a mechanism of bacterial Resistance to phages in which the phage is killed but so too is the phage-Infected bacterium. This latter definition I use as a synonym for bacterial 'Self-Sacrifice'. In any case, an abortive infection always results in bacterial death. Contrast this latter definition with both Negation and Reduced Infection Vigor as well as with Delay.

Abortive infection system Evolved bacterial mechanism of phage Resistance that is associated both with reductions (often severe) in the Fitness of infecting phages but so too in the death of the phage Infected bacterium. Traditionally, though, an abortive infection system does not necessarily result in reduction of phage Fitness to zero but instead can include Reduce Infection Vigor, i.e., as can result in lower phage Efficiencies of Plating.

Abortive transduction Donor DNA fragment that is carried by a phage into a recipient bacterium but then nothing happens, i.e., neither Recombination into the recipient's genome nor replication of the fragment.

Absolute fitness A measure of the Reproductive Success of a Genotype particularly in terms of numbers of offspring produced that produce offspring themselves. Thus, absolute fitness takes into account reproductive output, subsequent offspring Survival, and also whether those offspring succeed in reproducing. An absolute fitness of 2 means that on average a Genotype produces two offspring which survive and reproduce per generation, while an absolute fitness of one indicates simply parental organism replacement. Contrast with Relative Fitness and see also Darwinian Fitness.

Accessory gene Genetic Locus that generally is not essential for the normal functioning an organism, at least across all Environments in which the organism might be found. Accessory genes therefore typically are not present in the genomes of

S. T. Abedon, *Bacteriophages as Drivers of Evolution*,
https://doi.org/10.1007/978-3-030-94309-7

all of the members of a given species and also often are carried by Mobile Genetic Elements. Lysogenic Converting Genes are examples of bacterial accessory genes. Accessory genes can also be described as variable Genes and they are to be contrasted with the Core Genes of a Lineage.

Acquired resistance Mechanism of interference with the action of any type of antagonistic entity, such as of Antibiotics but also of phages, that requires genetic changes to the otherwise potentially negatively impacted organism. These changes can be Mutational (and these generally are chromosomally encoded, i.e., Chromosomal Resistance) or instead can be acquired via Horizontal Gene Transfer (and thus typically are extrachromosomal, i.e., Extrachromosomal Resistance). A number of bacterial Alleles and systems are known which confer acquired resistance to various phages.

Adaptation Referring to changes to an organism, or a Population of organisms, that are beneficial to the harboring organisms in some manner. With physiological or morphological adaptations, these are genetically encoded changes that are stimulated in response to some Environmental change and which have evolved because ancestral organisms that responded equivalently displayed a higher Evolutionary Fitness than those which did not. Evolutionary adaptations, which can include physiological and morphological adaptations, are by definition products of Natural Selection. Thus, a beneficial genetic change that is retained within a Population due to that benefit can be described as an Evolutionary adaptation. Generally, here, the term adaptation is being used in an Evolutionary rather than physiological or morphological sense.

Adaptive immunity Various anticipatory and, in terms of patterns recognized, highly specific immunological mechanisms. Contrasting Innate Immunity, adaptive immune responses can change in their specificity in response to exposure to new patterns. CRISPR-Cas Systems provide bacteria with adaptive immunity.

Adsorption Process of Virion Attachment especially to a host bacterium. Infection is something that largely or entirely follows adsorption so should not be equated with adsorption. The Extracellular Search, by contrast, precedes adsorption.

Adsorption resistance Bacterial phage-Resistance mechanism that acts after Virion encounter with a bacterium but prior to Irreversible Virion Attachment. That is, given adsorption resistance then free phages are prevented from becoming Adsorbed phages. This Resistance typically is due to either an absence of necessary bacterial surface Receptor molecules or instead mechanisms preventing phages from reaching those Receptor molecules, such as via bacterial display of Capsules. Adsorption resistance also has been described as Envelope-Level Resistance, though this latter term in principle might also include as well post-Irreversible Attachment mechanisms, such as Superinfection Exclusion, that should not be equated with adsorption resistance.

Adsorptive host range Those bacterial Strains that a given phage type is able to successfully Irreversibly Attach to.

All the world's a phage Concept implying that bacteriophage Mosaic Evolution involves the indirect genetic interaction of a majority of the world's phages via partially overlapping phage Host Ranges, or at least indirect genetic interaction of a majority of Tailed, Temperate Phages.

Allele Variant of a Gene possessing a Nucleotide Sequence which is distinct though typically not highly different from the sequence of occupants of the same Locus as found in different genomes or on different chromosomes.

Allele frequency Fractions of those Alleles of a specific type as found within a Population. The frequency of all Alleles found at a given genetic Locus add up collectively to one (1.0). Within haploid Populations, an allele's frequency is equal to the fraction of individuals within that Population carrying that Allele.

Allelic variation Differences in the Genes encoded by different individuals that make up a single Population. See also the concept of Polymorphism.

Allelopathy Often used to refer to the ability of plants, such as trees, to inhibit the growth of competitor plants attempting to grow around them. This idea has been coopted to describe the ability of bacteria to inhibit competitor organisms also attempting to grow around them. This includes bacteria that may be inhibited by means of the release of Lytic, Temperate Phages by a neighboring bacterium.

Amixis Lack of sex, here meaning lack of Gene Exchange. Contrast with Panmixis.

Antagonist Something which acts against something else. Thus, both Antibiotics and Lytic Phages can serve as antagonists to bacteria.

Antagonistic coevolution Essentially the idea of a genetic Arms Race between two Species, typically between an Exploiter and a Victim, where Adaptations toward more effective exploitation or resistance to exploitation are countered by Adaptations conferring more effective Resistance or ability to overcome that Resistance. Contrast with Mutualistic Coevolution.

Antagonistic interaction Fighting between individuals, as typically is a consequence of some form of larger Competition between those individuals. See, alternatively, Exploitative Competition.

Antagonistic pleiotropy The existence of more than one Phenotype associated with a given Allele where these Phenotypes are in conflict, such that improvements in one Phenotype results in reductions in the effectiveness of one or more other Phenotypes. For example, an Adaptation that allows a phage to replicate more effectively while infecting one type of bacterial Host may result as well in that phage being less able to replicate effectively while infecting a different bacterial Host. See also simply Pleiotropy.

Antibiotic Non-living substance, traditionally one that is naturally occurring as produced by bacteria or fungi and which is relatively broadly acting as an antibacterial substance. Antibiotics, particularly those which have been commercialized, tend to affect more bacterial Species than just the producing Species, thus contrasting Bacteriocins, are also not living and are produced by microorganisms (bacteria), but instead tend to have very narrow activity spectra.

Archaea Domain-level Prokaryotic Lineage, constituting the other Prokaryotic Lineage besides *Bacteria*. So far as is known, *Archaea* and *Bacteria* do not share Viruses.

Archaeal virus The Viruses which infect members of domain *Archaea*. Contrast, for example, with Bacterial Viruses. Generally archaeal viruses should not be described as phages though many archaeal viruses are difficult to distinguish morphologically from Tailed Phages.

Arms race Coevolutionary process whereby especially Antagonistic Adaptations by one Species are countered by Antagonistic Adaptations by a second Species, and so on. Also described as Antagonistic Coevolution, such as seen between phages and the potential Host bacteria of these phages.

Assortative mating Gene Exchange that preferentially occurs, as within Species, between individuals that possess more similar Genotypes. Assortative mating is a form of Non-Random Mating. Contrast with Random Mating as well as with Disassortative Mating.

Assortative transduction Idea that phage-mediated Gene Exchange will more likely occur between more similar versus more dissimilar bacterial types. The concept of assortative transduction as used here is analogous to that of Assortative Mating.

Attachment Especially non-covalent, often highly specific bonding between Virion Particles and the surfaces of target cells. See more or less equivalently, Adsorption.

Attachment site Location of physical integration of a Prophage into the genome of its bacterial Host. These are abbreviated as *att* sites and specifically *attB* is associated with a bacterium's chromosome where as *attP* refers to the phage Sequence responsible for recognizing the *attB* attachment site.

Auxiliary metabolic genes As encoded by phages, these express products that serve to replace or augment Host bacterium functions especially toward assuring that bacterial basal metabolic processes are favorable especially to Lytic Phage Infections.

Avoidance Consequence of phage-Resistance mechanisms that affect a phage's ability to deliver its genome into a bacterium's cytoplasm. A bacterium thus is 'avoiding' becoming phage Infected, here with the phage Infection defined as starting at the point of phage genome entrance into the cytoplasm (Uptake). Avoidance can act either prior to Virion Attachment (Adsorption Resistance) or following Virion Attachment (phage genome Uptake blocks, i.e., such as especially Superinfection Exclusion).

Bacteria Prokaryotic organisms which are not members of domain *Archaea*.

Bacterial virus Also known as a bacteriophages or phages, these are the Viruses that infect members of domain *Bacteria*.

Bactericidal host range Those bacterial Strains that a specific phage type is able to kill. Killing especially means able to complete a Lytic Cycle though operationally this also should include an ability to induce an Abortive Infection. The bactericidal host range of a phage in either case is a subset of that phage's Adsorptive Host Range. It is also a subset of a phage's Penetrative Host Range.

Bacteriocin A non-living substance produced by bacteria which displays antibacterial activity against especially different Strains of the producing bacterial Species.

Bacteriophage Viruses whose Hosts consist of members of domain *Bacteria*. Also commonly described as phages but also as Bacterial Viruses.

Bacteriophage exclusion (BREX) Mechanism of neutralization of phage Infections that involves recognition of phage DNA as foreign if that DNA has not been appropriately methylated.

Base sequence Standing for nucleic acid base sequence, this is equivalent to Nucleotide Sequence or simply a nucleic acid's sequence, i.e., as indicated using the abbreviations A, T, G, and C for DNA or A, U, G, and C for RNA.

Biofilm Association of multiple microbial cells with each other, not necessarily all of the same Species, often in association with a surface and with the cells embedded in some form of Extracellular Polymeric Substance (biofilm matrix). Biofilms often consist of conglomerations of Clonal Microcolonies and this is rather than conglomerations of individual cells that are spatially isolated from others of their kind.

Biogeochemical cycling Movement of nutrients through Biotic and Abiotic components of Ecosystems in combination with nutrient movement between Ecosystems. This can include nutrient movement on global scales. What sorts of things that these nutrients are found in association with often are recurrent (hence, 'cycling'), with this recurrence sometimes occurring over long, i.e., geological, spans of time. Note that 'bio' stands for Biotic, 'geo' refers to Abiotic, and 'chemical' is equivalent to nutrient.

Biogeography Study of what organisms are located where, geographically, and why they are located where they are.

Biological species concept Idea that Species are distinguished from each other especially in terms of likelihoods of Gene Exchange, where Gene Exchange occurs within Species more often but between Species relatively rarely. Species thus represent somewhat genetically coherent entities. Contrast the biological species concept especially with Mosaic Evolution that is associated with either Horizontal Gene Transfer or with Hybridization.

Biotic Referring to either a living organism or organism-like but dead remains of an organism. Certainly at the point where an organism has been fully mineralized (converted to inorganic components) it is no longer a biotic component of an Environment but instead an Abiotic component. At some point prior to then, however, it can still be considered to be biotic. Thus, organisms that are scavengers, detritivores, or decomposers obtain biotic aspects of Environments as nutrients as do also Predators, herbivores, and Parasites, etc., and indeed this is true for heterotrophic organisms generally, while autotrophic organisms mostly do not consume biotic aspects of Environments.

Burst size Number of Progeny Virions released especially at the end of a Lytic Cycle. The release of Virions from a phage-Infected bacterium at the end of a Lytic Cycle's Latent Period is known as a burst.

***Caudovirales* (order)** The Tailed Viruses, as collectively infect in substantial numbers and types both domains *Bacteria* and *Archaea*. Tails can be short (Podoviruses) or long (Myoviruses or Siphoviruses) and either contractile (Myoviruses) or not.

Capsid That proteinaceous aspect of a Virion that surrounds the Virus nucleic acid genome. In Tailed Phages, the head represents the capsid whereas the tail represents an appendage to the capsid.

Capsule Carbohydrate polymers that are located outside of a bacterium's cell envelope but which are not easily washed away. See similarly Extracellular Polymeric Substance.

Cellular arrangement Clonal association of bacteria that have remained physically attached following binary fission rather than separating. This lack of physical separation results in patterns of arrangements that are typical for a given bacterial type. This differs from Microcolonies or Biofilms because there daughter bacteria remain closely associated due to their being mutual embedded within an extracellular matrix (see Extracellular Polymeric Substance) or instead because they simply have not moved themselves apart, rather than daughter cells remaining attached cell-to-cell to each other. Typical cellular arrangements are diplococci, streptococci, staphylococci, and streptobacilli.

Chromosomal resistance Especially Mutations in targets of antibacterial agents. As these targets tend to be essential or otherwise highly useful to the carrying bacterium, they generally are not encoded by accessory genes but instead are the products of Genes found on bacterial chromosomes. Chromosomal resistance, as it is mutationally attained, nonetheless is a form of acquired resistance. See simply Resistance. Contrast with Extrachromosomal Resistance.

Chronic infection/release Transition of Virions from intracellular to extracellular locations via processes that both are not simultaneous for a given Infection's Progeny Virions nor directly resulting in the death or Lysis of the Infected bacterium. Thus, these Infections can be ongoing with both continued Virion production and continued Release over time.

Circular permutation Referring to a linear chromosome that is redundant at its ends and for which these physical ends can be located anywhere within the phage's genetic map. As a consequence of circular permutation, the genetic map is circular rather than linear and this is even though the chromosome itself is physically linear. Circular permutation occurs as a consequence of Headful Packaging of phage Concatemeric DNA.

Clone/clonal A group of organisms having been relatively recently generated by non-Sexual means from a single common ancestor. That is, clonally related cells or organisms have not individually incorporated Genetic Material from other, especially non-clonally related organisms. Clonally related organisms are either genetically identical or instead differ only as a consequence of Mutation-driven Divergence.

Coadsorption Especially the Attachment of more than one Virion to a single bacterial cell. Coadsorptions can lead to Coinfections, but not necessarily. That is, if the first Adsorbing phage encodes Superinfection Exclusion mechanisms and Adsorbs sufficiently sooner than the Secondarily Adsorbing phage or phages, then Superinfection Exclusion may be sufficiently expressed that genome Translocation into the bacterial cytoplasm by the Secondarily Adsorbed phage or phages (Uptake) is blocked.

Coevolution Genetic changes in one entity (i.e., one Species) that select for genetic changes in a second entity (such as a second Species). This results in the two entities becoming increasingly specialized for interacting with each other, especially in either Mutualistic or Antagonistic manners.

Coinfection More than one Parasitic organism colonizing a Host organism at the same time. For instance, this can be two different phages Infecting the same bacterium at the same time. Included in the concept of coinfection can be phage Infection of a bacterial Lysogen where the resident Prophage thereby represents the other Parasitic organism. Coinfection, particularly as including Prophages, can involve more than two different phages together Infecting the same bacterium.

Coliphage A bacteriophage that Infects especially the bacterium *Escherichia coli* and/or infects coliform bacteria more generally.

Colloid Highly wetted substance that technically is not also dissolved, often due to its large size. Both proteins and phages are described as colloidal.

Colony Among bacteria, these are groups of spatially associated cells that are found in the same location especially because they have not left this location after being formed. This is because the cells inherently lack motility, because the cells are trapped within an extracellular matrix (i.e., Extracellular Polymeric Substance, either of their own making or instead as imposed upon them, e.g., such as in the laboratory by agar), or because the cells are replicating within an only minimally aqueous Environment.

Colony-forming unit A cell or Cellular Arrangement that, upon growth in association with solid or semisolid media, will form a single Colony.

Commensal Especially Symbiotic organisms that neither help nor hinder the functioning or Evolutionary fitness of their Host organism. See also Commensalism.

Commensalism The relationship between Commensal organisms and their Host organisms.

Community Collection of the different Species found within an Ecosystem. For example, all of the different types of bacteriophages and the different types of bacteria that are found in the same Environment together make up a community, in this case a microbial community. A Population, by contrast, is a group of organisms consisting of only a single Species.

Community dynamics Study and description of changes in the number of individuals making up different Species as determined especially within individual Ecosystems over Ecological time spans. That is, study and description of changes in the number of individuals of different Species as found within specific Communities.

Community ecology Study of the interactions that occur between different Species within a single Environment. See also simply Community.

Community resistance Existence of phage Resistance other than due to known exposure to a selecting phage. Community here refers to human communities, such as towns or cities, with community resistance meaning that a bacterial Strain that a person is exposed to may not be susceptible to a given phage type especially for reasons other than that the bacterial Strain had previously been treated with Phages Therapeutically. Contrast with Treatment Resistance.

Compensating mutation A change in Genetic Material that helps to correct a Phenotypic defect resulting from a different, previous, independently occurring genetic change. Compensating mutations thus compensate for changes in Phenotype made by other Mutations. Those other Mutations need not have been overall detrimental to the carrying organism, however, as a compensating mutation can also compensate for Epistatic modifications of Phenotype, i.e., with the first Mutation providing benefits in one aspect of Phenotype but costs in another, and with the compensating mutation improving Fitness by compensating for the latter.

Competition Direct or indirect interactions between two or more organisms in which one organism or set of organisms is harmed and another, the 'winner' of the competitive interaction, either gains or is harmed less. Exploitative Competition is an example of indirect competition between organisms whereas phage-encoded anti-phage mechanisms, e.g., such as Superinfection Exclusion and Superinfection Immunity, play roles in more direct competition between phages, i.e., as Antagonistic Interactions.

Concatemer DNA double helix that contains multiple, duplicate genomes that are covalently linked end to end as well as, for some phages, linked together also by Holliday junctions. The latter are points of crossover-type Recombination events between two double helices. Concatemers often represent the substrate for phage genome Packaging into Capsids, in which case the concatemer must be locally cut after filling the Capsid with DNA.

Conspecific Individual of the same Species. See also Homospecific. Contrast with Heterospecific.

Converting gene Short for lysogenic converting gene, which is a Prophage Gene capable of giving rise to Lysogenic Conversion.

Conjugation Pilus-mediated mechanism (i.e., conjugation pili or sex pili) of transfer of a Plasmid from one bacterium (the donor) to another (the recipient).

Core gene Genetic Sequence that tends to be consistently found across Strains within a Lineage, e.g., such as within a bacterial Species. By contrast are Accessory Genes, which tend to be only inconsistently encoded by the different Strains making a single Species.

***cos* site** Short Nucleotide Sequences that come in pairs, e.g., as separated by a complete phage genome, which together serve as a Packaging signal. Contrast with *pac* Sites.

CRISPR Stands for Clustered Regularly Interspersed Short Palindromic Repeats. See CRISPR-Cas System.

CRISPR-Cas system Mechanism of bacterial Resistance to phages and other forms of infectious DNA that involves acquisition of 'spacer' sequences which correspond to DNA Sequences (proto-spacers) that are found as part of infectious DNA. These spacer sequences are then used as the basis of recognition of this infectious DNA given a bacterium's future exposure to the same DNA Sequence. CRISPR-Cas systems can cause the Negation of phage Infections but also may be associated with bacterial Self-Sacrifice and phage Delay.

CRISPR adaptation Acquisition of CRISPR spacer sequences from proto-spacer sequences. Proto-spacer sequences can be found, for example, as components of phage genomes. See also CRISPR-Cas System.

Cumulative evolution Multiple changes to a single Genotype. This contrasts with multiple changes just to a single Gene Pool. This generally implies for mostly asexual Populations that new Mutations are occurring within Genotypes containing old Mutations, thereby resulting in a Lineage in which individuals contain more than one newer Mutation.

Curing Loss of Prophage or Plasmid from a bacterium. Degrees of loss can range from complete elimination to only functional loss such as, for a Prophage, being no longer Inducible to produce Virions.

Cryptic prophage A Defective Prophage that, strictly speaking, is unable to display Superinfection Immunity. Often used, however, to mean simply Defective Prophage.

Darwinian fitness Measure of the capacity of a Genotype to increase its numbers going from one generation to the next. Thus, if there is a capacity to double in number per generation, then that would represent a Darwinian fitness of 2, which is an Absolute Fitness. Experimentally, Darwinian fitness often is determined by comparing one Genotype to another, e.g., with a Relative Fitness of 1.1 indicating a 10% greater capacity to increase numbers per generation than a Relative Fitness of 1.0.

Defective interfering particle (DI particle) Evolutionary product of excessive Coinfection by Viruses of the same type, where genetic variants evolve to be obligate Parasites of Infections caused by the wild-type Viruses. That is, the wild-type Virus serves as a Helper Virus for the defective interfering particle and the ability of the wild-type Virus to replicate while serving in this Helper-Virus role is reduced due to the intracellular Parasitic actions of the Parasitic DI particles.

Defective prophage Latently Infecting Virus that, due to Mutation, is unable to produce functional Virions or indeed necessarily even Virions at all. Defective prophages, depending on the location or extent of their Mutations, may continue to display Superinfection Immunity functions. Contrast therefore with Cryptic Prophage.

Delay Description of the potential impact of certain bacterial mechanisms of phage Resistance that can result in a slowing of phage Population Growth, e.g., through (1) Burst Size reductions, (2) extending phage Latent Periods, or (3) interfering with phage Virion movement to new bacteria to infect, etc. See also Reduced Infection Vigor.

Deletion mutation Numerical loss of nucleotides from the Genetic Material of an individual organism. Losses can range from a single nucleotide (or base pair) to large blocks of nucleotides, though the larger the deletion then the more likely that it will be detrimental. Also, within reading frames deletions in multiples of three tend to be less detrimental than non-multiples of three.

Destructive infection Phage Infection in which the phage is inactivated. For example, the phage may be restricted (see Restriction) or it may be subject to an Abortive Infection. See equivalently the concepts of Negation (Restriction equivalent) and Self-Sacrifice (Abortive Infection equivalent) as well as Avoidance (specifically in this latter case as a Superinfection Exclusion equivalent). Contrast with phage-Productive Infection and phage-Reductive Infection.

Determinism Idea that future events are predictable in a non-probabilistic manner. Thus, given determinism then knowledge of the values of all variables and constants at time zero should allow prediction of all variable values some time in the future. A properly functioning and otherwise well-engineered mechanical device generally will display deterministic behavior, e.g., you flip a light switch and the light goes on. Alternatively, more complex systems such as organisms and their Populations might still act deterministically even if it can be difficult to model those systems in high mechanistic detail, that is, have knowledge of all variables and their values. Contrast determinism with Stochasticism.

Deterministic evolution Another name for the consequences of Natural Selection. See also Determinism.

Direct reciprocal transduction Phage-mediated movement of bacterial DNA where movement from one bacterium to another and then back again requires the existence of only a single phage type, i.e., one which is able to Infect, Transduce from, and Transduce to both bacteria. Contrast with Indirect Reciprocal Transduction where this back-and-forth movement of DNA requires the existence of more than one phage type. See also simply Reciprocal Transduction.

Directional selection Change in Allele Frequencies toward Deterministic dominance by a specific Allele or some extreme in Phenotype. Also known as Positive Selection. Contrast with Stabilizing Selection, or Stabilizing Frequency-Dependent Selection.

Disassortative mating Reduced tendencies by more similar individuals to exchange Genes. Contrast with Assortative Mating as well as with Random Mating.

Disperse Movement apart and away.

Dispersion Description of the spatial distribution of a Population across an Environment. Dispersions can be, e.g., random, even, or clumped. These different manifestations of dispersion result from individuals not interacting with each other either positively or negatively (random dispersion), individuals interacting with each other negatively such that there is a tendency for individuals to stay as far apart from one another within a given Environment as they can (even dispersion), or individuals are attracted to each other, or at least not repelled (clumped dispersion). Bacteria within broth cultures, especially that are well mixed, display random dispersions whereas bacteria growing within Colonies or Microcolonies display clumped dispersions.

Disruptive frequency-dependent selection Increasing Evolutionary Fitness associated with Alleles as those Alleles become more common. The result is a tendency for such Alleles to become Fixed within at least local Populations. Contrast with Stabilizing Frequency-Dependent Selection.

Divergent evolution Genetic changes that serve to increasingly differentiate two Populations over time.

Diversifying selection Circumstances that motivate the Evolution of two or more distinctly different forms across a Population. See, for example, Disruptive Frequency-Dependent Selection.

Domestication (of genes) Transition of the coding context of Genetic Material from that especially of a Symbiont to instead that of a Host. For example, this could be the genomic transfer of Genes from Endosymbionts to nuclear DNA, or the stripping of a Gene from a Prophage (as a Symbiont) to become instead the Gene of a bacterium (as a Host).

Eclipse Period during phage Productive Infections spanning from Adsorption until the maturation of the first Progeny Virion. The eclipse, or eclipse period, is a pre-reproductive aspect of Virion-productive phage Infections.

Ecological species concept The idea that Species can be defined in terms of the unique Niches that they occupy rather than strictly in terms of reproductive or Horizontal Gene Transfer isolation from other Populations. Contrast especially with Biological Species Concept.

Ecology Study of interactions between organisms and their Environments. Environments consist of both Biotic and Abiotic aspects. Biotic aspects are either Conspecifics or instead are different Species (Heterospecifics).

Ecophysiology Study of the impact of Environmental conditions on an organism's functioning.

Ecosystem Reasonably well delineated aspect of an Environment in which both Communities (different Species) and Abiotic aspects are taken into account. This is basically everything except for interactions with other so-defined Environments. Ecosystems furthermore have characteristic properties. Thus, a pond represents an ecosystem, and so would a surrounding field or forest represent separate Ecosystems.

Ecosystem Ecology Study especially of the interactions between Biotic and Abiotic components of Environments. This is with greater emphasis on the impact on Abiotic components than is the case for Organismal Ecology, which instead focuses on impact on organisms rather than changes to the Abiotic aspects of Environments. Ecosystem ecology is concerned especially with nutrient cycling and energy flow within Ecosystems.

Ecotype Species subset that is distinguishable from other members of its Species due to specific Adaptations to specific Environmental aspects but is not so genetically distinct as to represent a separate Subspecies. Divergence of Species into different ecotypes could serve as a nascent basis of subsequent formation of Ecological Species.

Efficiency of plating (EOP) Measure of the ability of Viruses to form Plaques, either per plated Virion Particle or relative to under different conditions. Low efficiencies of plating can be due to low potentials for Infected bacteria to support the production of new phage Virions, low potential for Virions to form visible Plaques despite Infecting indicator bacteria Productively, or both.

Encounter Physical contact between a Virion and the surface of a potential Host bacterium. A Virion's Extracellular Search thus potentially ends with encounter with a potential Host bacterium, i.e., if ultimately resulting in Irreversible Attachment.

Endosymbiosis Cellular organism that lives within the cells of another organism over multiple generations for both. Mitochondria and chloroplasts are the best known and understood endosymbionts.

Environment What an organism exists within and interacts with. Environments consist of both Biotic and Abiotic components.

Environmental DNA (eDNA) Nucleic acid that is found especially outside of the boundaries of producing organisms. See also Environment.

Endolysin Phage protein that enzymatically degrades bacterial cell walls toward effecting especially phage-mediated bacterial Lysis from Within.

Envelope-level resistance Used equivalently to Adsorption Resistance, though arguably could also include mechanisms that could act at the level of the bacterial cell envelope but after irreversible phage attachment, i.e., such as via Superinfection Exclusion. Envelope-level resistance is an aspect of bacterial mechanisms of Avoidance of phage Infection.

Epistasis Gene-Gene Phenotypic interactions where the Phenotype associated with one Gene can be affected by what Allele is present of the other Gene. In terms of phage impact on bacterial Evolution, epistasis can be a concern to the extent that Resistance to two phages negatively affects bacterial fitness more than would be expected based on Resistance to each of the phages individually. Contrast epistasis with Pleiotropy but see also Compensating Mutation and Pseudoreversion.

Eukaryote Organisms whose cells possess nuclei. Contrast with Prokaryotes, such as bacteria, whose cells lack nuclei. Eukaryotes are members of domain *Eukarya*.

Evolution Change in the Frequency of Alleles within Populations as a function of time. These changes may or may not result in evolutionary Adaptation, where Adaptation generally is a product only of Natural Selection. Changes in Allele Frequencies over time, that is, can also result from Mutation, Genetic Migration, or Genetic Drift, though these latter processes only by chance are expected to directly result in evolutionary Adaptation.

Evolutionary biology Study of how organisms change genetically over time, how organisms are related, and why genetic changes have occurred. See also Evolution.

Evolutionary ecology Study of organism Adaptations, i.e., products especially of Natural Selection, as those Adaptations exist within an Ecological context.

Evolutionary fitness Measure of the reproductive potential of an organism, and especially the reproductive potential of a specific Genotype. Evolutionary fitness, or Fitness for short, can be measured Relatively or Absolutely, that is, in comparison to other Genotypes or instead in terms of absolute numbers of successful progeny produced. Generally, to be counted toward fitness totals, progeny should go on to produce progeny of their own. Equivalent to Darwinian Fitness. See also Reproductive Success as well as Survival.

Evolvability Potential for a Population to change genetically, especially as a function of Natural Selection. Higher levels of evolvability for a Population are associated with higher diversities of potentially Adaptive Alleles.

Exploitative competition What occurs when two organisms, especially as representing different Species or at least as representing separate Gene Pools are utilizing the same limiting Resources. Exploitative competition is particularly seen when units of Resource that are used by one organism are thereby no longer available for use by the other organism. Such Competition often does not involve Antagonistic Interactions and further is not a component of Exploiter-Victim interactions.

Extended phenotype Projection of an organism's properties beyond the body of the organism itself. Thus, a beaver's dam can be considered to be not just a product of a beaver's Phenotype, i.e., a product of a beaver's behaviors and physical capabilities, but so too an aspect of a beaver's Phenotype itself, e.g., just as would be the case for a spider's web. That is, *Phenotype extends* beyond the body. Here this concept is most relevant to the release of soluble factors such as often phage-encoded exotoxins by bacteria (see Virulence Factor). These have the effect of presumably adaptively modifying the bacterium's immediate Environment, as equivalent to a beaver adaptively modifying its immediate Environment by building a dam and creating a pond.

Extracellular polymeric substance (EPS) Materials that are molecularly large in size, are secreted by cells, are retained in the vicinity of cells, and which typically play more structural than enzymatic roles. Bacterial Capsules, slime layers, and Biofilm matrix all are examples or consist of extracellular polymeric substances.

Extracellular search Virion-associated process that begins with Release from its parental Infection and ends, ideally for the Virion, with encounter and then Attachment to a phage-permissive Host bacterium. Extracellular searches by phages are primarily diffusion driven but can include movement in association with the surrounding Environment as well, e.g., such as within the flow of water in a stream. Contrast the extracellular search with the phage infecting state (Virocell) but also with Virions that have been 'banked', that is, as found especially in a physically sequestered state during which encounter with a potential bacterial Host is unlikely.

Extrachromosomal resistance Genes associated with blocks on the action of antibacterial agents, especially as encoded by Plasmids. Extrachromosomal resistance is a form of Acquired Resistance. See also simply Resistance. Contrast with Chromosomal Resistance.

Extinction Loss of specific Alleles, Genotypes, or instead all of the organisms making up a Population. Specifically, these are reductions in numbers or frequencies to zero.

Exploiter Individual whose actions reduce the Fitness of other individuals, typically toward increasing the Fitness of the first individual. The individuals whose Fitness has been reduced by the action of an exploiter can be described as Victims. Predators as well as Parasites are exploiters whereas Prey along with Hosts of Parasites are Victims.

Fitness Description of the Reproductive Success of an organism, especially as a function of its Genotype. Also known as Darwinian Fitness, fitness can be attributed to individual Alleles, to an organism's whole Genotype, or even to a Population of organisms, e.g., see Unit of Selection. Fitness also can be differentiated into Relative Fitness versus Absolute Fitness. See equivalently, Evolutionary Fitness.

Fixation Increase of the Frequency of an Allele within a Population to approximately 1.0. That is, with fixation of an Allele there is a loss, i.e., Extinction within a Population of all of the other Alleles found at its Locus. An Allele that has 'gone to fixation' is said to be Fixed and the Locus can be described as now being monomorphic rather than Polymorphic.

Fixed Description of an Allele that has a Frequency of 1.0 within a Population. See also Fixation.

Focus organism Individual that is currently under discussion. This may be an actual individual or instead a category of individuals, e.g., for the latter, 'infecting phages'.

Founder effect Probable occurrence of Sampling Error resulting from the establishment of a new Population that consists initially of few individuals. This is a scenario for potentially substantial impact of Genetic Drift on Allele Frequencies as a Population is split off from a larger Population. A founder effect also can be described as a single round of Genetic Bottlenecking occurring at the start of the existence (establishment) of a Population.

Free DNA This is DNA that is unencapsidated and thereby available for Transformation. See especially Environmental DNA for equivalent.

Free phage Virions as they are found following their Release from Infected bacteria. I use 'free virion' equivalently.

Frequency-dependent selection Deterministic Evolution as motivated by Antagonists to a Population if these Antagonists are less effective against rare individuals than they are against common types. This is equivalent to Killing the Winner but involving Populations rather than Communities. Frequency-dependent selection also can be described as Stabilizing Frequency-Dependent Selection, contrasting Disruptive Frequency-Dependent Selection.

Gene Sequence of nucleic acid that is able to give rise to an RNA product. This RNA product is particularly other than an entire genome, i.e., as is the case for monopartite RNA Viruses. For many but certainly not all Genes, and most notably historically as well as in terms of our understanding of the formation of Phenotype, the resulting RNA product is a messenger RNA that encodes the production of a polypeptide.

Gene exchange Movement of Genetic Material from one organism to another. This gene exchange need not be reciprocal. Thus, Genetic Material can move from one organism to another organism without any requirement for simultaneous reverse movement. With bacteria, mechanisms of gene exchange traditionally are classified in terms of how the recipient bacterium acquires the new DNA, and these include via Conjugation, Transduction, or Transformation. Gene exchange, unlike Genetic Migration as strictly defined, can occur also within Populations rather than just between Populations.

Gene flow See equivalently Gene Exchange.

Gene pool Collection Alleles that are present within a Population. Gene pools generally imply some degree of Gene Exchange between individuals such that Alleles found in one individual have some reasonable potential to Recombine with Alleles found in another individual found within the same gene pool.

Gene transfer agent (GTA) Phage Virion-like structure that is capable of Transducing bacterial DNA but without having a biased potential for carrying its own Genes to new Hosts. Thus, unlike Temperate Phages, gene transfer agents lack mechanisms that result in the transfer of their encoding Genes, except as such transfer occurs by chance.

Generalized transduction Movement of bacterial DNA by phages to new bacteria but without also moving phage DNA. Generalized transduction has the property of being able to transfer large quantities of bacterial DNA, i.e., many tens of kilobases, and at the same time is not limited in terms of what DNA is transferred, neither of which are properties of Specialized Transduction. The process of generalized transduction resembles the action of Gene Transfer Agents but generalized transduction generally is attributed to phages which are able to preferentially Package their own encoding DNA. See also simply Transduction.

Genetic bottleneck Description of the impact of small Population sizes on Allele Frequencies. The primary impact is that of Genetic Drift. See also Founder Effect.

Genetic diversity Here referring especially to the number of types of Alleles found within a Population. A monomorphic Locus displays no genetic diversity whereas a Polymorphic Locus at a minimum displays greater than no genetic diversity.

Genetic drift Microevolutionary process, thus impacting Allele Frequencies within Populations, which is both Stochastic and a function of Sampling Error. Genetic drift especially operates more strongly the smaller a Population's size, and given small enough Population sizes can drive to Extinction even beneficial Alleles. See in particular Genetic Bottlenecking as well as Founder Effects. Genetic drift also underlies though is not identical to Muller's ratchet.

Genetic hitchhiking Increase in the Fitness of an Allele due to its Linkage to an actual higher-fitness Allele, with this other Allele typically associated with a different Gene. Hitchhiking has Stochastic aspects to the extent that the Linkage of one Allele with the second Allele occurs purely by chance. Also, keep in mind that in asexual lineages, all Genes are genetically Linked and thus hitchhiking can be common, though that Linkage breaks down the more that Gene Exchange occurs.

Genetic material DNA or less commonly RNA. The latter is the case for RNA viruses. The DNA content of chromosomes is genetic material and it is genetic material that is transferred in the course of Horizontal Gene Transfer as well as during vertical transmission from parent to offspring.

Genetic migration Movement of Genetic Material from one location to another, here used explicitly as movement from one Population to another. This especially is movement that results in the stable incorporation of Genetic Material into the genome of a member of the recipient Population. Note that Genetic Migration can occur even if the migrated Genetic Material fails to be retained by the recipient Population, i.e., as due to loss that is a consequence of Natural Selection or Genetic Drift. The impact of Genetic Migration is obviously more lasting, however, if the migrated Genetic Material is indeed retained in some form within the recipient Population.

Genetic sequence See Nucleotide Sequence.

Genetic structure Description of both what Alleles are present within a Population and how those Alleles are distributed across individuals, i.e., as within different Genotypes.

Genetic transduction See Transduction.

Genetic variation Nucleotide Sequence differences between individuals. These differences are in terms of what is present, in what order, Genes are present in, or to what extent these Genes are Linked.

Genomic island Continuous segments of DNA that both contain multiple Genes and are thought to be subject to Horizontal Gene Transfer between bacteria. The mobility of genomic islands is often phage associated and a subset of genomic islands are described as Pathogenicity Islands. See also Phage-Inducible Chromosomal Island.

Genotype An organism's hereditary material as encoded by a Sequence of nucleic acid bases.

Geographical range See Range.

Group selection Idea that collections of organisms, typically of the same Species, can together make up a Unit of Selection. Contrast group selection especially with Individual Selection, which typically is considered to be a much more powerful Evolutionary force than group selection. See also the related concept of Kin Selection.

Hard selection Removal of less-fit individuals from Populations especially via the death of those individuals. Lytic Phages, for example, impose hard selection on bacterial Populations, killing those bacteria that are phage-sensitive.

Hardy-Weinberg Referring to the equation or equilibrium, this is a description of circumstances in which both Allele and Genotype frequencies are not changing over time. That is, Hardy-Weinberg equilibriums is a no-Evolution Population Genetics default state.

Head-tailed As used especially in the Archaeal Virus literature, these are members of Virus order *Caudovirales*. Head-tailed Viruses are known among Bacterial Viruses as Tailed Phages.

Headful packaging Control of what amount of Virus genomic material is moved into a Capsid that is determined by a Capsid's volume. Headful packaging generally is associated with Circularly Permuted Virus genomes in combination with a bit of terminal redundancy in what viral Genetic Material is Packaged. Alternatively, in non-headful packaging, Virus genomic material available for Packaging is that found between specific Sequences that signal genome cutting, i.e., *cos* Sites.

Helper phage Coinfecting Virus that supplies Genes and functions missing from a second Coinfecting Virus. Helper phages can be naturally occurring, helping either unrelated viral Parasites of phage Infections or instead helping related, evolved Parasites, i.e., for the latter especially Defective Interfering Particles. In addition, helper phages can be used to allow the replication of phages that have been genetically engineered to be deficient in their ability to replicate on their own.

Heteroimmunity Description of a Temperate Phage that is not subject to the Superinfection Immunity effected by a Prophage associated with a Lysogen that the Temperate Phage is Infecting. That Prophage and the Superinfecting phage together are said to be heteroimmune. Contrast with Homoimmunity.

Heterospecific Two individuals if of different Species (not Conspecifics) can be said to be heterospecific. Contrast with Homospecific as well as Conspecific.

Historical contingency Difficulties in improving upon specific aspects of the Fitness of Genotypes owing to the complexity and therefore unlikeliness of Mutations occurring that are necessary to achieve those improvements. Especially, single Mutational steps are relatively easy to achieve while the likelihood of two Mutational steps occurring is somewhat lower and three Mutational steps somewhat lower still, with historical contingencies defined as impediments to the step-wise accumulation of these multiple mutations within individuals or Populations. This unlikeliness is particularly so if intermediate Mutations are neutral or especially detrimental to Fitness. Historical contingencies, however, can be overcome through Horizontal Gene Transfer.

Hitchhiking See Genetic Hitchhiking.

Homoimmunity Description of a Temperate Phage that is subject to the Superinfection Immunity effected by a Prophage associated with a Lysogen that the Temperate Phage is Infecting. That Prophage and the Superinfecting phage together are said to be homoimmune. Contrast with Heteroimmunity.

Homologous recombination Swapping of regions of nucleic acid that is driven by substantial Sequence similarity between the swapping molecules. Homologous recombination contrasts with Illegitimate Recombination which instead, sensu stricto, requires no Sequence similarity. Homologous recombination is equivalent to the molecular processes associated with crossing over during Meiosis in Eukaryotes. See also simply Recombination.

Homospecific Same Species, i.e., Conspecifics. Contrast with Heterospecific.

Horizontal gene transfer (HGT) Movement of Genetic Material from one individual to another, other than in the course of sexual production of offspring. Phage-mediated horizontal gene transfer is termed Transduction. Horizontal gene transfer otherwise is synonymous with lateral gene transfer.

Host Organism with which a Symbiotic organism is associated. The Symbiont usually is physically smaller than as well as dependent on (helped by) the host, e.g., such as is provided with Resources or protection. In a Parasite-host relationship, the host is harmed whereas in a Mutualistic relation the host is helped by the Symbiont.

Host range Specific Strains within a Species or collection of Species that a Parasite is able to infect. Note that it can be both possible and in many cases desirable to differentiate host ranges in terms of different Phenotypes, i.e., rather than just ability for a Parasite to successfully reproduce (i.e., as a Productive Host Range), and the range of Hosts impacted by a Parasite particularly can vary depending on what host-range determining assay is used. For phages, in alphabetical order, host range thus instead might be defined in terms of ability to Adsorb (Adsorptive Host Range), ability to kill (Bactericidal Host Range), ability to Lysogenize (Lysogenic Host Range), ability to deliver the phage genome to the bacterial cytoplasm (Penetrative Host Range), ability to Plaque (also a Productive Host Range), and ability to Transduce DNA (Transductive Host Range), with much though not complete overlap between many of these concepts. Contrast, however, Geographical Range, which at best is only minimally related to host range.

Hybridization Formation of an organism that is a product of Recombination between two somewhat genetically dissimilar parents. Classically, this would be the hybridization between two different Species, e.g., a mule as the product of a mating between a male donkey and a female horse. With phages or as mediated by phages, hybridization is what gives rise to Mosaic Evolution.

Illegitimate recombination Molecular mixing of nucleic acids that is not driven by Sequence similarity between the formerly two nucleic acid molecules. This contrasts especially Homologous Recombination. It is possible that at least some illegitimate recombination, however, involves small amounts of Sequence similarity. See also simply Recombination.

In situ Within or at a specific location, specifically as referring to a natural or experimental Environment.

In vitro Within a specific experimental Environment, especially one that is contained within glass, e.g., a test tube, or equivalent.

In vivo Within an individual organism. As a complication on this concept, while biochemists consider phages Infecting bacteria within test tubes to be in vivo, contrasting chemical reactions occurring acellularly, many phage biologists would consider this circumstance to be In Vitro instead, e.g., as compared to the phage Infection occurring in vivo within an animal. In either example, the same Infection might also be described as taking place In Situ.

Inclusive fitness Idea that the Reproductive Success associated with an Allele is a function not only of the interaction of its associated Phenotype with the Environment but also its Phenotype's ability to enhance the Reproductive Success of other carriers of the same Allele.

Indirect reciprocal transduction Phage-mediated movement of bacterial DNA where movement from one bacterium to another and then back again requires the existence of at least two different phage types. Contrast with Direct Reciprocal Transduction where this back-and-forth movement of DNA requires the existence of only one phage type. See also simply Reciprocal Transduction.

Individual selection Deterministic Evolution as conventionally considered, i.e., as acting on Populations of individual organisms rather than on Populations consisting of multiple groups of organisms. Individual selection, that is, is measured in terms of the Darwinian Fitness of individual organisms rather than Darwinian Fitness associated with groups of organisms. Contrast therefore with Group Selection and consider also the concept of Unit of Selection.

Induction Transition of a phage Lysogenic Cycle to a phage Productive Cycle. It is a Prophage that is induced. That induction might be either spontaneous or instead associated with exposure to an inducing agent (such as a DNA damaging agent). Typically the resulting Productive Cycle is Lytic, though this will vary with the type of phage.

Infection By a phage of a bacterium, here defined as starting with phage genome entrance into a bacterium's cytoplasm, i.e., as following genome Translocation/ Uptake. Phage Infections then follow well-choreographed pathways that, if successful, result in Lysogenic cycles, Lytic Cycles, or Chronic-Productive cycles.

Innate immunity Diverse and typically more generally acting mechanisms by which the actions of Parasites are thwarted by expressing Hosts. Innate immunity consists of fixed functions of an organism's Genotype. Contrast with Adaptive Immunity.

Insertion mutation Numerical gain of nucleotides in the Genetic Material of an individual organism. Gains can range from a single nucleotide (or base pair) to large blocks of nucleotides, though the latter especially tend to be derived from existing Genetic Material rather than being due to replication errors. Within reading frames, small insertions in multiples of three tend to be less detrimental than non-multiples of three. Integrated Prophages, upon their Integration, also will by definition cause insertion mutations, though often these phages supply replacement Sequence to prevent Gene functions from being disrupted. See also Insertional Mutagenesis.

Insertion (prophage) See Integration.

Insertion site Location within a bacterial genome that an Integrated Prophage comes to reside. That is, this is the site into which the phage genome is inserted via Site-Specific Recombination. See also Integration.

Insertional mutagenesis Disruption of a Nucleotide Sequence by the addition of new Sequence, that is, other than replacement of existing Sequence. Such insertions, especially if they occur at random locations without compensation, can be detrimental to cell functioning. See also Insertion Mutation.

Integrase Site-Specific Recombinase enzyme that is used by Temperate Phages to insert Prophages into chromosomes at specific locations. Integrases thus give rise to Prophage Integration (see Integrated Prophage). Not all Temperate Phages encode integrases, however, as not all Temperate Phages possess Integrating Prophages, i.e., some persist instead during Lysogenic Cycles as Plasmid Prophages.

Integrated prophage Latently Infecting Virus that inserts its genome into the chromosome of its bacterial Host. This can though typically does not result in a deleterious Insertion Mutation in the bacterial chromosome due to compensation by phage Sequence. The now genetic Linkage of Prophage and Host DNA assures Prophage retention in the course of Host cell division. Contrast with Plasmid Prophage.

Integration Insertion of a Prophage into existing DNA. Referred to equivalently here thus as insertion. Integration (or insertion) for many Prophages occurs at preferred (primary) locations on bacterial genomes, as is the case, e.g., for phage λ. For other Integrating Prophages, integration instead can occur more or less randomly, e.g., as for phage Mu.

Interspecific Referring to between Species. Contrast with Intraspecific.

Intraspecific Referring to within a Species. Contrast with Interspecific.

Irreversible attachment/adsorption Part of Virus acquisition of a Host cell in which the Virion *commits* to attempting to Infect a single, specific cell. Contrast with Reversible Attachment/Adsorption.

Invasive Referring to penetration into tissues, such as that of a patient by a pathogenic bacterium. Non-pathogenic bacteria tend to not be invasive, at least on their own accord. Not all pathogenic bacteria are invasive, however, though it is possible also for bacteria to penetrate into tissues solely with their toxins.

Isogenic Having almost the same Genotype. Ideally Phenotypic comparisons between microorganisms are made between isogenic Strains, i.e., as possessing a single Allelic difference.

Iterative coevolution Accumulation within two interacting Populations of multiple reciprocal changes. These changes, however, are not necessarily Cumulative within individual genomes.

Iteroparity Capable of undergoing more than one reproductive episode per life time. Here, analogously, this can be used to imply an ability to generate more than one Productive Infection Cycle per Infection-initiating phage Adsorption, i.e., as can be associated with Lysogenic Cycles in combination with Lysogen binary fission and randomly occurring Lysogen Induction. Thus, one phage Adsorption but many resulting Bursts. Chronically Infecting phages may also be described as iteroparous as their Virion production and especially Virion Release occurs non-simultaneously over long periods of time. Contrast with Semelparity.

Killing the winner Concept from bacteriophage Community Ecology that posits that it is those bacterial types that are especially numerous, or fastest replicating, which are most vulnerable to phage-mediated bacterial decimation. This is equivalent to Stabilizing Frequency-Dependent Selection except as applied to bacterial Communities (many Species) rather than to bacterial Populations (one Species). This phrasing is equivalent to *kill* the winner.

Kin selection Enhancement of Fitness motivated via preferential interaction with closely related individuals. Kin, that is, are organisms that are more closely related, such as clonally related, than are other members of a larger Population. See also Inclusive Fitness.

Lambdoid phage Phage capable of readily recombining with phage λ over large stretches of genomes, resulting in a functional Virion. Note that lambdoid phages are not necessarily Siphoviruses.

Latent infection Not a Virion Productive Infection but still a replicative intracellular interaction between a Virus and Host cell. For phages, latent infections are described as Lysogenic Cycles. Contrast with the concept of Latent Period.

Latent period Duration of Infection especially for a Lytic Cycle. Latent Periods are of lengths that are characteristic of different phages and Infection conditions. Contrast with Latent Infection.

Life cycle Description of an organism's characteristics starting from birth and ending with reproduction and/or with death. Organisms generally change in their characteristics over the course of their life cycles, with bacteria growing larger before dividing or phages transitioning from a Virion stage to bacterium-infecting stage, and back again. Note that the term 'birth' is used loosely in Ecology to describe the generation of a new organism via reproductive processes.

Lineage Group of organisms as they may be traced back phylogenetically (linearly in terms of descent) to a common ancestor. A lineage especially is represented by a descendant Species or instead by a descendant Clonal grouping of organisms that can be traced back vertically (parent-to-offspring relationships) to a single ancestral Species or single ancestral individual, respectively.

Linkage Description of the physical location of genetic Loci such as on the same chromosome, or same plasmid, along with the likelihood of Recombination occurring between individual nucleotides. Given the existence of substantial genetic Recombination, i.e., as seen with Meiotic cell division, then linkage typically will refer to a relatively close physical association on the same chromosome. When there is no Gene Exchange, however, then two Alleles can be genetically Linked even if they are found only in the same cell, just so long as in the course of reproduction there is a bias toward co-inheritance of the two Alleles.

Locus/Loci The specific physical region where a Gene is located on a chromosome or plasmid. Different Alleles of the same Gene generally are located at the same locus, though located on different copies of DNA.

Lotka-Volterra Eponymous description of Predator-Prey dynamics particularly where greater numbers of Predator individuals give rise to fewer Prey individuals within a Community while fewer Prey individuals are unable to support as many Predator individuals. The result is a cycling of Population densities especially where Predator densities follow Prey densities up and down.

Lysin Short for Endolysin.

Lysis Destruction of a cell's 'envelope', which for most bacteria means breaching of both the cell wall and plasma membrane. The result is that cytoplasmic contents mix with the external Environment. Phage Lytic Cycles end with a phage-induced lysis of the phage-Infected bacterium, a process that more precisely can be referred to as a Lysis from Within. Contrast phage-induced bacterial lysis with Chronic Release.

Lysis from within Phage-induced bacterial Lysis that is associated with the termination of a normal phage Lytic Cycle. Contrast with Lysis From Without as well as with Chronic Release.

Lysis from without Phage mechanism of bacterial cell-envelope destruction that is associated with anti-cell wall enzymes that are displayed by various phage Virions. Adsorption of these Virions in excessive numbers results in a weakening of the cell wall to the point of Lysis, which can occur prior to substantial progression of the initial phage Infection. Contrast with Lysis from Within.

Lysis-lysogeny decision Upon Infection of a bacterium by a Lytic, Temperate Phage, that phage must either display a Lytic Cycle or a Lysogenic Cycle, with the choice a product of a molecular decision that is controlled by what otherwise is known as a switch. This choice is termed a lysis-lysogeny decision. In many cases, the decision seems to be biased especially toward Lytic Cycles, though there are various circumstances such as high-Multiplicity Infections or Infection of physiologically older bacteria that bias the decision instead more toward Lysogenic Cycles. Note that for Chronically infecting Temperate Phages, the concept of a lysis-lysogeny decision at a minimum is a misnomer since these phages do not display Lysis; thus, lysis-lysogeny decisions might be generalized to 'Productive Infection-Latent Infection' decisions.

Lysogen Symbiosis consisting of an established Prophage and a bacterial Host. That is, a lysogen is a bacterium that is hosting a Temperate Phage Lysogenic Cycle. See also Sister Lysogen.

Lysogen allelopathy Antagonism by bacterial Lysogens against other, closely located and often also closely related bacteria, with the Antagonistic Interaction mediated by Prophage Induction and subsequent Release of Lytically Infecting Phage Virions. The released Virions would be acting as what might be thought of as the equivalent of replication-competent Bacteriocins.

Lysogeny The carriage by a bacterium of one or more Prophages.

Lysogenic conversion Modification of a bacterium's Phenotype by Prophage Gene expression and in ways that are not a direct consequence of the Prophage's Lysogenic Cycle. Thus, Superinfection Immunity is not considered to be an example of lysogenic conversion.

Lysogenic cycle Phage ongoing existence as an uninduced Prophage, thereby resulting in the carrying bacterium existing as a Lysogen. The 'cycle' of lysogenic cycles refers at least in part to the Life Cycle of the carrying bacterium, i.e., as involving repeated rounds of binary fission.

Lysogenic host range Those bacterial Strains that a given Temperate Phage type is able to display Lysogenic Cycles in. This assumes that these bacteria also fall within the same phage's Adsorptive Host Range and Penetrative Host Range, though not necessarily also the phage's Productive Host Range. That is, these are the bacteria that a Temperate Phage can succeed in Infecting and then establishing a Latent Infection in, without consideration of whether they necessarily can also precede to or otherwise effect a Productive Infection.

Lytic conversion Phage-induced blocks on Secondary Adsorption, i.e., on Virion Attachment rather than on Secondary Infection or Translocation. This mechanism should not contribute to lasting changes in bacterial properties unless it is associated with a Lysogenic Cycle or with a Chronic Infection.

Lytic cycle Phage Productive Cycle that both leads to and ends with Lysis of the phage-Infected bacterium. Note that it is also possible for a bacterium to be lysed by a phage without associated Virion Production or Release, e.g., as via Lysis From Without or with certain forms of Abortive Infection, but these generally are not referred to as lytic cycles.

Lytic phage Bacterial Virus with a Productive Cycle that ends in Lysis, whether or not that phage is Temperate. That is, most Temperate Phages are also Lytic Phages. See also the concept of Obligately Lytic or Strictly Lytic as well as Professionally Lytic.

Macroevolution Evolution as it occurs above the level of Species, specifically as involving differential new Species formation and differential Species extinction. The divergence of Species away from each other (Divergent Evolution) impacts macroevolution, and phylogenies typically will depict our understanding of the consequences macroevolutionary processes.

Mass action Interactions between particles that are a consequence of random motion by at least one of the particle types. Thus, the likelihood of Encounter of a bacterium with a phage Virion can be modeled as a product of mass action, in this case as particularly a consequence of Virion diffusion rates in combination with bacterial target size.

Meiosis Eukaryotic nuclear division that results in four offspring cells each possessing half of the chromosome set of the original parental cell. That is, diploid cells are converted to haploid cells. Often the newly formed cells, especially in animals, are gametes, which are then subject to fertilization. Even if the latter is not directly the case, meiotic cell divisions still are central to the sexual reproduction of Eukaryotic organisms.

Microcolony Group of microbial cells such as bacterial cells that are descended from a single parent cell and in which daughter cells have not moved away from each other once formed. The result is a Colony of cells, but if that Colony is extremely small or microscopic then it is referred to as a microcolony. Biofilms often are made up of associations of individual Microcolonies.

Microevolution Consequences of Mutation, Genetic Migration, Genetic Drift, or Natural Selection on Allele Frequencies within Populations.

Migration (genetic) See Genetic Migration.

Mobile genetic element Nucleic acid Sequence that readily moves either between organisms or instead from place to place within an organism, the latter especially to new locations within genomes. Examples of Mobile Genetic Elements are Phages, Plasmids, and Transposons.

Moron (broad definition) Horizontally acquired DNA Sequence that has been inserted into a phage's genome. This literally is 'more' phage DNA and particularly this is DNA that has ultimately been obtained from non-phage sources via Illegitimate Recombination. Morons are not limited in where they are located within a phage's genome nor in their functions there. They also may be sub-optimally incorporated, either Genotypically or Phenotypically or both. See especially Moron Accretion Hypothesis for use of this *broader* definition.

Moron (narrow definition) Horizontally acquired DNA Sequence that is inserted into a phage's genome that provides an additional Phenotype particularly during Lysogenic Cycles. This definition of moron is less-well associated with the concept of Moron Accretion but instead has significant overlap with the concept of Lysogenic Conversion, i.e., as converting Genes often also can be classified as morons.

Moron accretion hypothesis Idea that phage genomes have been built up over Evolutionary time via sequential acquisition, e.g., one Gene at a time, of relatively small DNA insertions, with these insertions subsequently coopted toward functioning during a phage Lysogenic Cycle or Lytic Cycle (or Productive Cycle more generally). Insertions would have (1) occurred mostly via Illegitimate Recombination, (2) been relatively rare in their occurrence, (3) consisted of random lengths and locations within genomes, and (4) been retained within recipient Populations mostly if they were Selectively beneficial (or at least not detrimental). Here the Broad Definition of Moron is being used, i.e., as simply Horizontally Transferred additions to phage genomes.

Mosaic evolution Variation in the Evolutionary history of different aspects of an organism, as seen in an organism's genome. These differences are a consequence of Recombination combining together parts of relatively unrelated organisms into a single organism, such as can occur following the Coinfection of a single bacterium by two somewhat genotypically different phages, or given Horizontal Gene Transfer from a phage to a bacterium. Mosaic evolution in phages appears to occur especially among Temperate Phages rather than phages that are Strictly Lytic, though nevertheless mosaic evolution is commonly seen among the latter as well.

Multiplicity of infection (MOI) Ratio of phages to phage-susceptible bacteria. Strictly, this should be the ratio of *Adsorbed* phages to adsorbable bacteria. Many authors, however, instead use multiplicity of infection to describe the ratio of *added* phages to bacteria. While often in practice added phages readily transition into being Adsorbed phages, thus often resulting in an experimental equivalence of the two concepts, when densities of adsorbable bacteria are low then numbers of added phages can greatly outnumber numbers of Adsorbed phages, thereby making the two alternative concepts of multiplicity of infection less quantitatively equivalent.

Muller's ratchet Essentially irreversible loss of wild-type Genotypes from asexual Populations due to Genetic Drift, and/or loss of least-Mutated Genotypes, with assumptions that lost Genotypes are not recoverable though Mutation alone. Instead, only through Recombination (matings or Horizontal Gene Transfer) can individuals possessing fewer mutations be reestablished within Populations that have been subject to Muller's ratchet. Note that while Muller's ratchet is driven by Genetic Drift, it is not identical to Genetic Drift. Similarly, loss of

Alleles from Populations as due to Genetic Drift is not Muller's ratchet since Muller's ratchet is premised on the idea that while Genotypes may be lost from Populations, their underlying Alleles nevertheless are retained.

Mutant Individual organism that carries a Mutation, especially a newly observed Mutation. The terms 'Mutation' and 'mutant' should not be used interchangeably as one is a change in genome Sequence while the other is an organism, respectively.

Mutation Change in the Sequence of Genetic Material, typically as occurring due to random modification of existing Sequence rather than due to Horizontal Gene Transfer or Homologous Recombination. Mutations especially give rise to new Alleles or instead can remove Genetic Material from organisms.

Mutation, compensating See Compensating Mutation.

Mutation, deletion See Deletion Mutation.

Mutation, insertion See Insertion Mutation.

Mutation, point See Point Mutation.

Mutation rate Number of discrete changes in Nucleotide Sequence as occur per round of replication, per generation, or per some unit of time.

Mutualism When the interactions between two Species result in net advantages for both of them. This, for example, may be seen with Lysogeny giving benefit to the Host bacterium via Lysogenic Conversion in combination with the Prophage being supported and replicated by the hosting bacterium.

Mutualistic coevolution Two Species interacting in a manner that is beneficial to both, where genetic changes to one organism are followed by changes in the other, all toward improving mutual benefits rather than resulting in increasingly Antagonistic or instead more communalistic (i.e., Selectively neutral) interactions. Mutualistic coevolution is the means by which Mutualisms can improve Evolutionarily toward improved benefits for one or both participants. Contrast with Antagonistic Coevolution.

Myovirus Viruses possessing long, contractile tails. See also Tailed Phage and *Caudovirales* as well as contrast with Podovirus and Siphovirus.

Natural selection Deterministic Microevolutionary force measured in terms of differences in the Reproductive Success of different organism Genotypes. Natural selection, that is, represents a biased contribution of Genotypes to the next generation: some Genotypes, in other words, are more effective at producing offspring than others. Natural selection occurs as a consequence of Phenotype-Environment interactions and can be measured in terms of Genotype Fitness.

Negation Here used synonymously with Restriction to mean a bacterial mechanisms of phage Resistance that results in loss of phage viability in conjunction with Survival of the phage-Infected bacterium. Negation is associated with Restriction-Modification Systems, CRISPR-Cas Systems, Bacteriophage Exclusion (BREX) Systems, and also Superinfection Immunity.

Negative selection Here referring to means of increasing the prevalence of Alleles that nevertheless are detrimental to their carriers. In microbiology, this typically is accomplished by replica plating, which involves growth in association with solid or semi-solid media under two conditions, selective and not selective. Those bacteria that form colonies under non-selective conditions but not under selective conditions are, in the case of negative selection, associated with sought Alleles, e.g., such as auxotrophic Mutations. Here, however, this concept is being used more generally to describe Alleles that require indirect means of Selection, especially of Alleles conferring phage Resistance but in which the expressing bacterium nevertheless dies upon phage Infection. See Abortive Infection, Self-Sacrifice, and Kin Selection. Contrast with Positive Selection.

Niche Environmental factors that determine a given Species' ability to Survive and reproduce. If essential niche-defining factors are missing from an Environment, e.g., necessary specific carbon and energy sources, then an organism will not be able to Survive and reproduce, or at least not do so to the degree required for that Species to persist in that location, other than its serving as a sink Environment, i.e., where a Species is present only because it has migrated in from a different location.

Non-Darwinian evolution Evolutionary processes other than as mediated by Natural Selection.

Non-homologous recombination See Illegitimate Recombination.

Non-random mating Biases in terms of what Genetic Material is available for Recombining into the Genetic Material associated with a different organism. Non-random mating is especially seen when mating is confined to within a single Species, i.e., as consistent with the Biological Species Concept, but generally this term is used instead to describe biases in what matings take place within rather than between Populations or Species. For non-meiotic Gene Exchange, non-random mating can be interpreted as meaning that the transfer of Genetic Material from an organism is more likely to one set of organisms than to a different set of organisms. Still, transfers tend to be more likely to Conspecifics than to Heterospecific organisms. See similarly Assortative Mating as well as Assortative Transduction. Contrast with Random Mating.

Non-reciprocal transduction Idea that phage-mediated movement of DNA may occur between two distinct bacterial Strains only in one direction, i.e., from Strain A to Strain B but not also from Strain B to Strain A, particularly as via the action of only a single phage type. Contrast with Reciprocal Transduction. See also simply Transduction.

Non-synonymous substitution Point Mutation that modifies a codon while changing what amino acid is encoded. Fixation of non-synonymous substitutions within Populations is considered to be a possible indicator of Natural Selection having acted upon the associated Genes, i.e., as a due to possible Directional Selection toward retention of these Mutations. Contrast with Synonymous Substitution.

Nucleotide sequence Order of nitrogenous bases making up DNA or RNA. This is, that is, the order of As, Ts, Gs, and Cs or As, Us, Gs, and Cs, respectively. Differences in Genotypes are a function of differences in genome nucleotide sequences, and indeed Genotypes consist of nucleotide sequences.

Numerical refuge Bacterial Population Survival, given phage presence, that is a consequence of the density of bacteria supporting phage Population growth being insufficient to allow phages to reach high enough Titers to result in substantial reductions in bacterial numbers. This is equivalent to Predator Species numbers being a function of Prey Species numbers, where lower Prey numbers support lower Predator numbers. As a result of there being fewer Predators, fewer Prey in absolute terms are lost to predation. See Lotka-Volterra and contrast with Numerical Vulnerability.

Numerical vulnerability Bacterial densities that are sufficiently high as to result in support of phage Population growth to Titers that are high enough to substantially reduce bacterial densities. A numerically vulnerable bacterial Population could be viewed as a 'winner' Population in Killing the Winner. See also the concept of Spatial Vulnerability and contrast with Numerical Refuge.

Obligately lytic Equivalent to Strictly Lytic.

Obligatorily sexual Referring especially to Meiosis-associated mechanisms that are engaged in relatively frequently by a Species, most familiarly in terms of an association between the transfer of Genes between individuals and reproduction of those individuals. Many obligatorily sexual Species thus obligatorily participate in sexual processes in order to reproduce, or at least must periodically engage in fertilization and Meiosis. Contrast Horizontal Gene Transfer which, though potentially evolutionarily useful to all Lineages, is not required over shorter times scales for the normal functioning or reproduction of these organisms.

One-step growth Equivalent to Single-Step Growth

Organismal ecology Study especially of an organism's interaction with Abiotic aspects of its Environment, as considered particularly from the perspective of the organism rather than from the perspective of the Ecosystem as a whole (that is, as contrasting Ecosystem Ecology). Study of phage Virion requirements for adsorption cofactors or variation in Burst Size as a function of temperature could be viewed as aspects of phage organismal ecology.

Orthologous replacement Swapping via Homologous Recombination of one similar Allele for another. With orthologous replacement, the genetic Locus containing a given Gene is not disrupted, though what Allele is present that represents that Gene is changed. With orthologous replacement, the synteny of a genome (order of genetic elements) does not change. Contrast with the insertion of Genetic Material (i.e., as via Illegitimate Recombination) which can generate new Genes rather than simply changing the Sequence of existing Genes, the latter, i.e., being the consequence of orthologous replacement.

***pac* site** Sequence on a phage genome that signals the location of a specific cut site for Packaging. An alternative form of signaling for Packaging is the use of *cos* Sites.

Packaging Movement of a phage genome into a phage Capsid, i.e., the encapsidation process. In the case of generalized transduction, it would instead be bacterial DNA that is packaged rather than phage DNA, but the packaging process in this case is otherwise equivalent. Genome packaging is an aspect of Virion maturation.

Panmixis Another name for Random Mating. Panmixis is also known as panmixia. Contrast with Non-Random Mating.

Parasite Organism that lives in intimate association with another organism, i.e., a Host organism, but that harms that Host organism and does not otherwise provide to the Host organism sufficient benefits to make up for that harm. Many Parasites will not directly or intentionally kill their Host organisms, though they certainly can weaken their Hosts. It is an open question, therefore, whether phages during Lytic Cycles truly are Parasites rather than Predators. Chronically Infecting phages, by contrast, are more traditionally parasite-like in that their Life Cycles to a degree harm their Host cells but don't intentionally kill those cells. It is an open question as well whether Prophages serve as Parasites versus instead as Commensalistic or Mutualistic Symbionts.

Parasitism Process by a Symbiotic organism of harming its Host organism. Parasitism is a Symbiotic relationship where the Parasite, as Symbiont, gains while the Host loses from the interaction. Compare to Commensalism as well as Mutualism.

Pathogenicity island Continuous segment of DNA that both contains multiple Genes and contributes to the ability of a bacterium to cause disease. Pathogenicity Islands typically are thought to be subject to Horizontal Gene Transfer between bacteria (hence the Linkage of their associated Genes) and otherwise represent a subset of a broader collection of such elements known as Genomic Islands. The mobility of pathogenicity islands is often phage associated, i.e., see Phage-Inducible Chromosomal Island.

Penetrative host range Those bacterial Strains that a phage is able to Adsorb and then Translocate its genome into. That is, if a phage can deliver its genome into a bacterium's cytoplasm then that bacterium is found within that phage's penetrative host range. The significance of successful genome penetration is that this is the minimum requirement for Transduction of DNA to a recipient bacterium from a Transducing Particle. The penetrative host range could be broader than the Productive Host Range of a phage, thereby potentially allowing a broader Transductive Host Range than a successful phage Infection (Productive Cycle) Host Range.

Periodic selection Sweeping of specific Genotypes through Populations toward Fixation. An example would be the rise in Allele Frequency of phage-Resistant bacterial Genotypes in response to phage exposure.

Phage Short for Bacteriophage. A Bacterial Virus.

Phage-inducible chromosomal island Multi-Gene DNA Sequence that actively Parasitizes one or more Helper Viruses toward its Packaging and transmission to new bacterial Hosts. See also Satellite Virus as well Genomic Island.

Phage-susceptibility type Collection of bacterial Strains that a specific Bacterial Virus is able to Infect. Thus, if the Titers of a lytic phage are high enough within a given Environment, then all bacteria found within the associated phage-susceptibility group will be subject to phage-associated reductions in numbers.

Phage λ The canonical Temperate Phage. Phage λ is a Coliphage, Infecting *E. coli.*

Phage therapy Use of Bacterial Viruses to reduce the prevalence of nuisance bacteria. This includes to combat bacterial infections, including ones that are antibiotic Resistant. Note that phage therapy also is referred to as bacteriophage therapy.

Phagocytosis Uptake of particles by Eukaryotic cells into cytoplasm-suspended vesicles. These vesicles are known as phagosomes. Typically phagocytosis is followed by phagosome fusion with a lysosome, forming a phagolysosome and ideally resulting in intracellular digestion of the particle. See also Phagotrophic.

Phagotrophic Organisms, especially single-celled Eukaryotes (protists), that obtain their nutrients by endocytic engulfment. That is, via Phagocytosis.

Phase variation Differences in Gene expression across an otherwise Clonal bacterial Population, generally involving a turning off or on of individual Gene expression. A key characteristic of phase variation is that the changes involved are reversible without the occurrence of simply random Mutation. The expression of phage Receptors can be affected by phase variation, resulting in bacterial Adsorption Resistance to those phages that use a Receptor that is not currently being expressed in the course of phase variation.

Phenotype The characteristics of an organism that are other than explicitly the Nucleotide Sequence of its genome. Phenotypes can vary Allelically and Natural Selection acts upon phenotypic differences among Conspecifics when those differences vary Allelically, that is, vary as a function of Genotype.

Physical Referring to aspects of Environments that are neither strictly chemical nor Biotic. Examples of physical aspects of Environments include temperature, amounts of light, radiation, osmolarity, and also pH.

Plaque Phage-mediated elimination or substantial reduction in number of spatially associated bacterial cells and Microcolonies. These are observed in the laboratory as circular clearings in bacterial lawns that are growing in or on agar-containing media.

Plaque-forming unit (PFU) Operational stand-in for number of Virions especially in determining Titers. That is, rather than actual numbers of phage Virions, instead numbers of phages able to produce Plaques upon plating are described as plaque-forming units. Note that numbers of plaque-forming units often will vary with phage characteristics, Host differences, and plating conditions, thereby giving rise to the concept of Efficiency of Plating.

Plasmid Extrachromosomal, somewhat autonomously replicating DNA. Typically plasmids are closed circular though some instead are linear. Some Prophages exist as plasmids, both closed circular or linear depending on the phage (see Plasmid Prophage).

Plasmid prophage Latently Infecting Virus that does not Integrate its genome into the chromosome of its bacterial Host. Such Prophages must possess strategies for successful Segregation, i.e., as do Plasmids generally. Contrast with Integrated Prophage.

Pleiotropy Single Gene underlying more than one Phenotype. Thus, different underlying Alleles have associated with them two or more definable Phenotypes rather than just one. The concept of pleiotropy can be extended to Phenotypes that are other than those associated with a multicellular organism, such as Resistance to more than one phage type as a consequence of a single Mutation (also known as cross resistance). See also the concept of Antagonistic Pleiotropy. See as a contrast the concept of Epistasis.

Podovirus Tailed Viruses possessing short, non-contractile tails. See also Tailed Phage and *Caudovirales* as well as contrast with Myovirus and Siphovirus.

Point mutation Modification of a single nucleotide within Genetic Material. If this modification is neither a Deletion Mutation or Insertion Mutation then it is described as a substitution.

Polymorphism Existence of more than one Allele at a single genetic Locus within a Population. The existence of polymorphisms is not dependent on the ploidy of organisms as, e.g., Allele *A* can exist in one haploid organisms while Allele *a* can exist in a Conspecific, and together this would represent a polymorphism. A more general name for polymorphism is simply genetic diversity within a Population. Contrast with monomorphism, or Fixed Allele.

Population Members of the same Species, i.e., Conspecifics, especially as existing within some reasonably well defined local area.

Population ecology Study of organism-Environment interactions that can be observed particularly from the perspective of groups of Conspecifics but not at the level of just individuals. Population ecology especially is the study of changes in Population sizes and age structures (demographics) as occurs as a consequence of Environment interactions as well as study of interactions between Conspecifics. Contrast with Organismal Ecology as well as Community Ecology.

Population genetics Study of the diversity of Genes (Allelic Variation) and changes in their frequency (Allele Frequency) within Populations as well as among related Populations.

Population growth Increases in the number of Conspecifics as a consequence of reproduction by those organisms. Population growth of bacteria for example occurs as a consequence of binary fission. Note that for population growth to positive, however, then per capita rates of reproduction (quantified as 'births') must exceed per capita rates of death. See also simply Population.

Positive selection Process of biased increases in the prevalence of specific Alleles or Genotypes within a Population. Selection for a phage-Resistant bacterial variant thus is an example of positive selection. As a method, positive selection is straightforward, as application of a Selection scheme will result in increases in the frequency of desired variants, e.g., again, such as via the application of phages to bacteria to Select for phage-Resistant bacteria. See, similarly, Directional Selection and contrast here with Negative Selection. See, too, Hard Selection as a specific as well as somewhat extreme category of positive selection.

Postzygotic isolation Mechanisms of prevention of successful mating that act after the point of fertilization, though here referring instead to inhibitors of Gene Exchange that act following the point of new DNA delivery into a bacterium's cytoplasm. This can include destruction of that DNA immediately upon entry, failure of the DNA to properly Segregate to daughter cells, or a lack of Natural Selection favoring the retention of that DNA.

Predator Organisms that kill Prey organisms either toward or in the course of consuming those Prey. Parasites, by contrast, do not necessarily intentionally kill their Victim organisms. Phages serve as predators in Lotka-Volterra-type predator-Prey interactions.

Prezygotic isolation Mechanisms of prevention of successful mating which act prior to the point of fertilization, though here referring instead to inhibitors of Gene Exchange that act prior to the point of new DNA delivery into a bacterium's cytoplasm. Prezygotic isolation mechanisms for bacteria thus would be barriers to Transducing Particle Adsorption or instead to genome Translocation, i.e., as due to bacterial Avoidance as a mechanism of phage Resistance. See also Adsorption Resistance as well as Superinfection Exclusion plus the concept of Transductive Host Range.

Prey Organism that is killed by a Predator prior to or in the course of that organism being consumed by the Predator. Bacteria serve as prey in Lotka-Volterra-type Predator-prey interactions.

Productive cycle/infection Assembly, maturation, and Release of new Virions. Productive Cycles/Infections can result in either Lytic or Chronic Virion Release, depending on the Virus. Contrast especially with phage-Reductive Infections but also with phage-Destructive Infections.

Productive host range Those bacterial Strains that are able to support a phage's Infection and subsequent Release of Progeny Virions. Productive host ranges are a subset of Adsorptive Host Ranges and Penetrative Host Ranges as phages must be able to Infect bacteria in order to produce new Virions. Productive host ranges are also a subset of Bactericidal Host Ranges, at least for Lytic Phages, since Lysis is bactericidal, though there exist other mechanisms by which phage Infections can result in the death of Host bacteria other than productive lytic infections, i.e., Abortive Infections. See also simply Host Range.

Professionally lytic Description of a phage that is not Temperate, does not Release Virions Chronically, and is not closely related to a Temperate Phage. That is, these are Strictly Lytic Phages that have not descended in recent Evolutionary time from Temperate Phages. Ideally this means that not only are professionally lytic phages Strictly Lytic, thus unable to display Lysogenic Cycles, but also are unlikely to encode bacterial Virulence Factors.

Progeny, Virion See Virion Progeny.

Prokaryote Cellular organism that is a member of either domain *Bacteria* or domain *Archaea*. Prokaryotes generally lack cell nuclei. Contrast with Eukaryotes. Note that not all modern researchers are comfortable with the term, prokaryote, because taxonomically a prokaryote would be polyphyletic, meaning that members of domain *Bacteria* are no more closely related to members of domain *Archaea* than they are to members of domain Eukarya. Alternatively, the concept of prokaryote can be used to mean a cellular organism that is not a member of domain Eukarya, as is the case here. That is, at least in this author's opinion, 'prokaryote' is a legitimate as well as useful term when used in a non-phylogenetic context.

Prophage Virus genome as it exists during a Lysogenic Cycle. Prophages can be found either integrated into already existing bacterial DNA (Integrated prophage) or can exist as Plasmids (Plasmid Prophage). More generally, a Prophage is a kind of provirus.

Prophage depletion Concept as used here to mean loss of Genetic Material associated with integrated Viruses infecting bacteria. These losses are such as due Deletion Mutations and can be accompanied by selective retention by bacteria of certain, not deleted Prophage Genes. That is, this is a description of one process of Prophage Gene Domestication as can occur in conjunction with Prophage genetic decay.

Prophage, integrated See Integrated Prophage.

Prophage, plasmid See Plasmid Prophage.

Pseudogene Nucleic acid Sequence that is not able to generate a functional product though resembles a Sequence that *is* potentially able to generate a functional product. Pseudogenes particularly are descended from such Genes that had been able to generate functional gene products. Typically pseudogenes are able to ascend in Allele Frequency within a Population, relative to the functional version of a Gene, when Selection for retention of the latter is minimal. This can occur either because Environmental change renders a functional Allele less useful or instead due to Gene duplication such that one copy of the duplicated Gene is now dispensable.

Pseudolysogeny Though different authors often use this term to mean different things, here this is a description of a stalled phage Infection, such as due to a bacterial Host's exposure to low nutrient densities, that can be restarted such as by adding nutrients.

Pseudoreversion Mutational recovery of an approximation of a wild-type Phenotype, not through direct reversal of a Mutation but rather as a consequence of Mutation in another location in an organism's genome. A pseudoreversion thus is a change in Genotype that recovers to some degree a previous but Mutationally lost Phenotype, but does not recover the original, wild-type Genotype. Pseudoreversions can also be described as products of Compensating Mutations and can occur either within Genes or between Genes, with the latter an example of Epistasis.

Random mating Direct intracellular interaction between the genomes of different organisms, particularly organisms from the same Species, where there are no biases against any possible interactions. Contrast random mating, for example, with Assortative Mating, where instead biases exist and therefore mating likelihoods within a Population are not random. Random mating within a bacterial

Population would mean that transfer of DNA from one cell or Strain type to another occurs with equal likelihood no matter the status of the recipient organism, and ideally no matter the status of the donor organism either. The existence of random mating is also described as Panmixis.

Range Geographical concept referring to where an organism is able to live and generally also where an organism is also able to successfully reproduce. Typically this is the naturally occurring geographical area within which a Species is endemic rather than an area that is occupied as a consequence of artificial introduction. Contrast with Host Range.

Receptor, adsorption Cell-surface molecule to which a Virion attaches in the course of its Adsorption.

Reciprocal transduction Idea that phage-mediated movement of bacterial DNA can occur between two distinct bacterial Strains in both directions, i.e., from Strain A to Strain B but also from Strain B to Strains A, and especially as via the action of the same phage type. See in addition Direct Reciprocal Transduction and Indirect Reciprocal Transduction. Contrast with Non-Reciprocal Transduction. See also simply Transduction.

Recombination Combining of two nucleic acid polymers into a single entity. For example, this occurs during Meiosis in which the genomes of two parent organisms are combined molecularly to contain Sequences in different chromosomal locations obtained from different parents. Molecular recombination often readily occurs between the genomes of phages Coinfecting the same cell. Recombination also represents a common outcome of Horizontal Gene Transfer, particularly when the donor Sequence in some manner is integrated into the recipient organism's DNA. Without the molecular component, recombination occurs as well simply when nucleic acid polymers from two different parents enter the same cell and particularly so if the new, literally recombinant organism persists. See also Homologous Recombination, Illegitimate Recombination, and Site-Specific Recombination.

Reduced infection vigor Mechanisms that lead especially to extended phage Infections and/or reduced Infection productivity. This, as the concept was originally formulated, was particularly as these could lead to a reduced Efficiency of Plating without simply being a consequence of phage inviability. Reduced infection vigor also serves as a basis of phage Delay as a form of bacterial Resistance to phages.

Reductive infection Term that minimally may be used to describe a Lysogenic Cycle but also latent infections more generally as well as latent-like viral Infections such as some forms of Pseudolysogeny. The application of the term to Lysogenic Cycles is derived from the writing of André Lwoff. See by contrast phage-Productive Infections and what I describe as phage-Destructive Infections.

Relative fitness Reproductive Success of a given Genotype as compared to the Reproductive Success of a different Genotype. Relative fitness is often easier to determine than Absolute Fitness as it involves Competition experiments between different Genotypes. That Genotype which wins these competitions is considered to have a relative fitness that is greater than their losing competitor's, at least under the conditions of the determination, with relative fitness quantified in terms of how much one Genotype increases in frequency, as equivalent to Allele Frequency given Isogenic comparisons, relative to the other Genotype.

Release Movement of a Virion from within its parental Infection to the extracellular Environment, thus beginning a Virion's Extracellular Search. Contrast Lysis with Chronic Release.

Reproductive success Definitive measure of an organism's Evolutionary fitness, typically defined as the number of offspring produced in combination with the potential for those offspring to also successfully reproduce. Reproductive success is also dependent on Survival as one has to survive to reproduce. See also Absolute Fitness.

Resistance Ability to reduce the harmful effects of exposure to an Antagonistic entity, especially where the underlying mechanism has a genetic basis. That is, resistance stems from one or more specific Alleles which distinguish the resistant organism from an Isogenic sensitive one. See also Acquired Resistance, Chromosomal Resistance, and Extrachromosomal Resistance. For phage-resistance, see as well Abortive Infection, Adsorption Resistance, Avoidance, CRISPR, Delay, Negation, Restriction, Restriction-Modification, and Self-Sacrifice.

Resistance, acquired See Acquired Resistance.

Resistance, chromosomal See Chromosomal Resistance.

Resistance, extrachromosomal See Extrachromosomal Resistance.

Restriction Destruction of a targeted individual, such as a phage, such that the targeted individual is no longer able to replicate. This concept is especially associated with the action of Restriction Endonucleases, which kill phages but at the same time are generally thought to spare the phage-Infected bacterium. Though I tend to use 'restriction' also to describe more generally phage-Resistance systems which kill phages but not the phage Infected bacteria, here to avoid confusion with solely the actions of Restriction Endonucleases, I use as a synonym instead the word 'Negation'. Contrast restriction among phage Destructive Infections especially with Abortive Infections, the latter which instead are associated with bacterial 'Self-Sacrifice'.

Restriction-modification Mechanism within Prokaryotic organisms in which 'self' DNA is tagged in such a manner (modification) such that it resists cleavage (restriction), thereby allowing non-self DNA to be recognized as not having been modified and thereby cleaved (restricted). Restriction-modification systems are associated with the Negation of phages among phage-Destructive Infections. See also Restriction Endonuclease.

Restriction endonuclease DNA cleavage-effecting enzyme that targets non-self DNA. Restriction endonucleases effect the Restriction in Restriction-Modification systems, with 'endo' referring to this cleavage occurring within rather than at the ends of targeted DNA. In particular, it is well-defined, short, unmethylated DNA sequences that are recognized by restriction endonucleases and which guide the positioning of cleavage.

Resource Something obtained by an organism which, once obtained, potentially results in an increase in that organism's Fitness. Failure by an organism to obtain adequate quantities or qualities of a resource will, by contrast, tend to result in reduced organism Fitness.

Reversible attachment/adsorption Specific interactions of a Virion with a cell surface Receptor that are not permanent. Thus, the Virion remains able to Irreversibly Adsorb to the reversibly attached cell or, instead, is able to separate from that cell. Contrast thus with Irreversible Attachment/Adsorption.

Reversion mutation Change in Genotype that recovers a previous but Mutationally lost Genotype. Thus, if the original Mutation were A $\rightarrow$ C, then the reversion mutation, at the same position, would be C $\rightarrow$ A. Reversion mutations tend to be rare relative to Mutation Rates because they represent highly specific Mutational events. Compare with Pseudoreversion as well as Compensating Mutation.

Sampling error Failure of a subgroup of entities to fully especially quantitatively represent a larger group. Implicit to this idea is that with the larger group there exists variation of some sort and it is this variation which the subgroup fails to fully represent given sampling error. Sampling error tends to be greater the smaller the sample, and sampling error represents the statistical underpinning of the concept of Genetic Drift.

Satellite plasmid Plasmid that has Mutationally lost an ability to express an Accessory Gene. Satellite plasmids thereby impose a reduced metabolic Gene-expression burden on the carrying cell, which could be Selectively advantageous under conditions in which Selection for the Accessory Gene is reduced.

Satellite virus Genetic entity that Parasitizes cells that are Infected by unrelated Viruses, particularly toward satellite virus transmission to new Hosts. Among satellite viruses are Phage-Inducible Chromosomal Islands.

Secondary adsorption Irreversible Attachment of a Virion to an already Virion-Infected cell.

Secondary infection Meaning not the first Infection, but still with competing meanings. The first meaning, dating back to at least 1948 for phages and much earlier otherwise, is phage Infection of an already phage-Infected bacterium, i.e., as equivalent to Superinfection. The second comes from epidemiology, which is the initiation of a new Infection by a phage that had been produced within the same local Environment. For the latter, a secondary infection is rather than an Infection by a phage, as an 'index case', which had newly entered that same Environment. For both, contrast with primary infection.

Secondary phage Phage that Adsorbs after a primary phage has Adsorbed and/or Infected. 'Secondary phage' is shorthand for Secondarily Adsorbing or Secondarily Infecting phage.

Segregation Retention of Genetic Material in both daughter cells following cell division. Plasmids, especially those found in low copy numbers, will possess partitioning systems to assure this segregation. A Temperate Phage genome that has been subject to Superinfection Immunity should lack segregation due to its retention within a bacterium's cytoplasm but without phage genome replication.

Selection Biased Reproductive Success as occurring in any context. This includes within the natural world in the form of Natural Selection but also while under human direction in the form of artificial selection. See also Unit of Selection.

Selective sweep Rise in prevalence within a Population of an Allele that bestows higher Fitness on its carriers. See also Natural Selection as well as Directional Selection and Positive Selection.

Self-sacrifice Bacterial death that occurs in association with successful display of phage Resistance. Self-sacrifice by bacteria as a phage-Resistance mechanism is seen most notably with display by bacteria of Abortive Infection Systems. Bacterial death is also an aspect of bacterial Resistance mechanisms resulting in phage Delay, as Virion production by a Lytic Phage, even if reduced in extent, is still associated with bacterial death. With Delay, though, the bacterial death would be phage- rather than self-inflicted and therefore technically would not represent a form of bacterial self-sacrifice.

Semelparity Not capable of undergoing more than one reproductive episode per life time. Thus, obligately Lytic Phages are semelparous, with their single burst per Infected bacterium the equivalent of a single reproductive episode per life time. Contrast with Iteroparity.

Sequence See Nucleotide Sequence.

Sequence space Possible changes to genetic material that may be achieved especially mutationally. *Exploration* of sequence space therefore is the creation (via Mutation) as well as testing (via Selection) of Genetic Variation within a Population.

Sex Not referring here to the act of mating as toward sexual reproduction but instead referring to the incorporation of Genetic Material associated with one organism into the genome of a second organism. That second organism may be an otherwise mature individual (e.g., a bacterium) or instead may be an individual which has been newly formed in association with the mixing of Genetic Material from more than one parent (i.e., as a product of sexual reproduction). With bacteria, sex is a product of Horizontal Gene Transfer.

Sexual reproduction Production of offspring (new individuals) whose genomes are a combination of Genetic Material sourced from different parents, e.g., a mother and a father. A key word in this term is 'reproduction' since Sex in non-obligately sexual organisms is not required for reproduction, e.g., for binary fission in bacteria. In Eukaryotic organisms, sexual reproduction is associated with Meiosis.

Single-nucleotide polymorphism (SNP) Mutational variation between individuals or Strains that consists of a change to only a one base or base pair. For example, this single-nucleotide change could be from an A to a C. The term Polymorphism refers simply to the existence of such differences across a Population. Thus, a Mutation affecting a single base pair (or base for single-stranded entities) will be found in some individuals within a Population, but not found in others, thereby implying Genetic Variation at that Locus and therefore the existence of a Polymorphism.

Single-step growth Experimental approach in which phages are allowed to initiate Infections of bacteria via Adsorption but any phages Released are prevented from Infecting other bacteria, at least within the experimental vessel. This assures that there is only a single round of phage Infection and thus a single *step* in terms of increases in phage Titers due to Productive Infections occurs. Single-step growth experiments typically will also be done starting with low phage Multiplicities so that only a single phage (on average) is Infecting individual bacteria and are then followed via ongoing determinations of Virus Titers until the accumulation of released Virions plateaus. These experiments are also done with some degree of Adsorption synchronization so that a phage-Infected Population can be studied as a whole. From single-step growth experiments both Virus Burst Size and Latent Period can be determined. With modification, Virus Eclipse Period lengths can be determined as well. Synonymous with one-step growth.

Siphovirus Tailed Viruses possessing long, non-contractile tails. See also Tailed Phage and *Caudovirales* as well as contrast with Myovirus and Podovirus.

Sister lysogen Relationship of two bacteria possessing the same Prophage, especially to the extent that those two bacteria are clonally related. See also simply Lysogen.

Site-specific recombination Mechanism of Integration used by many Temperate Phages into the genome of their Host bacteria. Not all integrating Temperate Phages recombinogenically Integrate at specific, sequence-defined sites (Insertion Sites), however, such as phage Mu, which inserts randomly into genomes and therefore does not Integrate in a site-specific manner. See also simply Recombination.

Spatial structure Impediments within Environments to mixing and potentially also to diffusion. Within spatially structured Environments, entities tend to persist nearer to their points of origin than they do in non-spatially structured Environments. In microbiology, compare well-mixed broth (not spatially structured) to agar-based media (spatially structured). The formation of phage Plaques or bacterial colonies in association with the latter, as localized phenomena, is a consequence of the imposition of spatial structure on bacterial and phage Populations by the agar matrix. This results in a combination of little or no Environmental mixing, hindrance of at least bacterial diffusion, and, if otherwise present, interference with bacterial motility.

Specialized transduction Virion-mediated movement of non-phage DNA in association with movement of phage DNA. A sensu stricto as well as canonical definition of specialized transduction, however, is the inadvertent carriage by a Temperate Phage of Insertion-Site adjacent bacterial Genes, and this is certainly included in the broader definition offered in the sentence found at the start of this paragraph. Also potentially included among products of specialized transduction, at least at the point of their acquisition, are Morons, whether or not this more DNA is derived from insertion site-adjacent bacterial Genes.

Spatial vulnerability Referring to higher localized susceptibility to Exploiter organisms as resulting from existence of Populations within clumps. For bacteria these clumps could be, for example, Clonal Microcolonies. The idea is not that clumps of organisms are always more vulnerable than isolated organisms to exploiters, but instead that for organisms found in clumps to be as Resistant or more Resistant to exploiters then they need to display additional Resistance mechanisms than do isolated (not clumped) individual organisms. For example, in Microcolonies this additional Resistance might be derived from existing within Extracellular Polymeric Substances.

Speciation Process of separation of a single Gene Pool into separate Gene Pools, particularly when that separation is more than geographical but instead is a function of Phenotypes associated with the members of the now separated Gene Pools. Or, in other words, speciation is the formation of a new Species from a parental Population, particularly as resulting from some mechanism of especially Prezygotic Isolation.

Species Collection of organisms that are more similar Genotypically and more likely to exchange Genes with individuals of their type than with individuals not of their type. Individuals of the same type in this case we would call Conspecifics (or Homospecifics), contrasting Heterospecifics. A collection of Conspecifics is called a Population whereas a collection of Heterospecifics is called a Community.

Stable polymorphism Longstanding retention across a Population of more than one Allele at a given Locus. Contrast with Allele Fixation. Directional Selection, by contrast, tends to destabilize polymorphisms.

Stabilizing frequency-dependent selection Mechanism in which increasing Allele rarity is associated with increasing Allele fitness. Thus, increasingly common Alleles are Selected against, resulting in Stable Polymorphisms. Stabilizing frequency-dependent selection is typically described as just Frequency-Dependent Selection, or instead as negative frequency-dependent selection, and is also a form of balancing selection. Contrast with Disruptive Frequency-Dependent Selection.

Stabilizing selection Deterministic Evolution favoring common Alleles and Phenotypes such that they retain their frequency dominance within Populations. Populations subject to stabilizing selection thus tend to not substantially change over time in terms of Allele Frequencies. This is not equivalent to saying that these Populations are not subject to Evolution, however, but instead that exiting Evolutionary processes result in Genotypic stasis (retention of current Allele Frequencies) rather than Genotypic change over time.

Stochastic Having a random component; that is, as cannot be precisely defined. For example, random Mutation is a stochastic process. Contrast with deterministic (see Determinism).

Strain (taxonomy) A distinct Genotype found within a Species, making up a sub-Population of the larger Species. Strains can be more closely or instead more distantly related. For example, Isogenic strains of two bacteria can in principle differ by only a single Mutation whereas naturally occurring strains within bacterial Species can differ by multiple percentages in terms of their overall Nucleotide Sequences.

Strictly lytic Phage for which successful Infections of bacteria obligately end with Lysis of the phage-Infected bacterium. A strictly lytic phage, or Virus more generally, inherently is never able to display Latent Infections nor pass through a Prophage stage. Obligately lytic is used synonymously and see also Professionally Lytic.

Subspecies Population that displays substantially reduced Gene Flow from outside of the Population, but still within the same Species, relative to inside of the Population. A subspecies, in other words, is a sub-Population found within a larger Species that does not display substantial Genetic Migration with other so-defined sub-Populations making up the same Species.

Superinfection Strictly the Infection by a phage of an already phage-Infected bacterium, thereby resulting in Coinfection. Many, however, use the term superinfection to describe what instead can be only the Secondary Adsorption of an already phage-Infected bacterium. This, though, is imprecise since Secondary Adsorption by a phage does not guarantee Secondary Infection.

Superinfection exclusion Especially a genome Translocation-level anti-phage Resistance mechanism, i.e., that interferes with the transition from Irreversible Adsorption to Infection. Superinfection exclusion often is expressed by already Infecting phages. Contrast with Superinfection Immunity.

Superinfection immunity Anti-phage bacterial Resistance mechanism that acts after phage-genome Translocation into a bacterium's cytoplasm. This is expressed by a resident Prophage and is a consequence of the action of phage-produced repressor proteins that interfere with Temperate Phage gene expression toward retaining Lysogenic Cycles by the Prophage. Superinfecting phages against which superinfection immunity is effective are said to be Homoimmune relative to the superinfection immunity-expressing Prophage. Contrast with Superinfection Exclusion.

Superspreader Individual that is especially effective at transmitting something, especially an infectious disease, to other individuals. Acquisition of an infection by a superspreader can result in dramatic acceleration of the spread of an infectious disease through a Population of susceptible individuals.

Surface receptor Molecules associated with the extracellular side of a cell envelope and to which Virions can specifically bind to in the course of Adsorption.

Survival Ability to continue a Life Cycle by avoiding death, thus allowing at least attempts to reproduce. Survival thereby serves as an important component of Evolutionary Fitness.

Symbiont The non-Host component of a Symbiosis. A Prophage, for example, can be viewed as the symbiont, and the bacterium the Host, in the overall Symbiosis known as a Lysogen.

Symbiosis Especially longer-term (relative to generation time) intimate interactions between different Species, often with one Species serving as the Host and the other as the Symbiont. These interactions can take on various forms known as Mutualisms, Commensalisms, or Parasitisms.

Synonymous substitution Point Mutation that has modified a codon without changing what amino acid is encoded. Biases in Fixation within Populations toward synonymous substitutions are considered to be possible indicators of Genetic Drift having impacted the associated Genes, that is to the extent that these mutations can have little impact on Fitness. Contrast with Non-Synonymous Substitution.

Tailed phage Virions possessing highly exaggerated, single appendages for Adsorption. Tailed Phages are members of Virus order *Caudovirales* and among the Archaeal Viruses, Tailed Phages instead are described as Head-Tailed.

Tailocin Bacteriocin resembling the tail of a Tailed Phage.

Temperate phage Bacterial Virus that is capable of displaying a Latent Infection. Latent Infections for temperate phages also are known as Lysogenic Cycles. Temperate phages should not be described as 'lysogenic phages' since historically it is bacteria that are considered to be lysogenic rather than phages and also most temperate phages also display Lytic Cycles, with many temperate phages seeming to preferentially display Lytic Cycles following Lysis-Lysogeny Decisions. See also Lysogen as well as Prophage.

Titer Term commonly used to denote the concentration of phage Virions within Environments. Usually titer is expressed in phages per ml or Plaque-Forming Units per ml units.

Tradeoff Association of costs with benefits. A tradeoff associated with Resistance by a bacterium to a bacteriophage could be slower bacterial growth than is seen with Isogenic phage-sensitive bacteria. See also Antagonistic Pleiotropy.

Transducing particle Phage Virion which carries bacterial DNA, i.e., as potentially effecting Horizontal Gene Transfer of bacterial DNA from one bacterium to another. Transducing particles are generated by Transducing Phages. See Transduction.

Transducing phage Bacterial Virus capable of transferring bacterial DNA between bacteria. In particular, not all phage types are capable of Transducing bacterial DNA. Transducing phages are phages that can generate Transducing Particles.

Transduction Virus-mediated transfer of cellular DNA from one cell to another, e.g., as effected by phages from one bacterium to another. See also Transducing Phage, Transducing Particle, Generalized Transduction, Specialized Transduction, Reciprocal Transduction, Direct Reciprocal Transduction, Indirect Reciprocal Transduction, Non-Reciprocal Transduction, Assortative Transduction, and Abortive Transduction.

Transductive host range Collection of those bacteria that a phage is able to directly effect Gene Exchange to. That is, the potential bacterial recipient Strains of DNA from a Transducing Particle. The transductive host range is a subset of the Penetrative Host Range, which in turn is a subset of the Adsorptive Host Range of a phage.

Transductive species Bacterial grouping defined by the potential for Gene Flow between members as mediated by one or more bacteriophages. Bacteria found outside of the normal Transductive Host Range of a Transducing Phage would not be considered to be members of that transductive species. More narrowly, we can consider only those bacteria for which Reciprocal Transduction as mediated by a given Transducing Phage is possible.

Transformation Acquisition by a cell of Free DNA from the Environment and subsequent incorporation of that DNA into the genome of that recipient cell. Bacteria which are capable of being so transformed outside of the laboratory are said to be naturally competent. Contrast with Transduction as well as with Conjugation. See also Environmental DNA.

Translocation Movement from one place to another but here used primarily synonymously with phage genome Uptake into an Adsorbed cell's cytoplasm.

Transposon Short for transposable element, this is a sequence of DNA that can move from place to place within a genome by inserting into new locations. Transposons may or may not carry the Genes required for their transposition. Certain phages also act as transposable elements, e.g., phage Mu.

Treatment resistance Conversion of sensitive targets of antimicrobial therapy to no longer sensitive targets, as occurs over the course of that therapy. Treatment resistance is Acquired Resistance that Evolves in the course of treatment of bacterial infections such as with phages, i.e., with infections initially consisting of phage-sensitive bacteria. Treatment resistance in particularly can be contrasted with Community Resistance.

Unit of selection Entity to which an Evolutionary Fitness measure can be assigned, whether an individual organism or, less frequently, an individual Allele or instead a group of organisms. See also Natural Selection, Individual Selection, and Group Selection as well as simply Selection.

Uptake Translocation of Virus Genetic Material from outside of a cell to inside of a cell. Given successful Virion Adsorption, then uptake of the phage genome usually will occur, i.e., as following Virion Attachment. Also described for some phages as injection.

Victim Organism that is subject to Antagonistic actions by an Exploiter organism. Typically if these Antagonistic actions are successful, then this will result in reductions in victim Fitness, potentially in association with increases in Exploiter Fitness. Prey are examples of victims as too are the Hosts of Parasites. Bacteria are victims of phages especially that are displaying Lytic Cycles.

Virion Especially mature, infectious Viruses, i.e., the viral transmission-to-new-cells stage. See also Free Phage.

Virion particle Equivalent to simply Virion, though emphasizing the discrete, particle nature of Virions.

Virion progeny Encapsidated products of Virus-Infected cells. Virion progeny do not include daughter Lysogens produced by Prophage-Infected bacteria since those are not Virions.

Virocell The Infection stage of a Virus Life Cycle. Contrast with the free Virion or Free Phage stage.

Virulence (pathogen) Capacity to cause disease. For example, *E. coli* serotype O157:H7 (the 'bad' *E. coli*) is more virulent than typical Commensal *E. coli* Strains living in your gut.

Virulence (phage) Capacity to clear bacterial cultures. This is generally a relatively ambiguously defined concept, but the lower the initial phage starting Titer needed to clear a given bacterial culture (meaning Lyse most of the bacteria present), then the more virulent the phage. In addition, and related, phages that are more virulent tend to not be Temperate Phages and many use the term 'virulent' as synonymous with Strictly Lytic. Note that phage virulence as so-defined has no bearing on a phage's encoding of bacterial Virulence Factors, and in fact non-Temperate Phages tend to *not* encode bacterial Virulence Factors.

Virulence factor Especially proteins that bacteria use to directly effect disease. Virulence factors are generally other than simply bacterial Fitness factors, thereby with virulence factors modifying bacterial Phenotypes particularly within a context of causing an infection of some Host organism. Many bacterial virulence factors are encoded by Prophages.

Virus Acellular obligately intracellular Parasite of cellular organisms, as typically possessing a proteinaceous Capsid enclosing a nucleic acid genome, though lipid materials may be present as well. Phages are viruses.

Index

S. T. Abedon, *Bacteriophages as Drivers of Evolution*,
https://doi.org/10.1007/978-3-030-94309-7

B

D

F

N

T

The manufacturer's authorised representative in the EU is Springer Nature Customer Service Centre GmbH, Europaplatz 3, 69115 Heidelberg, Germany. If you have any concerns regarding our products, please contact ProductSafety@springernature.com

Printed and bound by CPI Group (UK) Ltd, Croydon, CR0 4YY
30/09/2026
02225233-0009